BIRKHÄUSER

Progress in Probability
Volume 62

Spin Glasses: Statics and Dynamics

Summer School, Paris 2007

Anne Boutet de Monvel
Anton Bovier
Editors

Birkhäuser
Basel · Boston · Berlin

Editors:

Anne Boutet de Monvel
Institut de Mathématique de Jussieu
Université Paris Diderot Paris 7
175 rue du Chevaleret
75013 Paris
France
e-mail: aboutet@math.jussieu.fr

Anton Bovier
Institut für Angewandte Mathematik
Rheinische Friedrich-Wilhelms-
Universität
Wegelerstrasse 6
53115 Bonn
Germany
e-mail: bovier@uni-bonn.de

2000 Mathematics Subject Classification: 82-06, 82B44

Library of Congress Control Number: 2009933494
Additional material to this book can be downloaded from http://extras.springer.com
Bibliographic information published by Die Deutsche Bibliothek.
Die Deutsche Bibliothek lists this publication in the Deutsche Nationalbibliografie; detailed bibliographic data is available in the Internet at http://dnb.ddb.de

ISBN 978-3-7643-8999-4 Birkhäuser Verlag AG, Basel · Boston · Berlin

Basel · Boston · Berlin
P.O. Box 133, CH-4010 Basel, Switzerland
Part of Springer Science+Business Media
Printed on acid-free paper produced from chlorine-free pulp. TCF ∞
Printed in Germany

ISBN 978-3-7643-8999-4 e-ISBN 978-3-7643-9891-0

9 8 7 6 5 4 3 2 1 www.birkhauser.ch

Contents

Preface

Over the last decade, spin glass theory has turned from a fascinating part of theoretical physics to a flourishing and rapidly growing subject of probability theory as well. These developments have been triggered to a large part by the mathematical understanding gained on the fascinating and previously mysterious "Parisi solution" of the Sherrington–Kirkpatrick mean field model of spin glasses, due to the work of Guerra, Talagrand, and others. At the same time, new aspects and applications of the methods developed there have come up.

The present volume collects a number of reviews as well as shorter articles by lecturers at a summer school on spin glasses that was held in July 2007 in Paris. These articles range from pedagogical introductions to state of the art papers, covering the latest developments. In their whole, they give a nice overview on the current state of the field from the mathematical side.

The review by Bovier and Kurkova gives a concise introduction to mean field models, starting with the Curie–Weiss model and moving over the Random Energy models up to the Parisi solution of the Sherrington–Kirkpatrik model. Ben Arous and Kuptsov present a more recent view and disordered systems through the so-called local energy statistics. They emphasize that there are many ways to look at Hamiltonians of disordered systems that make appear the Random Energy model (or independent random variables) as a universal mechanism for describing certain rare events. An important tool in the analysis of spin glasses are correlation identities. This aspect is taken up in the articles by Contucci, Giardinà, and Nishimori on the one hand, and Franz and de Sanctis on the other. In some sense closely related to this is the question on how the Gibbs states of a spin glass react to small changes in the parameters, in particular the temperature. This question of "chaos" is discussed in the contribution by Rizzo. Two articles by Hanen discuss limit laws for covariances and mean magnetization in the SK model at high temperatures.

One of the main unresolved issues in spin glass theory is the relevance of the mean field model for more realistic short range models. Unfortunately, the study of short range models seems to require very different methods, and no satisfactory results are available. In the paper by Machta, Newman, and Stein an approach to this problem via percolation methods and the FK representation is presented, and some rigorous and numerical results are shown.

From the practical point of view, dynamical properties of spin glasses are even more important than equilibrium properties. Two articles in this volume address this elusive issue. Chamon and Cugliandolo give a brief discussion of the theory of out-of-equilibrium fluctuations in mean field and finite-dimensional models, whereas Černý presents recent rigorous results on aging in the dynamics of the random energy model.

The two final articles of this collection treat another lively area of the field of disordered systems. Giacomin gives an overview on recent progress on the behavior of random walks in the presence of a pinning interface and disorder, while Lacoin and Toninelli treat a specific hierarchical Verizon of this model where they discuss in particular the smoothing effect of disorder in the phase transition.

We would like to express our gratitude to all the contributers to this volume. Special thanks are due to Jean-Jacques Sansuc for his help and support in the editing process.

April 2009

Anne Boutet de Monvel (Paris)
Anton Bovier (Bonn)

Acknowledgement

The International Summer School "Spin Glasses" was held in Paris from June 25 to July 6, 2007. This was the annual Summer School organized by the Institute of Mathematics of Jussieu.

The subject of the school was chosen at the suggestion of the Director of the Institute, Gilles Godefroy. It was largely due to the presence of Michel Talagrand as member of the Institute.

The Summer School was formally organized by Anne Boutet de Monvel and Michel Talagrand, but with the very effective help of Anton Bovier.

We are very grateful to Gilles Godefroy for his constant and unwavering support. For the preparation of the School we received the help of the administrative staff of the Institute, in particular of Rosita Monchanin. We also had the complicity of Alberto Arabia for all technical problems related to web and mail communication.

Monique Douchez was in charge of the technical organization of this two-week series of lectures, in two places in Paris, the Institute Henri Poincaré and the Center Javelot. She was also responsible for welcoming participants and speakers and every day she has wonderfully solved all technical problems. We thank her very warmly.

The School was co-sponsored by the Institute of Mathematics of Jussieu (IMJ), the National Center for Scientific Research (CNRS), the Pierre-et-Marie Curie University, the Paris-Diderot University, the Region Ile-de-France, the Ministry of Higher Education and Research, and the Ministry of Foreign Affairs (MAE). We thank all these institutions for their support.

List of Contributors

Gérard Ben Arous
Courant Institute of Mathematical Sciences, New York University, 251 Mercer Street, New York, NY 10012, USA, benarous@cims.nyu.edu

Anton Bovier
Institut für Angewandte Mathematik, Rheinische Friedrich-Wilhelms-Universität Bonn, Endenicher Allee 60, D-53115 Bonn, Germany, bovier@uni-bonn.de

Jiří Černý
Departement Mathematik, ETH Zürich, Rämistrasse 101, CH-8092 Zürich, Switzerland, jiri.cerny@math.ethz.ch

Claudio Chamon
Physics Department, Boston University, 590 Commonwealth Avenue, Boston, MA 02215, USA, chamon@bu.edu

Pierluigi Contucci
Dipartimento di Matematica, Università di Bologna, I-40127 Bologna, Italy, contucci@dm.unibo.it

Leticia F. Cugliandolo
Université Pierre et Marie Curie – Paris VI, Laboratoire de Physique Théorique et Hautes Énergies, 5ème étage, Tour 25, 4 Place Jussieu, F-75252 Paris Cedex 05, France, leticia@lpthe.jussieu.fr

Luca De Sanctis
ICTP, Strada Costiera 11, I-34014 Trieste, Italy, lde_sanc@ictp.it

Silvio Franz
Université Paris-Sud Paris 11, F-91405 Orsay Cedex, France, silvio.franz@lptms.u-psud.fr

Giambattista Giacomin
U.F.R. de Mathématiques, case 7012 (site Chevaleret), Laboratoire de Probabilités et Modèles Aléatoires (LPMA, CNRS U.M.R. 7599), Université Paris Diderot (Paris 7), F-75205 Paris Cedex 13, France, giacomin@math.jussieu.fr

Cristian Giardinà
Eindhoven University of Technology & EURANDOM, P.O. Box 513, NL-5600 MB Eindhoven, The Netherlands, c.giardina@tue.nl

Albert Hanen
University Paris X Nanterre, 200 avenue de la République, F-92000 Nanterre, France, ahanen3752@aol.com

Alexey Kuptsov
Courant Institute of Mathematical Sciences, New York University, 251 Mercer Street, New York, NY 10012, USA, kuptsov@cims.nyu.edu

Irina Kurkova
Laboratoire de Probabilités et Modèles Aléatoires, Université Paris 6, 4, place Jussieu, B.C. 188, F-75252 Paris Cedex 05, France, irina.kourkova@upmc.fr

Hubert Lacoin
U.F.R. de Mathématiques, case 7012 (site Chevaleret), Laboratoire de Probabilités et Modèles Aléatoires (LPMA, CNRS U.M.R. 7599), Université Paris Diderot (Paris 7), F-75205 Paris Cedex 13, France, lacoin@clipper.ens.fr

Jonathan Machta
Physics Department, University of Massachusetts, Amherst, MA 01003, USA, machta@physics.umass.edu

Charles M. Newman
Courant Institute of Mathematical Sciences, New York University, 251 Mercer Street, New York, NY 10012, USA, newman@courant.nyu.edu

Hidetoshi Nishimori
Department of Physics, Tokyo Institute of Technology, Oh-okayama, Meguru-ku, Tokyo 152-8551, Japan, nishimori@phys.titech.ac.jp

Tommaso Rizzo
"E. Fermi" Center, Via Panisperna 89 A, Compendio Viminale, I-00184 Roma, Italy & Dipartimento di Fisica, Università di Roma "La Sapienza", P.le Aldo Moro 2, I-00185 Roma, Italy, tommaso.rizzo@inwind.it

Daniel L. Stein
Physics Department & Courant Institute of Mathematical Sciences, New York University, 251 Mercer Street, New York, NY 10012, USA, daniel.stein@nyu.edu

Fabio Lucio Toninelli
Laboratoire de Physique (CNRS U.M.R. 5672), École Normale Supérieure de Lyon, 46 Allée d'Italie, F-69364 Lyon Cedex 07, France, fltonine@ens-lyon.fr

Part I

Mean Field

Progress in Probability, Vol. 62, 3–44

A Short Course on Mean Field Spin Glasses

Anton Bovier and Irina Kurkova

Abstract. We give a brief introduction to the theory of mean field models of spin glasses. This includes a concise presentation of the Random Energy model and the Generalized Random Energy model and the connection to the corresponding asymptotic models based on Poisson cascades. We also explain the nature of the Parisi solution of the Sherrington–Kirkpatrick model and its derivation via Gaussian interpolation methods.

Mathematics Subject Classification (2000). 82B44.

Keywords. Mean field models, spin glasses, random energy model, Sherrington–Kirkpatrick model, Gaussian processes, interpolation, Parisi solution

1. Preparation: The Curie–Weiss model

The main topic of this lecture series are disordered mean field spin systems. This first section will, however, be devoted to ordered spin systems, and more precisely essentially to the Curie–Weiss model. This will be indispensable to appreciate later the much more complicated Sherrington–Kirkpatrick spin glass.

1.1. Spin systems

In his Ph.D. thesis in 1924, Ernst Ising [18, 19] attempted to solve a model, proposed by his advisor Lenz, intended to describe the statistical mechanics of an interacting system of magnetic moments. The setup of the model proceeds from a lattice, $\mathbb{Z}^d$, and a finite subset, $\Lambda \subset \mathbb{Z}^d$. The lattice is supposed to represent the positions of the atoms in a regular crystal. Each atom is endowed with a magnetic moment that is quantized and can take only the two values $+1$ and -1, called the *spin* of the atom. This spin variable at site $x \in \Lambda$ is denoted by σ_x. The spins are supposed to interact via an interaction potential $\phi(x, y)$; in addition, a magnetic field, h, is present. The energy of a *spin configuration* is then

$$H_\Lambda(\sigma) \equiv - \sum_{x \neq y \in \Lambda} \phi(x, y)\sigma_x\sigma_y - h \sum_{x \in \Lambda} \sigma_x. \tag{1}$$

The spin system with the Hamiltonian (1) with the particular choice

$$\phi(x,y) = \begin{cases} J, & \text{if } |x-y| = 1, \\ 0, & \text{otherwise,} \end{cases} \tag{2}$$

is known as the Ising spin system or *Ising model.* This model has played a crucial rôle in the history of statistical mechanics.

The essential game in statistical mechanics is to define, once a Hamiltonian is given, a probability measure, called the *Gibbs measure*, on the space of spin-configurations. This entails some interesting subtleties related to the fact that we would really want to do this in infinite volume, but we will not enter into these here. For finite volumes, Λ, we can easily define this probability as

$$\mu_{\beta,h,\Lambda}(\sigma) \equiv \frac{\exp\left(-\beta H_\Lambda(\sigma)\right)}{Z_{\beta,h,\Lambda}}, \tag{3}$$

where $Z_{\beta,h.\Lambda}$ is a normalizing factor called the *partition function*,

$$Z_{\beta,h,\Lambda} \equiv \sum_{\sigma\in\mathcal{S}_\Lambda} \exp\left(-\beta H_\Lambda(\sigma)\right). \tag{4}$$

An interesting fact of statistical mechanics is that the behavior of the partition function as a function of the parameters β and h contains a lot of information.

We will call

$$F_{\beta,h,\Lambda} \equiv -\frac{1}{\beta} \ln Z_{\beta,h,\Lambda} \tag{5}$$

the *free energy* of the spin system.

1.2. Sub-additivity and the existence of the free energy

The main concern of statistical mechanics is to describe systems in the limit when their size tends to infinity. The first question one asks is whether quantities defined for finite Λ have limits as $\Lambda \uparrow \mathbb{Z}^d$. The free energy (5) is expected to grow linearly with the volume of the system, and thus one may ask whether the *specific* free energy,

$$f_{\beta,h,\Lambda} \equiv \frac{1}{|\Lambda|} F_{\beta,h,\lambda}, \tag{6}$$

converges. Since these questions will recur, it will be useful to see how that can be proven.

It will be useful to note that we can express the Hamiltonian in the equivalent form

$$\widehat{H}_\Lambda(\sigma) = \sum_{x,y\in\Lambda} \phi(x,y)\left(\sigma_x - \sigma_y\right)^2 - h\sum_{x\in\Lambda}\sigma_x, \tag{7}$$

which differs from H_Λ only by a constant. Now let $\Lambda = \Lambda_1 \cup \Lambda_2$, where Λ_i are disjoint volumes. Clearly we have that

$$Z_{\beta,\Lambda} = \sum_{\sigma_x,\, x\in\Lambda_1} \sum_{\tau_y,\, y\in\Lambda_2} \exp\left(-\beta\left[H_{\Lambda_1}(\sigma) + H_{\Lambda_2}(\tau)\right]\right)$$
$$\times \exp\left(-\beta \sum_{x\in\Lambda_1} \sum_{y\in\Lambda_2} \phi(x,y)(\sigma_x - \tau_y)^2\right). \tag{8}$$

If $\phi(x,y) \geq 0$, this implies that

$$Z_{\beta,\Lambda} \leq Z_{\beta,\Lambda_1} Z_{\beta,\Lambda_2} \tag{9}$$

and therefore

$$-F_{\beta,\Lambda} \leq (-F_{\beta,\Lambda_1}) + (-F_{\beta,\Lambda_2}). \tag{10}$$

The property (9) is called *sub-additivity* of $(-F_{\beta,\Lambda})$ (as a function of the volume Λ). The importance of sub-additivity is that it implies convergence through the following elementary lemma:

Lemma 1.1. *Let a_n be a real-valued sequence that satisfies, for any $n, m \in \mathbb{N}$,*

$$a_{n+m} \leq a_n + a_m. \tag{11}$$

(i) *Then, $\lim_{n\uparrow\infty} n^{-1} a_n$ exists.*
(ii) *If, moreover, $n^{-1}a_n$ is uniformly bounded from below, then the limit is finite.*

By successive iteration, the lemma has an immediate extension to arrays:

Lemma 1.2. *Let $a_{n_1,n_2,\ldots,n_d}$, $n_i \in \mathbb{N}$ be a real-valued array that satisfies, for any $n_i, m_i \in \mathbb{N}$,*

$$a_{n_1+m_1,\ldots,n_d+m_d} \leq a_{n_1,\ldots,n_d} + a_{m_1,\ldots,m_d}. \tag{12}$$

(i) *Then, $\lim_{n\uparrow\infty}(n_1 n_2 \ldots n_d)^{-1} a_{n_1,\ldots,n_d}$ exists.*
(ii) *If $(n_1 n_2 \ldots n_d)^{-1} a_{n_1,\ldots,n_d} \geq b > -\infty$, then the limit is finite.*

Lemma 1.2 can be used straightforwardly to prove convergence of the free energy over rectangular boxes:

Proposition 1.1. *If the Gibbs free energy $F_{\beta,\Lambda}$ of a model satisfies the sub-additivity property* (10), *and if $\sup_\sigma H_\Lambda(\sigma)/|\Lambda| \geq C > -\infty$, then, for any sequence Λ_n of rectangles*

$$\lim_{n\uparrow\infty} |\Lambda_n|^{-1} F_{\beta,\Lambda_n} = f_\beta \tag{13}$$

exists and is finite.

Obviously this proposition gives the existence of the free energy for Ising's model, but the range of applications of Proposition 1.1 is far wider, and virtually covers all lattice spin systems with bounded and absolutely summable interactions. Further details can be found, e.g., in [28].

1.3. The Curie–Weiss model

Although the Ising model can be solved exactly in dimensions one (easy) and two (hard), exact solutions in statistical mechanics are rare. To get a quick insight into specific systems, one often introduces exactly solvable *mean field models*. It will be very instructive to study the simplest of these models, the *Curie–Weiss model* in some detail. All we need to do to go from the Ising model to the Curie–Weiss model is to replace the nearest-neighbor pair interaction of the Ising model by another extreme choice, namely the assumption that each spin variable interacts with each other spin variable at any site of the lattice with exactly the same strength. Since then the structure of the lattice becomes irrelevant, and we simply take $\Lambda = \{1, \dots, N\}$. The strength of the interaction should be chosen of order $1/N$, to avoid the possibility that the Hamiltonian takes on values larger than $O(N)$. Thus, the Hamiltonian of the Curie–Weiss model is

$$H_N(\sigma) = -\frac{1}{N} \sum_{1 \le i,j \le N} \sigma_i \sigma_j - h \sum_{i=1}^{N} \sigma_i. \tag{14}$$

A crucial notion is that of *macroscopic variables*. The most important one here is

$$m_N(\sigma) \equiv N^{-1} \sum_{i=1}^{N} \sigma_i \tag{15}$$

the *empirical magnetization*. Here we divided by N to have a specific magnetization.

Note that the particular structure of the Curie–Weiss model entails that the Hamiltonian can be written as a function of this single macroscopic function:

$$H_N(\sigma) = -\frac{N}{2} \left[m_N(\sigma)\right]^2 - hNm_N(\sigma) \equiv N\Psi_h(m_N(\sigma)). \tag{16}$$

This can be considered as a defining feature of *mean field models*.

Let us now compute the free energy of this model. Because of the interaction term, this problem looks at first sight complicated. But it is not. To solve the problem we change from the ensemble of fixed magnetic field to that of fixed magnetization. That is, we write

$$Z_{\beta,h,N} = \sum_{m \in \mathcal{M}_N} e^{N\beta(\frac{m^2}{2} + mh)} z_{m,N} \tag{17}$$

where $\mathcal{M}_N$ is the set of possible values of the magnetization, i.e.,

$$\begin{aligned} \mathcal{M}_N &\equiv \{m \in \mathbb{R} : \exists \sigma \in \{-1,1\}^N : m_N(\sigma) = m\} \\ &= \{-1, -1 + 2/N, \dots, 1 - 2/N, 1\} \end{aligned} \tag{18}$$

and

$$z_{m,N} \equiv \sum_{\sigma \in \{-1,1\}^N} \mathbb{1}_{m_N(\sigma) = m} \tag{19}$$

is a 'micro-canonical partition function'. Fortunately, the computation of this micro-canonical partition function is easy. In fact, all possible values of m are of the form $m = 1 - 2k/N$, where k runs from 0 to N and counts the number of spins that have the value -1. Thus, the computation of $z_{m,N}$ amounts the elementary combinatorial problem of counting the number of subsets of size k in the set of the first N integers. Thus,

$$z_{m,N} = \binom{N}{N(1-m)/2} \equiv \frac{N!}{[N(1-m)/2]![N(1+m)/2]!}. \tag{20}$$

The asymptotics of the logarithm of the binomial coefficients is given, to leading order, for $m \in \mathcal{M}_N$,

$$N^{-1} \ln z_{m,N} \sim \ln 2 - I(m) \tag{21}$$

where

$$I(m) = \frac{1+m}{2} \ln(1+m) + \frac{1-m}{2} \ln(1-m) \tag{22}$$

is called *Cramèr's entropy function* and worth memorizing.

Some elementary properties of I are useful to know: First, I is symmetric, convex, and takes its unique minimum, 0, at 0. Moreover $I(1) = I(-1) = \ln 2$. Its derivative, $I'(m) = \operatorname{arcth}(m)$, exists in $(-1, 1)$. While I is not uniformly Lipschitz continuous on $[-1, 1]$, it has the following property:

Lemma 1.3. *There exists $C < \infty$ such that for any interval $\mathbb{D} \subset [-1, 1]$ with $|\mathbb{D}| < 0.1$,*

$$\max_{x,y \in \mathbb{D}} |I(x) - I(y)| \leq C|\mathbb{D}|\,|\ln|\mathbb{D}||.$$

We would like to say that $\lim_{N\uparrow\infty} \frac{1}{N} \ln z_{m,N} = \ln 2 - I(m)$. But there is a small problem, due to the fact that the relation (21) does only hold on the N-dependent set $\mathcal{M}_N$. Otherwise, $\ln z_{m,N} = -\infty$. A precise asymptotic statement could be the following:

Lemma 1.4. *For any $\tilde{m} \in [-1, 1]$,*

$$\lim_{\epsilon\downarrow 0} \lim_{N\uparrow\infty} \frac{1}{N} \ln \sideset{}{'}\sum_{\substack{m \in \mathcal{M}_M: \\ |m-\tilde{m}|<\epsilon}} z_{m,N} = \ln 2 - I(\tilde{m}). \tag{23}$$

Proof. The proof is elementary from the properties of $z_{m,N}$ and $I(m)$ mentioned above and is left to the reader. □

The following formulation of Lemma 1.4 is known as *Cramèr's theorem.* It is the simplest so-called *large deviation principle* [12]:

Lemma 1.5. *Let $A \in \mathcal{B}(\mathbb{R})$ be a Borel-subset of the real line. Define a probability measure p_N by $p_N(A) \equiv 2^{-N} \sum_{m \in \mathcal{M}_N \cap A} z_{m,N}$, and let $I(m)$ be defined in (22). Then*

$$-\inf_{m \in \mathring{A}} I(m) \leq \liminf_{N\uparrow\infty} \frac{1}{N} \ln p_N(A) \leq \limsup_{N\uparrow\infty} \frac{1}{N} \ln p_N(A) \leq -\inf_{m \in \bar{A}} I(m). \tag{24}$$

Moreover, I is convex, lower-semi-continuous, Lipschitz continuous on $(-1, 1)$, bounded on $[-1, 1]$, and equal to $+\infty$ on $[-1, 1]^c$.

Remark 1.1. The classical interpretation of the preceding theorem is the following. The spin variables $\sigma_i = \pm 1$ are *independent, identically distributed* binary random variables taking the values ± 1 with equal probability. $m_N(\sigma)$ is the normalized sum of the first N of these random variables. p_N denotes the probability distribution of the random variable m_N, which is inherited from the probability distribution of the family of random variables σ_i. It is well known, by the law of large numbers, that p_N will concentrate on the value $m = 0$, as N tends to ∞. A large deviation principle states in a precise manner how small the probability will be that m_N take on different values. In fact, the probability that m_N will be in a set A, that does not contain 0, will be of the order $\exp(-Nc(A))$, and the value of $c(A)$ is precisely the smallest value that the function $I(m)$ takes on the set A.

The computation of the canonical partition function is now straightforward:

$$Z_{\beta,h,N} = \sum_{m \in \mathcal{M}_N} \binom{N}{N(1-m)/2} \exp\left(N\beta\left(\frac{m^2}{2} + hm\right)\right) \tag{25}$$

and, by the preceding lemma, one finds that:

Lemma 1.6. *For any temperature, β^{-1}, and magnetic field, h,*

$$\begin{aligned} \lim_{N\uparrow\infty} \frac{-1}{\beta N} \ln Z_{\beta,h,N} &= \inf_{m\in[0,1]} \left(-m^2/2 + hm - \beta^{-1}(\ln 2 - I(m))\right) \\ &= f(\beta, h). \end{aligned} \tag{26}$$

Proof. We first prove an upper bound for $Z_{\beta,h,N}$:

$$\begin{aligned} Z_{\beta,h,N} &\leq N \max_{m\in\mathcal{M}_N} \exp\left(N\beta\left(\frac{m^2}{2} + hm\right)\right) \binom{N}{N(1-m)/2} \\ &\leq N \max_{m\in[-1,1]} \exp\left(N\beta\left(\frac{m^2}{2} + hm\right) + N(\ln 2 - I(m)) - O(\ln N)\right). \end{aligned} \tag{27}$$

Hence

$$\begin{aligned} &N^{-1} \ln Z_{\beta,h,N} \\ &\quad \leq N^{-1}\ln N + \max_{m\in[-1,1]} \left(\beta\left(\frac{m^2}{2} + hm\right) + \ln 2 - I(m) - N^{-1}O(\ln N)\right) \\ &\quad \leq \ln 2 + \sup_{m\in[-1,1]} \left(\beta\left(\frac{m^2}{2} + hm\right) - I(m)\right) + N^{-1}O(\ln N) \end{aligned} \tag{28}$$

so that

$$\limsup_{N\uparrow\infty} N^{-1} \ln Z_{\beta,h,N} \leq \beta \sup_{m\in[-1,1]} \left(\frac{m^2}{2} + hm - \beta^{-1}I(m)\right) + \ln 2.$$

This already looks good. Now we need a matching lower bound. Using that a sum is bigger than its parts, we get

$$Z_{\beta,h,N} \geq \max_{m\in\mathcal{M}_N} \exp\left(N\beta\left(\frac{m^2}{2} + hm\right)\right) \binom{N}{N(1-m)/2}. \tag{29}$$

All that remains to be done is to pass from the max over $\mathcal{M}_N$ to the max over $[-1,1]$, after inserting the bound for the binomial coefficient in terms of $I(m)$. In fact,

$$N^{-1}\ln Z_{\beta,h,N} \geq \ln 2 + \beta \max_{m\in\mathcal{M}_N}\left(\frac{m^2}{2} + hm - \beta^{-1}I(m)\right) - O(\ln N/N) \tag{30}$$

for any N. Now, we can check that

$$\max_{m\in\mathcal{M}_N}\left|\left(\frac{m^2}{2} + hm - \beta^{-1}I(m)\right) - \sup_{\substack{m'\in[0,1]\\|m'-m|\leq 2/N}}\left(\frac{m^2}{2} + hm - \beta^{-1}I(m)\right)\right| \leq C\ln N/N \tag{31}$$

so that

$$\liminf_{N\uparrow\infty}\frac{1}{\beta N}\ln Z_{\beta,h,N} \geq \beta^{-1}\ln 2 + \sup_{m\in[-1,1]}\left(\frac{m^2}{2} + hm - \beta^{-1}I(m)\right) \tag{32}$$

and the assertion of the lemma follows. □

The function $g(\beta,m) \equiv -m^2/2 - \beta^{-1}(\ln 2 - I(m))$ is called the *Helmholtz free energy* for zero magnetic field, and

$$\lim_{\epsilon\downarrow 0}\lim_{N\uparrow\infty}\frac{-1}{\beta N}\ln\sum_{\tilde{m}:|\tilde{m}-m|<\epsilon}\widetilde{Z}_{\beta,\tilde{m},N} = g(\beta,m) \tag{33}$$

where

$$\widetilde{Z}_{\beta,\tilde{m},N} = \sum_{\sigma\in\{-1,1\}^N} \mathrm{e}^{\beta H_N(\sigma)}\mathbb{1}_{m_N(\sigma)=m} \tag{34}$$

for $h=0$. Thermodynamically, the function $f(\beta,h)$ is then called *Gibbs free energy*, and the assertion of the lemma would then be that the Gibbs free energy is the *Legendre transform* of the Helmholtz free energy. The latter is closely related to the rate function of a large deviation principle for the distribution of the magnetization under the Gibbs distribution. Namely, if we define the Gibbs distribution on the space of spin configurations

$$\mu_{\beta,h,N}(\sigma) \equiv \frac{\mathrm{e}^{-\beta H_N(\sigma)}}{Z_{\beta,h,N}} \tag{35}$$

and denote by $\tilde{p}_{\beta,h,N}(A) \equiv \mu_{\beta,h,N}\left(\{m_N(\sigma)\in A\}\right)$ the law of m_N under this distribution, then we obtain very easily

Lemma 1.7. *Let $\tilde{p}_{\beta,h,N}$ be the law of $m_N(\sigma)$ under the Gibbs distribution. Then the family of probability measures $\tilde{p}_{\beta,h,N}$ satisfies a large deviation principle, i.e.,* *for all Borel subsets A of* $\mathbb{R}$,

$$\begin{aligned} -\inf_{m\in A}(g(\beta,m)-hm)+f(\beta,h) &\leq \liminf_{N\uparrow\infty}\frac{1}{\beta N}\ln\tilde{p}_{\beta,h,N}(A) \\ &\leq \limsup_{N\uparrow\infty}\frac{1}{\beta N}\ln\tilde{p}_{\beta,h,N}(A) \\ &\leq -\inf_{m\in\bar{A}}(g(\beta,m)-hm)+f(\beta,h). \end{aligned} \tag{36}$$

We see that the thermodynamic interpretation of equilibrium emerges very nicely: the equilibrium value of the magnetization, $m(\beta,h)$, for a given temperature and magnetic field, is the value of m for which the rate function in Lemma 1.7 vanishes, i.e., which satisfies the equation

$$g(\beta,m(\beta,h))-hm(\beta,h)=f(\beta,h). \tag{37}$$

By the definition of f (see (26)), this is the case whenever $m(\beta,h)$ realizes the infimum in (26). If $g(\beta,m)$ is strictly convex, this infimum is unique, and, as long as g is convex, it is the set on which $\frac{\partial g(\beta,m)}{\partial m}=h$.

Note that, in our case, $g(\beta,m)$ is not a convex function of m if $\beta>1$.

In fact, it has two local minima at the values $\pm m^*_\beta$, where m^*_β is defined as the largest solution of the equation

$$m=\tanh\beta m. \tag{38}$$

Moreover, the function g is symmetric, and so takes the same value at both minima. As a consequence, the minimizer of the function $g(\beta,m)-mh$, the magnetization as a function of the magnetic field, is not unique at the value $h=0$ (and only at this value). For $h>0$, the minimizer is the positive solution of $m=\tanh(\beta(m+h))$, while for negative h it is the negative solution. Consequently, the magnetization has a jump discontinuity at $h=0$, where it jumps by $2m^*_\beta$. One says that the Curie–Weiss model exhibits a first-order phase transition.

1.4. A different view on the CW model

We will now have a slightly different look at the Curie–Weiss model. This will be very instructive from the later perspective of the Sherrington–Kirkpatrick model.

To get started, we may want to compute the distribution of the spin variables as such. The perspective here is that of the product topology, so we should consider a fixed finite set of indices which without loss we may take to be $\{1,\dots,K\}$ and ask for the Gibbs probability that the corresponding spin variables, $\sigma_1,\dots,\sigma_K$ take specific values, and then take the thermodynamic limit.

To do these computations, it will be useful to make the following choices. The total volume of the system will be denoted $K+N$, where K is fixed and N will later tend to infinity. We will write $\hat{\sigma}\equiv(\sigma_1,\dots,\sigma_K)$, and $\check{\sigma}\equiv(\sigma_{K+1},\dots,\sigma_{K+N})$.

We set $\sigma = (\hat{\sigma}, \check{\sigma})$. We re-write the Hamiltonian as

$$\begin{aligned} -H_{K+N}(\sigma) &= \frac{1}{2(N+K)} \sum_{i,j\leq K} \sigma_i \sigma_j \\ &+ \frac{1}{2(N+K)} \sum_{i,j>K} \sigma_i \sigma_j \\ &+ \frac{1}{N+K} \sum_{i=1}^{K} \sigma_i \sum_{j=K+1}^{N+K} \sigma_j . \end{aligned} \tag{39}$$

This can be written as

$$\begin{aligned} -H_{K+N}(\sigma) &= \frac{K^2}{2(N+K)} (m_K(\hat{\sigma}))^2 \\ &+ \frac{N^2}{2(N+K)} (m_N(\check{\sigma}))^2 \\ &+ \frac{N}{N+K} \sum_{i=1}^{K} \sigma_i m_N(\check{\sigma}). \end{aligned} \tag{40}$$

The first term in this sum is of order $1/N$ and can be neglected. Also $N/(N+K) \sim 1+O(1/N)$. But note that $N^2/(N+K) = N - K(1-K/N) \sim N-K$. Using these approximations, we see that, up to terms that will vanish in the limit $N \uparrow \infty$,

$$\begin{aligned} \mu_{\beta,N+K}(\hat{\sigma}) &= \frac{\mathbb{E}_{\check{\sigma}} e^{\beta(N-K)(m_N(\check{\sigma}))^2/2} e^{\beta m_N(\check{\sigma}) \sum_{i=1}^{K} \sigma_i}}{\mathbb{E}_{\check{\sigma}} e^{\beta(N-K)(m_N(\check{\sigma}))^2/2} \sum_{\hat{\sigma}} e^{\beta m_N(\check{\sigma}) \sum_{i=1}^{K} \sigma_i}} \\ &= \frac{\int \mathbb{Q}_{\beta,N}(dm) e^{-\beta K m^2/2} e^{\beta m \sum_{i=1}^{K} \sigma_i}}{\int \mathbb{Q}_{\beta,N}(dm) e^{-\beta K m^2/2} \prod_{i=1}^{K} 2\cosh(\beta m)}, \end{aligned} \tag{41}$$

where $\mathbb{Q}_{\beta,N}$ is the distribution of the magnetisation under the Gibbs measure,

$$\mathbb{Q}_{\beta,N} \equiv \mu_{\beta,N} \circ m_N^{-1}. \tag{42}$$

Now, we can easily show, using the computations of the preceding subsection, that $\mathbb{Q}_{\beta,N}$ converges to either a Dirac measure on m^* or the mixture of two Dirac measures on m^* and $-m^*$. Thus it follows from (41) that $\mu_{\beta,N+K}$ converges to a product measure.

But assume that we did not know anything about $\mathbb{Q}_\beta$. Could we find out about it?

First, we write (assuming convergence, otherwise take a subsequence)

$$\mu_\beta(\hat{\sigma}) = \frac{\int \mathbb{Q}_\beta(dm) e^{-\beta K m^2/2} e^{\beta m \sum_{i=1}^{K} \sigma_i}}{\int \mathbb{Q}_\beta(dm) e^{-\beta K m^2/2} \prod_{i=1}^{K} 2\cosh(\beta m)}, \tag{43}$$

for any K. (41) establishes that the Gibbs measure of our model is completely determined by a single probability distribution, $\mathbb{Q}_\beta$, on a scalar random variable. Thus the task of finding the Gibbs measure is reduced to finding this distribution. How could we do this? A natural idea would be to use the Gibbs variational principle that says that the thermodynamic state must minimize the free energy. For this we would just need a representation of the free energy in terms of $\mathbb{Q}_\beta$.

To get there, we write the analog of (41) for the partition function. This yields

$$\frac{Z_{\beta,N+K}}{Z_{\beta,N}} = \int \mathbb{Q}_{\beta,N}(\mathrm{d}m)\mathrm{e}^{-\beta K m^2/2} \prod_{i=1}^{K} 2\cosh(\beta m). \tag{44}$$

Now it is not hard to see that the free energy can be obtained as

$$\lim_{N\uparrow\infty} \frac{1}{K\beta} \ln \frac{Z_{\beta,N+K}}{Z_{\beta,N}} = -f_\beta.$$

Thus we get the desired representation of the free energy

$$-f_\beta = \frac{1}{K\beta} \ln \int \mathbb{Q}_{\beta,N}(\mathrm{d}m) \exp\left(-\beta K \left(m^2/2 - \beta^{-1} \ln 2\cosh(\beta m)\right)\right). \tag{45}$$

Thus the Gibbs principle implies that

$$-f_\beta = \sup_{\mathbb{Q}} \frac{1}{K\beta} \ln \int \mathbb{Q}(\mathrm{d}m) \exp\left(-\beta K \left(m^2/2 - \beta^{-1} \ln 2\cosh(\beta m)\right)\right), \tag{46}$$

where the supremum is taken over all probability measures on $\mathbb{R}$. It is of course not hard to see that the supremum is realized by any probability measure that has support on the minimizer of the function $m^2/2 - \beta^{-1}\ln\cosh(\beta m)$.

We will see later that a curious analog, with the sup replaced by an inf, of the formula (46) is the key to the solution of the Sherrington–Kirkpatrick model.

Should we not have known about the Gibbs principle, we could instead have observed that (45) can only hold for *all* K, if $\mathbb{Q}_\beta$ is supported on the minimizer of the function $m^2/2 - \beta^{-1}\ln\cosh(\beta m)$.

Remark 1.2. An other way to reach the same conclusion is to derive the consistency relation

$$\int \mathbb{Q}_\beta(\mathrm{d}m)m = \frac{\int \mathbb{Q}_\beta(\mathrm{d}m)\mathrm{e}^{-\beta K m^2/2}\left[\cosh(\beta m)\right]^K \tanh(\beta m)}{\int \mathbb{Q}_\beta(\mathrm{d}m)\mathrm{e}^{-\beta K m^2/2}\left[\cosh(\beta m)\right]^K} \tag{47}$$

for arbitrary K. But then it is clear that this can hold for all K only if $\mathbb{Q}_\beta$ is concentrated on the minimizers of the function $m^2/2 - \beta^{-1}\ln\cosh(\beta m)$, which happen also to solve the equation $m^* = \tanh(\beta m^*)$ so that in the end all is consistent.

2. Random mean field models

The naive analog of the Curie–Weiss Hamiltonian with random couplings would be

$$H_N[\omega](\sigma) = -\frac{1}{2N} \sum_{1\leq i,j\leq N} J_{ij}[\omega]\sigma_i\sigma_j \tag{48}$$

for, say, J_{ij} some family of i.i.d. random variables. Thus, we must estimate

$$\mathbb{P}[\max_{\sigma} H_N(\sigma) \geq CN].$$

But,

$$\begin{aligned} \mathbb{P}[\max_{\sigma} H_N(\sigma) \geq CN] &\leq \sum_{\sigma\in\mathcal{S}_N} \mathbb{P}[H_N(\sigma) \geq CN] \\ &= \sum_{\sigma\in\mathcal{S}_N} \inf_{t\geq 0} \mathrm{e}^{-tCN} \mathbb{E}\mathrm{e}^{t\frac{1}{2N}\sum_{i,j\in\Lambda_N\times\Lambda_N} J_{ij}[\omega]\sigma_i\sigma_j} \\ &= \sum_{\sigma\in\mathcal{S}_N} \inf_{t\geq 0} \mathrm{e}^{-tCN} \prod_{i,j\in\Lambda_N\times\Lambda_N} \mathbb{E}\mathrm{e}^{t\frac{1}{2N} J_{ij}[\omega]\sigma_i\sigma_j} \end{aligned} \tag{49}$$

where we assumed that the exponential moments of J_{ij} exist. A standard estimate then shows that, for some constant c, $\mathbb{E}\mathrm{e}^{t\frac{1}{2N} J_{ij}[\omega]\sigma_i\sigma_j} \leq \mathrm{e}^{c\frac{t^2}{2N^2}}$, and so

$$\mathbb{P}[\max_{\sigma} H_N(\sigma) \geq CN] \leq 2^N \inf_{t\geq 0} \mathrm{e}^{-tCN}\mathrm{e}^{ct^2/2} \leq 2^N \mathrm{e}^{-\frac{C^2N^2}{2c}} \tag{50}$$

which tends to zero with N. Thus, our Hamiltonian is never of order N, but at best of order $\sqrt{N}$. The proper Hamiltonian for what is called the *Sherrington–Kirkpatrick model* (or short SK-model), is thus

$$H_N^{\mathrm{SK}} \equiv -\frac{1}{\sqrt{2N}} \sum_{i,j\in\Lambda_N\times\Lambda_N} J_{ij}\sigma_i\sigma_j \tag{51}$$

where the random variables $J_{ij} = J_{ji}$ are i.i.d. for $i \leq j$ with mean zero (or at most $J_0 N^{-1/2}$) and variance normalized to one for $i \neq j$ and to two for $i = j$. In its original, and mostly considered, form, the distribution is taken to be Gaussian.

This model was introduced by Sherrington and Kirkpatrick in 1976 [27] as an attempt to furnish a simple, solvable mean-field model for the then newly discovered class of materials called *spin-glasses.*

2.1. Gaussian process

This point of view consists of regarding the Hamiltonian (51) as a *Gaussian random process* indexed by the set $\mathcal{S}_N$, i.e., by the N-dimensional hypercube. Its covariance function

$$\begin{aligned} \mathrm{cov}(H_N(\sigma), H_N(\sigma')) &= \frac{1}{2N} \sum_{1\leq i,j,l,k\leq N} \mathbb{E}J_{ij}J_{kl}\sigma_i\sigma_j\sigma'_k\sigma'_l \\ &= \frac{1}{N} \sum_{1\leq i,j\leq N} \sigma_i\sigma'_i\sigma_j\sigma'_j = NR_N(\sigma,\sigma')^2 \end{aligned} \tag{52}$$

where $R_N(\sigma,\sigma') \equiv N^{-1}\sum_{i=1}^N \sigma_i\sigma'_i$ is usually called the *overlap* between the two configurations σ and σ'. It is closely related to the *Hamming distance*, $d_{\mathrm{HAM}}(\sigma,\sigma') \equiv \#(i \leq N : \sigma_i \neq \sigma'_i)$, namely, $R_N(\sigma,\sigma') = (1 - 2N^{-1} d_{\mathrm{HAM}}(\sigma,\sigma'))$.

A more general class of models is obtained by choosing

$$\mathrm{cov}(H_N(\sigma), H_N(\sigma')) = N\xi(R_N(\sigma,\sigma')) \tag{53}$$

normalized such that $\xi(1) = 1$. In particular, the p-spin SK-models are obtained by choosing $\xi(x) = |x|^p$:

$$H_N^{p\text{-SK}}(\sigma) = \frac{-1}{\sqrt{N^{p-1}}} \sum_{1 \le i_1, \dots, i_p \le N} J_{i_1 \dots i_p} \sigma_{i_1} \dots \sigma_{i_p}. \tag{54}$$

As we will see later, the difficulties in studying the statistical mechanics of these models is closely linked to the understanding of the extremal properties of the corresponding random processes. While Gaussian processes have been heavily analyzed in the mathematical literature (see, e.g., [22, 1]), the known results were not enough to recover the heuristic results obtained in the physics literature.

2.2. The generalized random energy models

Further classes of models are obtained by choosing different distances on the hypercube.

Of special interest to us is the *lexicographic distance*. Here

$$d_N(\sigma, \tau) \equiv N^{-1} \left(\min(i : \sigma_i \neq \tau_i) - 1\right) \tag{55}$$

is analogous to the overlap $R_N(\sigma, \tau)$. The corresponding Gaussian processes are then characterized by covariances given by

$$\operatorname{cov}(H_N(\sigma), H_N(\tau)) = N A(d_N(\sigma, \tau)) \tag{56}$$

where A can be chosen to be any non-decreasing function on $[0, 1]$, and can be thought of as a probability distribution function. The choice of the lexicographic distance entails some peculiar features. First, this distance is an *ultrametric*, i.e., for any three configurations σ, τ, ρ,

$$d_N(\sigma, \tau) = \min\left(d_N(\sigma, \rho), d_N(\tau, \rho)\right). \tag{57}$$

3. The simplest example: the REM

We set

$$H_N(\sigma) = -\sqrt{N} X_\sigma \tag{58}$$

where X_σ, $\sigma \in \mathcal{S}_N$, are 2^N i.i.d. standard normal random variables.

3.1. Ground-state energy and free energy

Lemma 3.1. *The family of random variables introduced above satisfies*

$$\lim_{N \uparrow \infty} \max_{\sigma \in \mathcal{S}_N} N^{-1/2} X_\sigma = \sqrt{2 \ln 2} \tag{59}$$

both almost surely and in mean.

Proof. Since everything is independent,

$$\mathbb{P}\left[\max_{\sigma \in \mathcal{S}_N} X_\sigma \le u\right] = \left(1 - \frac{1}{\sqrt{2\pi}} \int_u^\infty e^{-x^2/2} dx\right)^{2^N} \tag{60}$$

and we just need to know how to estimate the integral appearing here. This is something we should get used to quickly, as it will occur all over the place. It will always be done using the fact that, for $u > 0$,

$$\frac{1}{u}\mathrm{e}^{-u^2/2}\left(1-2u^{-2}\right) \leq \int_u^\infty \mathrm{e}^{-x^2/2}\mathrm{d}x \leq \frac{1}{u}\mathrm{e}^{-u^2/2}. \tag{61}$$

Now we define a function $u_N(x)$ by

$$\frac{2^N}{\sqrt{2\pi}}\int_{u_N(x)}^\infty \mathrm{e}^{-z^2/2}\mathrm{d}z = \mathrm{e}^{-x}. \tag{62}$$

It turns out that (for $x > -\ln N/\ln 2$)

$$u_N(x) = \sqrt{2N\ln 2} + \frac{x}{\sqrt{2N\ln 2}} - \frac{\ln(N\ln 2)+\ln 4\pi}{2\sqrt{2N\ln 2}} + o(1/\sqrt{N}). \tag{63}$$

Thus

$$\mathbb{P}\left[\max_{\sigma\in\mathcal{S}_N} X_\sigma \leq u_N(x)\right] = \left(1-2^{-N}\mathrm{e}^{-x}\right)^{2^N} \to \mathrm{e}^{-\mathrm{e}^{-x}}. \tag{64}$$

In other terms, the random variable $u_N^{-1}\left(\max_{\sigma\in\mathcal{S}_N} X_\sigma\right)$ converges in distribution to a random variable with double-exponential distribution. □

Next we turn to the analysis of the partition function

$$Z_{\beta,N} \equiv 2^{-N}\sum_{\sigma\in\mathcal{S}_N} \mathrm{e}^{\beta\sqrt{N}X_\sigma}. \tag{65}$$

A first guess would be that a *law of large numbers* might hold, implying that $Z_{\beta,N} \sim \mathbb{E}Z_{\beta,N}$, and hence

$$\lim_{N\uparrow\infty}\Phi_{\beta,N} = \lim_{N\uparrow\infty}\frac{1}{N}\ln\mathbb{E}Z_{\beta,N} = \frac{\beta^2}{2},\quad \text{a.s.} \tag{66}$$

But this holds only for small enough values of β!

Theorem 3.1. *In the REM,*

$$\lim_{N\uparrow\infty}\mathbb{E}\Phi_{\beta,N} = \begin{cases} \frac{\beta^2}{2}, & \textit{for } \beta\leq\beta_c,\\ \frac{\beta_c^2}{2}+(\beta-\beta_c)\beta_c, & \textit{for } \beta\geq\beta_c, \end{cases} \tag{67}$$

where $\beta_c = \sqrt{2\ln 2}$.

Proof. We use the method of truncated second moments.

We will first derive an upper bound for $\mathbb{E}\Phi_{\beta,N}$. Note first that by Jensen's inequality, $\mathbb{E}\ln Z \leq \ln\mathbb{E}Z$, and thus

$$\mathbb{E}\Phi_{\beta,N} \leq \frac{\beta^2}{2}. \tag{68}$$

On the other hand we have that

$$\mathbb{E}\frac{\mathrm{d}}{\mathrm{d}\beta}\Phi_{\beta,N} = N^{-1/2}\mathbb{E}\frac{\mathbb{E}_\sigma X_\sigma \mathrm{e}^{\beta\sqrt{N}X_\sigma}}{Z_{\beta,N}} \leq N^{-1/2}\mathbb{E}\max_{\sigma\in\mathcal{S}_N} X_\sigma \leq \beta\sqrt{2\ln 2}(1+C/N) \tag{69}$$

for some constant C. Combining (68) and (69), we deduce that

$$\mathbb{E}\Phi_{\beta,N} \leq \inf_{\beta_0\geq 0}\begin{cases}\frac{\beta^2}{2}, & \text{for } \beta\leq\beta_0,\\ \frac{\beta_0^2}{2}+(\beta-\beta_0)\sqrt{2\ln 2}(1+C/N), & \text{for } \beta\geq\beta_0.\end{cases} \tag{70}$$

It is easy to see that the infimum is taken at $\beta_0=\sqrt{2\ln 2}$. This shows that the right-hand side of (67) is an upper bound.

It remains to obtain a matching lower bound. Note that, since $\frac{\mathrm{d}^2}{\mathrm{d}\beta^2}\Phi_{\beta,N}\geq 0$, the slope of $\Phi_{\beta,N}$ is non-decreasing, so that the theorem will be proven if we can show that $\Phi_{\beta,N}\to\beta^2/2$ for all $\beta<\sqrt{2\ln 2}$, i.e., that the law of large numbers holds up to this value of β. A natural idea to prove this is to estimate the variance of the partition function. One would compute

$$\begin{aligned}\mathbb{E}Z_{\beta,N}^2 &= \mathbb{E}_\sigma\mathbb{E}_{\sigma'}\mathbb{E}\mathrm{e}^{\beta\sqrt{N}(X_\sigma+X_{\sigma'})}\\ &= 2^{-2N}\left(\sum_{\sigma\neq\sigma'}\mathrm{e}^{N\beta^2}+\sum_\sigma \mathrm{e}^{2N\beta^2}\right)\\ &= \mathrm{e}^{N\beta^2}\left[(1-2^{-N})+2^{-N}\mathrm{e}^{N\beta^2}\right]\end{aligned} \tag{71}$$

where all we used is that for $\sigma\neq\sigma'$, X_σ and $X_{\sigma'}$ are independent. The second term in the square brackets is exponentially small if and only if $\beta^2<\ln 2$. For such values of β we have that

$$\begin{aligned}\mathbb{P}\left[\left|\ln\frac{Z_{\beta,N}}{\mathbb{E}Z_{\beta,N}}\right|>\epsilon N\right] &= \mathbb{P}\left[\frac{Z_{\beta,N}}{\mathbb{E}Z_{\beta,N}}<\mathrm{e}^{-\epsilon N} \text{ or } \frac{Z_{\beta,N}}{\mathbb{E}Z_{\beta,N}}>\mathrm{e}^{\epsilon N}\right]\\ &\leq \mathbb{P}\left[\left(\frac{Z_{\beta,N}}{\mathbb{E}Z_{\beta,N}}-1\right)^2>\left(1-\mathrm{e}^{-\epsilon N}\right)^2\right]\\ &\leq \frac{\mathbb{E}Z_{\beta,N}^2/(\mathbb{E}Z_{\beta,N})^2-1}{(1-\mathrm{e}^{-\epsilon N})^2}\\ &\leq \frac{2^{-N}+2^{-N}\mathrm{e}^{N\beta^2}}{(1-\mathrm{e}^{-\epsilon N})^2}\end{aligned} \tag{72}$$

which is more than enough to get (66). But of course this does not correspond to the critical value of β claimed in the proposition!

Instead of the second moment of Z one should compute a truncated version of it, namely, for $c\geq 0$,

$$\widetilde{Z}_{\beta,N}(c)\equiv\mathbb{E}_\sigma\mathrm{e}^{\beta\sqrt{N}X_\sigma}\mathbb{1}_{X_\sigma<c\sqrt{N}}. \tag{73}$$

An elementary computation using (61) shows that, if $c > \beta$, then

$$\mathbb{E}\widetilde{Z}_{\beta,N}(c) = \mathrm{e}^{\frac{\beta^2 N}{2}} \left(1 - \frac{\mathrm{e}^{-N\beta^2/2}}{\sqrt{2\pi N}(c-\beta)}(1 + O(1/N))\right) \tag{74}$$

so that such a truncation does not alter the averaged partition function. Now compute the average of the square of the truncated partition function:

$$\mathbb{E}\widetilde{Z}^2_{\beta,N}(c) = (1 - 2^{-N})[\mathbb{E}\widetilde{Z}_{\beta,N}(c)]^2 + 2^{-N}\mathbb{E}\mathrm{e}^{\beta\sqrt{N}2X_\sigma}\mathbb{1}_{X_\sigma < c\sqrt{N}}) \tag{75}$$

where

$$\mathbb{E}\,\mathrm{e}^{2\beta\sqrt{N}X_\sigma}\mathbb{1}_{X_\sigma<c\sqrt{N}} = \begin{cases} \mathrm{e}^{2\beta^2 N}, & \text{if } 2\beta < c, \\ 2^{-N}\frac{\mathrm{e}^{2c\beta N - \frac{c^2 N}{2}}}{(2\beta-c)\sqrt{2\pi N}}, & \text{otherwise.} \end{cases} \tag{76}$$

Combined with (74) this implies that, for $c/2 < \beta < c$,

$$\frac{2^{-N}\mathbb{E}\,\mathrm{e}^{2\beta\sqrt{N}X_\sigma}\mathbb{1}_{X_\sigma<c\sqrt{N}}}{\left(\mathbb{E}\,\widetilde{Z}_{N,\beta}\right)^2} = \frac{\mathrm{e}^{-N(c-\beta)^2 - N(2\ln 2 - c^2)/2}}{(2\beta - c)\sqrt{N}}. \tag{77}$$

Therefore, for all $c < \sqrt{2\ln 2}$, and all $\beta < c$,

$$\mathbb{E}\left[\frac{\widetilde{Z}_{\beta,N}(c) - \mathbb{E}\widetilde{Z}_{\beta,N}(c)}{\mathbb{E}\widetilde{Z}_{\beta,N}(c)}\right]^2 \le \mathrm{e}^{-Ng(c,\beta)} \tag{78}$$

with $g(c,\beta) > 0$. Thus, by Chebyshev's inequality

$$\mathbb{P}\left[|\widetilde{Z}_{\beta,N}(c) - \mathbb{E}\widetilde{Z}_{\beta,N}(c)| > \delta\mathbb{E}\widetilde{Z}_{\beta,N}(c)\right] \le \delta^{-2}\mathrm{e}^{-Ng(c,\beta)} \tag{79}$$

and so,

$$\lim_{N\uparrow\infty}\frac{1}{N}\mathbb{E}\ln\widetilde{Z}_{\beta,N}(c) = \lim_{N\uparrow\infty}\frac{1}{N}\ln\mathbb{E}\widetilde{Z}_{\beta,N}(c) \tag{80}$$

for all $\beta < c < \sqrt{2\ln 2} = \beta_c$. But this implies that for all $\beta < \beta_c$, we can choose c such that

$$\lim_{N\uparrow\infty}\frac{1}{N}\ln\mathbb{E}Z_{\beta,N} \ge \lim_{N\uparrow\infty}\frac{1}{N}\ln\mathbb{E}\widetilde{Z}_{\beta,N}(c) = \frac{\beta^2}{2}. \tag{81}$$

This proves the theorem. □

3.2. Fluctuations and limit theorems

Theorem 3.2. *Let $\mathcal{P}$ denotes the Poisson point process on $\mathbb{R}$ with intensity measure $\mathrm{e}^{-x}\mathrm{d}x$. Then, in the REM, with $\alpha = \beta/\sqrt{2\ln 2}$, if $\beta > \sqrt{2\ln 2}$,*

$$\mathrm{e}^{-N[\beta\sqrt{2\ln 2} - \ln 2] + \frac{\alpha}{2}[\ln(N\ln 2) + \ln 4\pi]}Z_{\beta,N} \xrightarrow{\mathcal{D}} \int_{-\infty}^{\infty}\mathrm{e}^{\alpha z}\mathcal{P}(\mathrm{d}z) \tag{82}$$

and

$$N\left(\Phi_{\beta,N} - \mathbb{E}\Phi_{\beta,N}\right) \xrightarrow{\mathcal{D}} \ln\int_{-\infty}^{\infty}\mathrm{e}^{\alpha z}\mathcal{P}(\mathrm{d}z) - \mathbb{E}\ln\int_{-\infty}^{\infty}\mathrm{e}^{\alpha z}\mathcal{P}(\mathrm{d}z). \tag{83}$$

Proof. Basically, the idea is very simple. We expect that for β large, the partition function will be dominated by the configurations σ corresponding to the largest values of X_σ. Thus we split $Z_{\beta,N}$ carefully into

$$Z_{N,\beta}^x \equiv \mathbb{E}_\sigma e^{\beta\sqrt{N}X_\sigma} \mathbb{1}_{\{X_\sigma \leq u_N(x)\}} \tag{84}$$

and $Z_{\beta,N} - Z_{\beta,N}^x$. Let us first consider the last summand. We introduce the random variable

$$\mathcal{W}_N(x) = Z_{\beta,N} - Z_{\beta,N}^x = 2^{-N} \sum_{\sigma\in\mathcal{S}_N} e^{\beta\sqrt{N}X_\sigma} \mathbb{1}_{\{X_\sigma > u_N(x)\}} \tag{85}$$

It is convenient to rewrite this as (we ignore the sub-leading corrections to $u_N(x)$ and only keep the explicit part of (63))

$$\begin{aligned}
\mathcal{W}_N(x) &= 2^{-N} \sum_{\sigma\in\mathcal{S}_N} e^{\beta\sqrt{N}u_N(u_N^{-1}(X_\sigma))} \mathbb{1}_{\{u_N^{-1}(X_\sigma)>x\}} \\
&= e^{N(\beta\sqrt{2\ln 2}-\ln 2)-\frac{\alpha}{2}[\ln(N\ln 2)+\ln 4\pi]} \sum_{\sigma\in\mathcal{S}_N} e^{\alpha u_N^{-1}(X_\sigma)} \mathbb{1}_{\{u_N^{-1}(X_\sigma)>x\}} &(86)\\
&\equiv \frac{1}{C(\beta,N)} \sum_{\sigma\in\mathcal{S}_N} e^{\alpha u_N^{-1}(X_\sigma)} \mathbb{1}_{\{u_N^{-1}(X_\sigma)>x\}} &(87)
\end{aligned}$$

where

$$\alpha \equiv \beta/\sqrt{2\ln 2} \tag{88}$$

and $C(b,N)$ is defined through the last identity. The key to most of what follows relies on the famous result on the convergence of the extreme value process to a Poisson point process (for a proof see, e.g., [21]):

Theorem 3.3. *Let $\mathcal{P}_N$ be the point process on $\mathbb{R}$ given by*

$$\mathcal{P}_N \equiv \sum_{\sigma\in\mathcal{S}_N} \delta_{u_N^{-1}(X_\sigma)}. \tag{89}$$

Then $\mathcal{P}_N$ converges weakly to a Poisson point process on $\mathbb{R}$ with intensity measure $e^{-x}dx$.

Clearly, the weak convergence of $\mathcal{P}_N$ to $\mathcal{P}$ implies convergence in law of the right-hand side of (86), provided that $e^{\alpha x}$ is integrable on $[x,\infty)$ w.r.t. the Poisson point process with intensity e^{-x}. This is, in fact, never a problem: the Poisson point process has almost surely support on a finite set, and therefore $e^{\alpha x}$ is always a.s. integrable. Note, however, that for $\beta \geq \sqrt{2\ln 2}$ the mean of the integral is infinite, indicating the passage to the low-temperature regime.

Lemma 3.2. *Let $\mathcal{W}_N(x), \alpha$ be defined as above, and let $\mathcal{P}$ be the Poisson point process with intensity measure $e^{-z}dz$. Then*

$$C(\beta,N)\mathcal{W}_N(x) \xrightarrow{\mathcal{D}} \int_x^\infty e^{\alpha z}\mathcal{P}(dz). \tag{90}$$

Next we show that the contribution of the truncated part of the partition function is negligible compared to this contribution. For this it is enough to compute the mean values

$$\begin{aligned}\mathbb{E}Z^x_{\beta,N} &\sim e^{N\beta^2/2}\int_{-\infty}^{u_N(x)-\beta\sqrt{N}}\frac{dz}{\sqrt{2\pi}}e^{-\frac{z^2}{2}}\\ &\sim e^{N\beta^2/2}\frac{e^{-(u_N(x)-\beta\sqrt{N})^2/2}}{\sqrt{2\pi}(\beta\sqrt{N}-u_N(x))}\\ &\sim \frac{2^{-N}e^{x(\alpha-1)}}{\alpha-1}e^{N(\beta\sqrt{2\ln 2}-\ln 2)-\frac{\alpha}{2}[\ln(N\ln 2)+\ln 4\pi]}\\ &= \frac{e^{x(\alpha-1)}}{\alpha-1}\frac{1}{C(\beta,N)}\end{aligned} \tag{91}$$

so that

$$C(\beta,N)\mathbb{E}Z^x_{\beta,N}\sim\frac{e^{x(\alpha-1)}}{\alpha-1}$$

which tends to zero as $x\downarrow-\infty$, and so $C(\beta,N)\mathbb{E}Z^x_{\beta,N}$ converges to zero in probability. The assertions of Theorem 3.2 follow. □

3.3. The Gibbs measure

We now want to describe the Gibbs measure of the REM. A nice way to do this consists in mapping the hypercube to the interval $(0,1]$ via

$$\mathcal{S}_N\ni\sigma\to r_N(\sigma)\equiv 1-\sum_{i=1}^{N}(1-\sigma_i)2^{-i-1}\in(0,1]. \tag{92}$$

Define the pure point measure $\tilde{\mu}_{\beta,N}$ on $(0,1]$ by

$$\tilde{\mu}_{\beta,N}\equiv\sum_{\sigma\in\mathcal{S}_N}\delta_{r_N(\sigma)}\mu_{\beta,N}(\sigma). \tag{93}$$

Our results will be expressed in terms of the convergence of these measures. It will be understood in the sequel that the space of measures on $(0,1]$ is equipped with the topology of weak convergence, and all convergence results hold with respect to this topology.

Let us introduce the Poisson point process $\mathcal{R}$ on the strip $(0,1]\times\mathbb{R}$ with intensity measure $\frac{1}{2}dy\times e^{-x}dx$. If (Y_k,X_k) denote the atoms of this process, define a new point process $\mathcal{M}_\alpha$ on $(0,1]\times(0,1]$ whose atoms are (Y_k,w_k), where

$$w_k\equiv\frac{e^{\alpha X_k}}{\int\mathcal{R}(dy,dx)e^{\alpha x}} \tag{94}$$

for $\alpha>1$.

With this notation we have that:

Theorem 3.4. *If $\beta > \sqrt{2\ln 2}$, with $\alpha = \beta/\sqrt{2\ln 2}$, then*

$$\tilde{\mu}_{\beta,N} \xrightarrow{\mathcal{D}} \tilde{\mu}_\beta \equiv \int_{(0,1]\times(0,1]} \mathcal{M}_\alpha(\mathrm{d}y, \mathrm{d}w)\delta_y w. \tag{95}$$

Proof. With $u_N(x)$ defined in (63), we define the point process $\mathcal{R}_N$ on $(0,1]\times\mathbb{R}$ by

$$\mathcal{R}_N \equiv \sum_{\sigma\in\mathcal{S}_N} \delta_{(r_N(\sigma), u_N^{-1}(X_\sigma))}. \tag{96}$$

A standard result of extreme value theory (see [21], Theorem 5.7.2) is easily adapted to yield that

$$\mathcal{R}_N \xrightarrow{\mathcal{D}} \mathcal{R}, \quad \text{as } N\uparrow\infty. \tag{97}$$

Note that

$$\mu_{\beta,N}(\sigma) = \frac{\mathrm{e}^{\alpha u_N^{-1}(X_\sigma)}}{\sum_\sigma \mathrm{e}^{\alpha u_N^{-1}(X_\sigma)}} = \frac{\mathrm{e}^{\alpha u_N^{-1}(X_\sigma)}}{\int \mathcal{R}_N(\mathrm{d}y,\mathrm{d}x)\mathrm{e}^{\alpha x}}. \tag{98}$$

Since $\int \mathcal{R}_N(\mathrm{d}y,\mathrm{d}x)\mathrm{e}^{\alpha x} < \infty$ a.s., we can define the point process

$$\mathcal{M}_{\alpha,N} \equiv \sum_{\sigma\in\mathcal{S}_N} \delta_{\left(r_N(\sigma), \frac{\exp(\alpha u_N^{-1}(X_\sigma))}{\int \mathcal{R}_N(\mathrm{d}y,\mathrm{d}x)\exp(\alpha x)}\right)} \tag{99}$$

on $(0,1]\times(0,1]$. Then

$$\tilde{\mu}_{\beta,N} = \int \mathcal{M}_{\alpha,N}(\mathrm{d}y,\mathrm{d}w)\delta_y w. \tag{100}$$

The only non-trivial point in the convergence proof is to show that the contribution to the partition functions in the denominator from atoms with $u_N(X_\sigma) < x$ vanishes as $x\downarrow-\infty$. But this we have shown already in the proof of Theorem 3.2. Standard arguments now imply that first $\mathcal{M}_{\alpha,N} \xrightarrow{\mathcal{D}} \mathcal{M}_\alpha$, and, consequently, (95) holds. □

The measure $\tilde{\mu}_\beta$ is closely related to a classical object in probability theory, the α-stable Lévy subordinator. To see this, denote by

$$\mathcal{Z}_\alpha(t) \equiv \int_0^t \int_{-\infty}^{+\infty} \mathrm{e}^{\alpha x}\mathcal{R}(\mathrm{d}y,\mathrm{d}x). \tag{101}$$

Clearly, the probability distribution function associated to the measure $\tilde{\mu}_\beta$ satisfies, for $t\in[0,1]$,

$$\int_0^t \mu_\beta(\mathrm{d}x) = \frac{\mathcal{Z}_\alpha(t)}{\mathcal{Z}_\alpha(1)}. \tag{102}$$

Lemma 3.3. *For any $\alpha > 1$, the stochastic process $\mathcal{Z}_\alpha(t)$ is the $1/\alpha$-stable Lévy process (subordinator) with Lévy measure $y^{-1/\alpha-1}\mathrm{d}y$.*

Proof. There are various ways to prove this result. Note first that the process has independent, identically distributed increments. It is then enough, e.g., to compute the Laplace transform of the one-dimensional distribution, i.e., one shows that

$$\mathbb{E}e^{-\lambda \mathcal{Z}_\alpha(t)} = \exp\left(\int_0^\infty (e^{-\lambda y} - 1) y^{-1/\alpha - 1} dy\right). \tag{103}$$

This can be done by elementary calculus and is left to the reader. □

Let us note that it is not difficult to show that the process

$$\sum_{\sigma \in \mathcal{S}_N} e^{\alpha u_N^{-1}(X_\sigma)} \mathbb{1}_{r_N(\sigma) \le t} \tag{104}$$

converges in the Skorokhod J_1-topology to the α-stable Lévy subordinator. Hence the distribution function of the measures $\tilde{\mu}_{\beta,N}$ to $\tilde{\mu}_\beta$ can be interpreted in the sense of the corresponding convergence of their distribution functions as stochastic processes on Skorokhod space.

3.4. The asymptotic model

In the case $\beta > 1$, we can readily interpret our asymptotic results in terms of a new statistical mechanics model *for the infinite volume limit.* It has the following ingredients:

- State space: $\mathbb{N}$;
- Random Hamiltonian: $x : \mathbb{N} \to \mathbb{R}$, where x_i is the ith atom of the Poisson process $\mathcal{P}$;
- Temperature: $1/\alpha = \beta_c/\beta$;
- Partition function: $\mathcal{Z}_\alpha = \sum_{i \in \mathbb{N}} e^{\alpha x_i}$;
- Gibbs measure: $\hat{\mu}_\sigma(i) = \mathcal{Z}_\alpha e^{\alpha x_i}$.

Our convergence results so far can be interpreted in terms of this model as follows:

- The partition function of the REM converges, after division by $\exp(\beta\sqrt{N} u_N(0))$, to $\mathcal{Z}_\alpha$;
- If we map the Gibbs measure $\hat{\mu}_\alpha$ to the unit interval via

$$\hat{\mu}_\alpha \to \check{\mu}_\alpha = \sum_{i \in \mathbb{N}} \delta_{U_i} \hat{\mu}_\alpha(i), \tag{105}$$

  where U_i, $i \in \mathbb{N}$, is a family of independent random variables that are distributed uniformly on the interval $[0, 1]$, then $\check{\mu}_\sigma$ has the same distribution as $\tilde{\mu}_\beta$.

This is a reasonably satisfactory picture. What is lacking, however, is a proper reflection of the geometry of the Gibbs measure on the hypercube. Clearly the convergence of the embedded measures on the unit interval is insufficient to capture this.

In the next section we will see how this should be incorporated.

3.5. The replica overlap

If we want to discuss the geometry of Gibbs measures, we first must decide on how to measure distance on the hypercube. The most natural one is the Hamming distance, or its counterpart, the *overlap*, $R_N(\sigma,\sigma')$. Of course we might also want to use the ultrametric distance d_N, and we will comment on this later.

To describe the geometry of $\mu_{\beta,N}$, we may now ask how much mass one finds in a neighborhood of a point $\sigma \in \mathcal{S}_N$, i.e., we may define

$$\phi_{\beta,N}(\sigma,t) \equiv \mu_{\beta,N}\left(R_N(\sigma,\sigma') > t\right). \tag{106}$$

Clearly this defines a probability distribution on $[-1,1]$ (as we will see, in reality it will give zero mass to the negative numbers).

Of course these 2^N functions are not very convenient. The first reflex will be to average over the Gibbs measures, i.e. to define

$$\psi_{\beta,N}(t) \equiv \sum_{\sigma\in\mathcal{S}_N} \mu_{\beta,N}(\sigma)\phi_{\beta,N}(\sigma,t) = \mu_{\beta,N}[\omega] \otimes \mu_{\beta,N}[\omega]\left(R_N(\sigma,\sigma') \in \mathrm{d}z\right). \tag{107}$$

The following theorem expresses the limit of ψ in the form we would expect, namely in terms of the asymptotic model.

Theorem 3.5. *For all $\beta > \sqrt{2\ln 2}$*

$$\psi_{\beta,N}(t) \xrightarrow{\mathcal{D}} \begin{cases} 0, & \text{if } t < 0, \\ 1 - \sum_{i\in\mathbb{N}} \hat{\mu}_\alpha(i)^2, & \text{if } 0 \le t < 1, \\ 1, & \text{if } t \ge 1. \end{cases} \tag{108}$$

Proof. The only new thing we have to show is that the function ψ increases only at 0 and at one; that is to say, we have to show that with probability tending to one the overlap R_N takes on only the values one or zero.

We write for any $\Delta \subset [-1,1]$

$$\psi_{\beta,N}(\Delta) \equiv \mu_{\beta,N}^{\otimes 2}(R_N \in \Delta) \equiv \psi_{\beta,N}(\Delta) = Z_{\beta,N}^{-2}\mathbb{E}_\sigma\mathbb{E}_{\sigma'} \sum_{\substack{t\in\Delta \\ R_N(\sigma,\sigma')=t}} \mathrm{e}^{\beta\sqrt{N}(X_\sigma+X_{\sigma'})}. \tag{109}$$

We use the truncation introduced in Section 3.2. Note first that, for any interval Δ,

$$\left| \psi_{\beta,N}(\Delta) - Z_{\beta,N}^{-2}\mathbb{E}_\sigma\mathbb{E}_{\sigma'} \sum_{\substack{t\in\Delta \\ R_N(\sigma,\sigma')=t}} \mathbb{1}_{X_\sigma,X_{\sigma'}\ge u_N(x)} \mathrm{e}^{\beta\sqrt{N}(X_\sigma+X_{\sigma'})} \right| \le \frac{2Z_{\beta,N}^x}{Z_{\beta,N}}. \tag{110}$$

We have already seen in the proof of Theorem 3.2 that the right-hand side of (110) tends to zero in probability, as first $N \uparrow \infty$ and then $x \downarrow -\infty$. On the other hand, for $t \neq 1$,

$$\begin{aligned}&\mathbb{P}\left[\exists\sigma,\sigma' : R_N(\sigma,\sigma') = t : X_\sigma > u_N(x) \wedge X_{\sigma'} > u_N(x)\right] \\ &\leq \mathbb{E}_{\sigma\sigma'} 1\!\!1_{R_N(\sigma,\sigma')=t}\, 2^{2N}\mathbb{P}\left[X_\sigma > u_N(x)\right]^2 = \frac{2\mathrm{e}^{-\phi(t)N}\mathrm{e}^{-2x}}{\sqrt{2\pi N}\sqrt{1-t^2}}\end{aligned} \tag{111}$$

by the definition of $u_N(x)$ (see (62)). This implies again that any interval $\Delta \subset [-1,1) \cup [-1,0)$ has zero mass. To conclude the proof it is enough to compute $\psi_{\beta,N}(1)$. Clearly

$$\psi_{\beta,N}(1) = \frac{2^{-N} Z_{2\beta,N}}{Z^2_{\beta,N}}. \tag{112}$$

By Theorem 3.2, one sees easily that

$$\psi_{\beta,N}(1) \xrightarrow{\mathcal{D}} \frac{\int \mathrm{e}^{2\alpha z}\mathcal{P}(\mathrm{d}z)}{\left(\int \mathrm{e}^{\alpha z}\mathcal{P}(\mathrm{d}z)\right)^2}. \tag{113}$$

It is now very easy to conclude the proof. □

The empirical distance distribution. Rather than just taking the mean of the functions $\phi_{\beta,N}$, we can naturally define their empirical distribution. It is natural to do this biased with their importance in the Gibbs measures. This lead to the object

$$\mathcal{K}_{\beta,N} \equiv \sum_{\sigma\in\mathcal{S}_N} \mu_{\beta,N}(\sigma)\delta_{\phi_{\beta,N}(\sigma,\cdot)}. \tag{114}$$

Here we think of the δ-measure as a measure on probability distribution functions, respectively probability measures on $[-1,1]$, and to $\mathcal{K}_{\beta,N}$ as a random probability measure on the same space. This measure carries a substantial amount of information on the geometry of the Gibbs measure and is in fact the fundamental object to study.

Note that we can define an analogous object in the asymptotic model. We just have to decide on how to measure distance, or overlap, between points in $\mathbb{N}$. In view of the results above, the natural choice is to say that the overlap between a point and itself is one, and is zero between different points. Then set

$$\mathcal{K}_\alpha = \sum_{i\in\mathbb{N}} \hat{\mu}_\alpha(i)\delta_{(1-\hat{\mu}_\alpha(i))\mathbf{1}\{\cdot\in[0,1)\}+\hat{\mu}_\alpha(i)\mathbf{1}\{\cdot\geq 1\}}. \tag{115}$$

A fairly simple to prove extension of Theorem 3.5 gives the strongest link between the REM and the asymptotic model.

Theorem 3.6. *With the standard relation between α and β,*

$$\mathcal{K}_{\beta,N} \to \mathcal{K}_\alpha, \tag{116}$$

where the convergence is in distribution with respect to weak topology of measures on the space of distribution functions equipped with the weak topology.

4. Derrida's generalized random energy models

We will now turn to the investigation of the second class of Gaussian models we have mentioned above, namely Gaussian processes whose covariance is a function of the lexicographic distance on the hypercube (see (56)). B. Derrida introduced these models in the case where A is a step function with finitely many jumps as a natural generalization of the REM and called it the *generalized random energy model* (GREM)[11, 13, 14, 15]. The presentation below is based on our results obtained in [6, 7, 8].

4.1. The GREM and Poisson cascades

A key in the analysis of the REM was the theory of convergence to Poisson processes of the extreme value statistics of (i.i.d.) random variables. In the GREM, analogous results will be needed in the correlated case.

We assume that A is the distribution function of a measure that is supported on a finite number, n, of points $x_1, \dots, x_n \in [0,1]$, as shown in Figure 1.

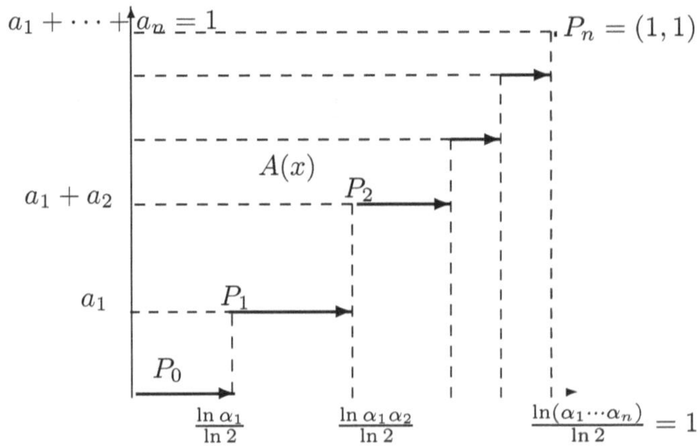

FIGURE 1. The function $A(x)$.

In that case we denote the mass of the atoms x_i by a_i, and we set

$$\ln \alpha_i = (x_i - x_{i-1}) \ln 2, \quad i = 1, \dots, n \tag{117}$$

where $x_0 \equiv 0$. We normalize in such a way that $\sum_{i=1}^n a_i = 1$, and $\prod_{i=1}^n \alpha_i = 2$.

It is very useful that there is an explicit representation of the corresponding process X_σ. We write $\sigma = \sigma_1 \sigma_2 \dots \sigma_n$ where $\sigma_i \in \mathcal{S}_{N \ln \alpha_i / \ln 2}$. Usually we will assume that $x_1 > 0$, $x_n = 1$, and all $a_i > 0$.

Then the Gaussian process X_σ can be constructed from independent standard Gaussian random variables $X_{\sigma_1}, X_{\sigma_1\sigma_2}, \dots, X_{\sigma_1 \dots \sigma_2}$, where $\sigma_i \in \{-1,1\}^{N \ln \alpha_i / \ln 2}$ as

$$X_\sigma \equiv \sqrt{a_1} X_{\sigma_1} + \sqrt{a_2} X_{\sigma_1\sigma_2} + \cdots + \sqrt{a_n} X_{\sigma_1\sigma_2\dots\sigma_n}, \quad \text{if } \sigma = \sigma_1\sigma_2 \dots \sigma_n. \tag{118}$$

4.2. Poisson cascades and extremal processes

Our first concern is to understand the structure of the extremes of such processes. The key ideas are easiest understood in the case where $n = 2$. Let us consider the set, S_x, of σ_1 for which $X_{\sigma_1} \sim \sqrt{a_1 2N \ln \alpha_1 x}$. We know that the cardinality of this set is rather precisely $\alpha_1^{N(1-x)}$ if $x < 1$. Now all the $\alpha_1^{N(1-x)}\alpha_2^N = 2^N \alpha_1^{-xN}$ random variables $X_{\sigma_1\sigma_2}$ with $\sigma_1 \in S_x$ are independent, so that we know that their maximum is roughly $\sqrt{2a_2 N(\ln 2 - x \ln \alpha_1)}$. Hence, the maximum of the X_σ with $\sigma_1 \in S_x$ is

$$\sqrt{a_1 2N \ln \alpha_1 x} + \sqrt{2a_2 N(\ln 2 - x \ln \alpha_1)}. \tag{119}$$

Finally, to determine the overall maximum, it suffices to find the value of x that maximizes this quantity, which turns out to be given by $x^* = \frac{a_1 \ln 2}{\ln \alpha_1}$, provided the constraint $\frac{a_1 \ln 2}{\ln \alpha_1} < 1$ is satisfied. In that case we also find that

$$\sqrt{a_1 2N \ln \alpha_1 x} + \sqrt{2a_2 N(\ln 2 - x^* \ln \alpha_1)} = \sqrt{2 \ln 2}, \tag{120}$$

i.e., the same value as in the REM. On the other hand, if $\frac{a_1 \ln 2}{\ln \alpha_1} > 1$, the maximum is realized by selecting the largest values in the first generation, corresponding to $x = 1$, and then for each of them the extremal members of the corresponding second generation. The value of the maximum is then (roughly)

$$\sqrt{a_1 2N \ln \alpha_1} + \sqrt{2a_2 N \ln \alpha_2} \leq \sqrt{2 \ln 2} \tag{121}$$

where equality holds only in the borderline case $\frac{a_1 \ln 2}{\ln \alpha_1} = 1$, which requires more care. The condition $\frac{a_1 \ln 2}{\ln \alpha_1} < 1$ has a nice interpretation: it simply means that the function $A(x) < x$, for all $x \in (0, 1)$.

In terms of the point processes, the above considerations suggest the following picture (which actually holds true): If $\frac{a_1 \ln 2}{\ln \alpha_1} < 1$, the point process

$$\sum_{\sigma \in \mathcal{S}_N} \delta_{u_N^{-1}(X_\sigma)} \to \mathcal{P} \tag{122}$$

exactly as in the REM, while in the opposite case this process would surely converge to zero. On the other hand, we can construct (in both cases) another point process,

$$\sum_{\sigma = \sigma_1\sigma_2 \in \{-1,+1\}^N} \delta_{\sqrt{a_1} u^{-1}_{\ln \alpha_1, N}(X_{\sigma_1} + \sqrt{a_2} u^{-1}_{\ln \alpha_2, N}(X_{\sigma_1\sigma_2})} \tag{123}$$

where we set

$$u_{\alpha,N}(x) \equiv u_{N \ln \alpha / \ln 2}(x). \tag{124}$$

This point process will converge to a process obtained from a *Poisson cascade*: The process

$$\sum_{\sigma_1 \in \{-1,+1\}^{\ln \alpha_1 N}} \delta_{u^{-1}_{\alpha_1, N}(X_{\sigma_1})} \tag{125}$$

converges to a Poisson point process, and, for any σ_1, so do the point process

$$\sum_{\sigma_2 \in \{-1,+1\}^{\ln \alpha_2 N}} \delta_{u_{\alpha_2,N}^{-1}(X_{\sigma_1\sigma_2})}. \tag{126}$$

Then the two-dimensional point process

$$\sum_{\sigma=\sigma_1\sigma_2 \in \{-1,+1\}^N} \delta_{(u_{\alpha_1,N}^{-1}(X_{\sigma_1}),u_{\alpha_2,N}^{-1}(X_{\sigma_1\sigma_2}))} \tag{127}$$

converges to a *Poisson cascade* in $\mathbb{R}^2$: we place the Poisson process (always with intensity measure $\mathrm{e}^{-x}\mathrm{d}x$) on $\mathbb{R}$, and then, for each atom, we place an independent PPP on the line orthogonal to the first line that passes through that atom. Adding up the atoms of these processes with the right weight yields the limit of the process defined in (123). Now this second point process does not yield the extremal process, as long as the first one exists, i.e., as long as the process (122) does not converge to zero. Interestingly, when we reach the borderline, the process (122) converges to the PPP with intensity $K\mathrm{e}^{-x}\mathrm{d}x$ with $0 < K < 1$, while the cascade process yields points that differ from those of this process only in the sub-leading order.

Having understood the particular case of two levels, it is not difficult to figure out the general situation.

The next result shows which Poisson point processes can emerge.

Theorem 4.1. *Let $0 < a_i < 1$, $\alpha_i > 1$, $i = 1, 2, \ldots, n$ with $\sum_{i=1}^n a_i = 1$. Set $\bar{\alpha} \equiv \prod_{i=1}^n \alpha_i$. Then the point process*

$$\sum_{\sigma=\sigma_1\ldots\sigma_n \in \{-1,+1\}^{N \ln \bar{\alpha}/\ln 2}} \delta_{u_{\bar{\alpha},N}^{-1}(\sqrt{a_1}X_{\sigma_1}+\sqrt{a_2}X_{\sigma_1\sigma_2}+\cdots+\sqrt{a_n}X_{\sigma_1\sigma_2\ldots\sigma_n})} \tag{128}$$

converges weakly to the Poisson point process $\mathcal{P}$ on $\mathbb{R}$ with intensity measure $K\mathrm{e}^{-x}\mathrm{d}x$, $K \in \mathbb{R}$, if and only if, for all $i = 2, 3, \ldots, n$,

$$a_i + a_{i+1} + \cdots + a_n \geq \ln(\alpha_i \alpha_{i+1} \cdots \alpha_n)/\ln \bar{\alpha}. \tag{129}$$

Furthermore, if all inequalities in (129) are strict, then the constant $K = 1$. If some of them are equalities, then $0 < K < 1$.

Remark 4.1. An explicit formula for K can be found in [6].

Remark 4.2. The conditions (129) can be re-expressed in the simple form "$A(x) \leq x$ for all $x \in (0, 1)$".

Theorem 4.2. *Let $\alpha_i \geq 1$, and $\bar{\alpha} \equiv \prod_{i=1}^k \alpha_i$. Let $Y_{\sigma_1}, Y_{\sigma_1\sigma_2}, \ldots, Y_{\sigma_1\ldots\sigma_k}$ be identically distributed random variables, such that the vectors*

$$(Y_{\sigma_1})_{\sigma_1 \in \{-1,1\}^{N \ln \alpha_1/\ln \bar{\alpha}}}, (Y_{\sigma_1\sigma_2})_{\sigma_2 \in \{-1,1\}^{N \ln \alpha_2/\ln \bar{\alpha}}}, \ldots$$
$$\ldots, (Y_{\sigma_1\sigma_2\ldots\sigma_k})_{\sigma_k \in \{-1,1\}^{N \ln \alpha_k/\ln \bar{\alpha}}}$$

are independent. Let $v_{N,1}(x),\dots,v_{N,k}(x)$ *be functions on* $\mathbb{R}$ *such that the following point processes*

$$\begin{aligned}&\sum_{\sigma_1}\delta_{v_{N,1}(Y_{\sigma_1})}\to\mathcal{P}_1\\&\sum_{\sigma_2}\delta_{v_{N,2}(Y_{\sigma_1\sigma_2})}\to\mathcal{P}_2\qquad\forall\sigma_1\\&\dots\dots\dots\dots\dots\dots\dots\dots\\&\sum_{\sigma_k}\delta_{v_{N,k}(Y_{\sigma_1\sigma_2\dots\sigma_k})}\to\mathcal{P}_k\quad\forall\sigma_1,\dots,\sigma_{k-1}\end{aligned}\tag{130}$$

converge weakly to Poisson point processes, $\mathcal{P}_1,\dots,\mathcal{P}_k$*, on* $\mathbb{R}$ *with intensity measures* $K_1\mathrm{e}^{-x}\mathrm{d}x,\dots,K_k\mathrm{e}^{-x}\mathrm{d}x$*, for some constants* $K_1,\dots,K_k$*. Then the point processes on* $\mathbb{R}^k$*,*

$$\mathcal{P}_N^{(k)}\equiv\sum_{\sigma_1}\delta_{v_{N,1}(Y_{\sigma_1})}\sum_{\sigma_2}\delta_{v_{N,2}(Y_{\sigma_1\sigma_2})}\cdots\sum_{\sigma_k}\delta_{v_{N,k}(Y_{\sigma_1\sigma_2\dots\sigma_k})}\to\mathcal{P}^{(k)}\tag{131}$$

converge weakly to point processes $\mathcal{P}^{(k)}$ *on* $\mathbb{R}^k$*, called* Poisson cascades *with* k *levels.*

Poisson cascades are best understood in terms of the following iterative construction. If $k=1$, it is just a Poisson point process on $\mathbb{R}$ with intensity measure $K_1\mathrm{e}^{-x}\mathrm{d}x$. To construct $\mathcal{P}^{(2)}$ on $\mathbb{R}^2$, we place the process $\mathcal{P}^{(1)}$ for $k=1$ on the axis of the first coordinate and through each of its points draw a straight line parallel to the axis of the second coordinate. Then we put on each of these lines independently a Poisson point process with intensity measure $K_2\mathrm{e}^{-x}\mathrm{d}x$. These points on $\mathbb{R}^2$ form the process $\mathcal{P}^{(2)}$. This procedure is now simply iterated k times.

Theorems 4.1 and 4.2 combined tell us which are the different point processes that may be constructed in the GREM.

Theorem 4.3. *Let* $\alpha_i\geq 1$*,* $0<a_i<1$*, such that* $\prod_{i=1}^n\alpha_i=2$*,* $\sum_{i=1}^n a_i=1$*. Let* $J_1,J_2,\dots,J_m\in\mathbb{N}$ *be indices such that* $0=J_0<J_1<J_2<\dots<J_m=n$*. We denote by* $\bar a_l\equiv\sum_{i=J_{l-1}+1}^{J_l}a_i$*,* $\bar\alpha_l\equiv\prod_{i=J_{l-1}+1}^{J_l}\alpha_i$*,* $l=1,2,\dots,m$*, and set*

$$\bar X^{\sigma_1\dots\sigma_{J_{l-1}}}_{\sigma_{J_{l-1}+1}\sigma_{J_{l-1}+2}\cdots\sigma_{J_l}}\equiv\frac{1}{\sqrt{\bar a_l}}\sum_{i=1}^{J_l-J_{l-1}}\sqrt{a_{J_{l-1}+i}}\,X_{\sigma_{J_1}\dots\sigma_{J_{l-1}+i}}.\tag{132}$$

To a partition $J_1,J_2,\dots,J_m$*, we associate the function* A_J *obtained by joining the sequence of straight line segments going from* $(x_{J_i},A(x_{J_i}))$ *to* $(x_{J_{i+1}},A(x_{J_{i+1}}))$*,* $i=0,\dots,m-1$*. A partition is* admissible *if* $A(x)\leq A_J(x)$ *for all* $x\in[0,1]$*. Then, for any admissible partition, the point process*

$$\begin{aligned}\mathcal{P}_N^{(m)}\;\equiv\;&\sum_{\sigma_1\dots\sigma_{J_1}}\delta_{u^{-1}_{\bar\alpha_1,N}(\bar X_{\sigma_1\dots\sigma_{J_1}})}\sum_{\sigma_{J_1+1}\dots\sigma_{J_2}}\delta_{u^{-1}_{\bar\alpha_2,N}(\bar X^{\sigma_1\dots\sigma_{J_1}}_{\sigma_{J_1+1}\dots\dots\sigma_{J_2}})}\cdots\\&\cdots\sum_{\sigma_{J_{m-1}+1}\dots\sigma_{J_m}}\delta_{u^{-1}_{\bar\alpha_m,N}(\bar X^{\sigma_1\dots\sigma_{J_{m-1}}}_{\sigma_{J_{m-1}+1}\dots\sigma_{J_m}})}\end{aligned}\tag{133}$$

converges weakly to the process $\mathcal{P}^{(m)}$ *on* $\mathbb{R}^m$ *defined in Theorem* 4.2 *with constants* $K_1, \ldots, K_m$. *If* $A_J(x) < A(x)$, *for all* $x \in (x_{J_i}, x_{J_{i+1}})$, *then*

$$K_l = 1. \tag{134}$$

Otherwise $0 < K_l < 1$.

Having constructed all possible point processes, we now find the extremal process by choosing the one that yields the largest values. It is easy to see that this is achieved if as many intermediate hierarchies as possible are grouped together. In terms of the geometrical construction just described, this means that we must choose the partition J in such a way that the function A_J has no convex pieces, i.e., that A_J is the *concave hull*, $\bar{A}$, of the function A (see Figure 2). (The concave

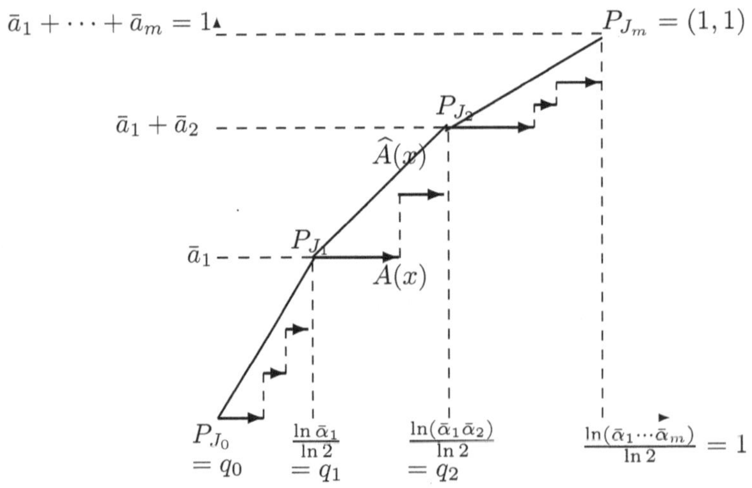

FIGURE 2. The concave hull of $A(x)$.

hull, $\bar{A}$, of a function A is the smallest concave function such that $\bar{A}(x) \geq A(x)$, for all x in the domain considered.) Algorithmically, this is achieved by setting $J_0 \equiv 0$, and

$$J_l \equiv \min\{J > J_{l-1} : A_{J_{l-1}+1,J} > A_{J+1,k} \ \ \forall k \geq J+1\} \tag{135}$$

where $A_{j,k} \equiv \sum_{i=j}^{k} a_i/(2\ln(\prod_{i=j}^{k} \alpha_i))$.

Set $\gamma_l \equiv \sqrt{\bar{a}_l}/\sqrt{2\ln\bar{\alpha}_l}$, $l = 1, 2, \ldots, m$. Clearly, by (135), $\gamma_1 > \gamma_2 > \cdots > \gamma_m$. Define the function $U_{J,N}$ by

$$U_{J,N}(x) \equiv \sum_{l=1}^{m} \left(\sqrt{2N\bar{a}_l \ln\bar{\alpha}_l} - N^{-1/2}\gamma_l(\ln(N(\ln\bar{\alpha}_l)) + \ln 4\pi)/2\right) + N^{-1/2}x \tag{136}$$

and the point process

$$\mathcal{E}_N \equiv \sum_{\sigma\in\{-1,1\}^N} \delta_{U_{J,N}^{-1}(X_\sigma)}. \tag{137}$$

Theorem 4.4.

(i) *The point process $\mathcal{E}_N$ converges weakly, as $N \uparrow \infty$, to the point process on* $\mathbb{R}$

$$\mathcal{E} \equiv \int_{\mathbb{R}^m} \mathcal{P}^{(m)}(\mathrm{d}x_1,\ldots,\mathrm{d}x_m)\delta_{\sum_{l=1}^m \gamma_l x_l} \tag{138}$$

where $\mathcal{P}^{(m)}$ is the Poisson cascade introduced in Theorem 4.3 *corresponding to the partition $J_1,\ldots,J_m$ given by* (135).

(ii) *$\mathcal{E}$ exists, since $\gamma_1 > \cdots > \gamma_m$. It is the cluster point process on $\mathbb{R}$ containing an a.s. finite number of points in any interval $[b,\infty)$, $b\in\mathbb{R}$. The probability that there exists at least one point of $\mathcal{E}$ in the interval $[b,\infty)$ is decreasing exponentially, as $b\uparrow\infty$.*

The proofs of these theorems can be found in [6].

4.3. Convergence of the partition function

We now turn to the study of the Gibbs measures. The main point will be to show that the infinite-volume limit of the properly rescaled partition function can be expressed as a certain functional of Poisson cascade processes, as suggested by Ruelle [25].

For *any* sequence of indices, J_i, such that the function A^J is concave, the partition function can be written as:

$$Z_{\beta,N} = \mathrm{e}^{\sum_{j=1}^m \left(\beta N\sqrt{2\bar{a}_j \ln\bar{\alpha}_j} - \beta\gamma_j[\ln(N\ln\bar{\alpha}_j)+\ln 4\pi]/2\right)} \times$$
$$\times\, \mathbb{E}_{\sigma_1\ldots\sigma_{J_1}} \mathrm{e}^{\beta\gamma_1 u_{\bar{\alpha}_1,N}^{-1}(\bar{X}_{\sigma_1\ldots\sigma_{J_1}})} \cdots \mathbb{E}_{\sigma_{J_{m-1}+1}\ldots\sigma_{J_m}} \mathrm{e}^{\beta\gamma_m u_{\bar{\alpha}_m,N}^{-1}(\bar{X}^{\sigma_1\ldots\sigma_{J_{m-1}}}_{\sigma_{J_{m-1}+1}\ldots\sigma_{J_m}})}. \tag{139}$$

These representations will only be useful, if the sums in the second line converge to a finite random variable. For this to happen, from what we learned in the REM, each of the sums should be at 'low temperature', meaning here that $\beta\gamma_\ell > 1$. Moreover, we should expect that there is a relation to the maximum process; in fact, this will follow from the condition that $\gamma_i > \gamma_{i+1}$, for all i that appear. Thus we will have to choose the partition J that yields the extremal process, and we have to cut the representation (139) at some temperature-dependent level, $J_{l(\beta)}$, and treat the remaining hierarchies as *high-temperature* REM's, i.e., replace them by their mean value. The level $l(\beta)$ is determined by

$$l(\beta) \equiv \max\{l \geq 1 : \beta\gamma_l > 1\} \tag{140}$$

and $l(\beta) \equiv 0$ if $\beta\gamma_1 \leq 1$.

From these considerations it is now very simple to compute the asymptotics of the partition function.

The resulting formula for the free energy was first found in [9]:

Theorem 4.5 ([9]). *With the notation introduced above,*

$$\lim_{N\to\infty} \Phi_{\beta,N} = \beta \sum_{i=1}^{l(\beta)} \sqrt{2\bar{a}_i \ln \bar{\alpha}_i} + \sum_{i=J_{l(\beta)}+1}^{n} \beta^2 a_i/2, \quad a.s. \tag{141}$$

The condition that for $\beta \leq \beta_c$, $l(\beta) = 0$, defines the critical temperature, $\beta_c = 1/\gamma_1$.

The more precise asymptotics of the partition function is as follows.

Theorem 4.6. *Let $J_1, J_2, \ldots, J_m \in \mathbb{N}$, be the sequence of indices defined by* (135) *and $l(\beta)$ defined by* (140)*. Then, with the notations introduced above,*

$$\begin{aligned} &\mathrm{e}^{-\beta \sum_{j=1}^{l(\beta)} \left(N\sqrt{2\bar{a}_j \ln \bar{\alpha}_j} - \gamma_j [\ln(N \ln \bar{\alpha}_j) + \ln 4\pi]/2\right) - N\beta^2 \sum_{i=J_{l(\beta)}+1}^{n} a_i/2} Z_{\beta,N} \\ &\xrightarrow{\mathcal{D}} C(\beta) \int_{\mathbb{R}^{l(\beta)}} \mathrm{e}^{\beta \sum_{i=1}^{l(\beta)} \gamma_i x_i} \mathcal{P}^{(l(\beta))}(\mathrm{d}x_1 \ldots \mathrm{d}x_{l(\beta)}). \end{aligned} \tag{142}$$

This integral is over the process $\mathcal{P}^{(l(\beta))}$ on $\mathbb{R}^{l(\beta)}$ from Theorem 4.2 *with constants K_j from Theorem* 4.3*. The constant $C(\beta)$ satisfies*

$$\begin{cases} C(\beta) = 1, & \text{if } \beta\gamma_{l(\beta)+1} < 1, \\ 0 < C(\beta) < 1, & \text{otherwise.} \end{cases} \tag{143}$$

Remark 4.3. An explicit formula for $C(\beta)$ is given in [6].

The integrals over the Poisson cascades appearing in Theorem 4.6 are to be understood as

$$\begin{aligned} &\int_{\mathbb{R}^m} \mathrm{e}^{\beta\gamma_1 x_1 + \cdots + \beta\gamma_m x_m} \mathcal{P}^{(m)}(\mathrm{d}x_1 \ldots \mathrm{d}x_m) \qquad (144) \\ &\equiv \lim_{x \downarrow -\infty} \int_{\substack{(x_1,\ldots,x_m)\in\mathbb{R}^m, \\ \exists i,\ 1\leq i \leq m: \\ \gamma_1 x_1 + \cdots + \gamma_i x_i > (\gamma_1 + \cdots + \gamma_i)x}} \mathrm{e}^{\beta\gamma_1 x_1 + \cdots + \beta\gamma_m x_m} \mathcal{P}^{(m)}(\mathrm{d}x_1 \ldots \mathrm{d}x_m). \end{aligned}$$

The existence of these limits requires the conditions on the γ_i mentioned before, and thus can be seen as responsible for the selection of the partition J and the cut-off level $l(\beta)$. Namely:

Proposition 4.1 ([6]). *Assume that $\gamma_1 > \gamma_2 > \cdots > \gamma_m > 0$, and $\beta\gamma_m > 1$. Then*

(i) *For any $a \in \mathbb{R}$ the process $\mathcal{P}^{(m)}$ contains a.s. a finite number of points $(x_1, \ldots, x_m)$ such that $\gamma_1 x_1 + \cdots + \gamma_m x_m > a$.*
(ii) *The limit in* (144) *exists and is finite a.s.*

4.4. The asymptotic model

As in the REM, we are able to reinterpret the convergence of the partition function in terms of an asymptotic statistical mechanics model.

This time, it has the following ingredients:

- State space: $\mathbb{N}^\ell$, that should be though of as an ℓ-level tree with infinite branching number.
- A sequence $\boldsymbol{\gamma} \equiv (\gamma_1 > \gamma_2 > \cdots > \gamma_\ell)$ of numbers.
- Random Hamiltonian: $\mathcal{H}_\gamma^\ell : \mathbb{N}^\ell \to \mathbb{R}$, where

$$\mathcal{H}_\gamma^\ell = \sum_{k=1}^{\ell} \gamma_k x_{\boldsymbol{i}_k}, \tag{145}$$

  and $x_{\boldsymbol{i}_k}$ is the $\boldsymbol{i}_k$th atom of the Poisson process $\mathcal{P}^{(k)}$.
- Temperature: $1/\beta$.
- Partition function: $\mathcal{Z}_\beta = \sum_{\boldsymbol{i}\in\mathbb{N}^\ell} \mathrm{e}^{\beta\mathcal{H}(\boldsymbol{i})}$.
- Gibbs measure:

$$\hat{\mu}_{\beta,\gamma}^\ell(\boldsymbol{i}) = \mathcal{Z}_\beta^\ell \mathrm{e}^{\beta\mathcal{H}_\gamma^\ell(\boldsymbol{i})}. \tag{146}$$

Our convergence results so far can be interpreted in terms of this model as follows:

- The partition function of the GREM converges, after multiplication with the correct scaling factor, to $\mathcal{Z}_\beta$.

A new feature compared to the situation in the REM is that the state space $\mathbb{N}^\ell$ of the asymptotic model carries a natural non-trivial distance, namely the hierarchical distance, respectively the corresponding hierarchical overlap

$$d(\boldsymbol{i},\boldsymbol{j}) = \frac{1}{\ell}(\min(k : i_k \neq j_k) - 1). \tag{147}$$

This allows to define in the asymptotic model the analog of the local mass distribution (see (106)) as

$$\phi_\beta(\boldsymbol{i},t) \equiv \hat{\mu}_\beta\left(d(\boldsymbol{i},\boldsymbol{j}) > t\right). \tag{148}$$

This allows also to write the empirical distance distribution function for the asymptotic model in the form

$$\mathcal{K}_\beta \equiv \sum_{\boldsymbol{i}\in\mathbb{N}^\ell} \hat{\mu}_\beta(\boldsymbol{i})\delta_{\phi_\beta(\boldsymbol{i},\cdot)}. \tag{149}$$

Clearly we expect K_β to be related to the analogous object in the GREM, i.e., to $\mathcal{K}_{\beta,N}$ defined as in (114). The only additional ingredient is a translation between the overlap on the hypercube and the tree-overlap (148). This is in fact given by the following Lemma:

Lemma 4.1. *Let*

$$q_\ell \equiv \sum_{n=1}^{\ell} \frac{\ln \bar{\alpha}_n}{\ln 2}, \tag{150}$$

and let $f(q) \equiv \sup\{k : q_k \leq q\}/\ell(\beta)$. *For any* β, *for* $q \leq q_{\max} \equiv q_{\ell(\beta)}$,

$$\lim_{N\uparrow\infty} \mu_{\beta,N}^{\otimes 2}\left(R_N(\sigma,\sigma') \leq q\right) = \hat{\mu}_\beta^{\otimes 2}\left(d(\boldsymbol{i},\boldsymbol{j}) \leq f(q)\right). \tag{151}$$

A nontrivial aspect of the lemma above is that the overlap defined in terms of the non-hierarchical R_N is asymptotically given in terms of the distribution of a hierarchical overlap, d. In fact, it would be quite a bit easier to show that

$$\lim_{N\uparrow\infty} \mu_{\beta,N}^{\otimes 2}\left(d_N(\sigma,\sigma') \leq q\right) = \hat{\mu}_{\beta}^{\otimes 2}\left(d(\boldsymbol{i},\boldsymbol{j}) \leq f(q)\right), \tag{152}$$

where d_N is defined in (55). The fact that the two distances are asymptotically the same on the support of the Gibbs measures is remarkable and the simplest instance of the apparent universality of ultrametric structures in spin glasses. Bolthausen and Kistler [5] (see also Jana [20]) have shown that the same occurs in a class of models where the covariance depends on several different hierarchical distances.

The main result on the limiting Gibbs measures can now be formulated as follows:

Theorem 4.7. *Under the assumptions and with the notation of Lemma* 4.1

$$\lim_{N\uparrow\infty} \mathcal{K}_{\beta,N} = \mathcal{K}_{\beta}^{f}, \tag{153}$$

where $f : [0, q_{\max}] \to [0,1]$ *is defined in Lemma* 4.1 *and*

$$\mathcal{K}_{\beta}^{f} = \sum_{\boldsymbol{i}\in\mathbb{N}^{\ell}} \hat{\mu}_{\beta}(\boldsymbol{i})\delta_{\phi_{\beta}(\boldsymbol{i},f(\cdot))}. \tag{154}$$

We see that in the asymptotic model, we have so far three ingredients:

1) the Poisson cascade;
2) the weights γ_i;
3) the mapping f that readjusts the tree-distance.

In fact, $\mathcal{K}_{\beta}$ as a probability distribution on distributions of distances contains a lot of *gauge invariance*. In particular, neither the measures $\hat{\mu}_{\beta}$ nor the underlying space $\mathbb{N}^{\ell}$ play a particular rôle. In fact, there is a canonical way to shift all the structure into the ultrametric and to choose as a canonical space the interval $[0,1]$ and as a canonical measure on it the Lebesgue measure. To do this, choose a one-to-one map,

$$\theta\colon \mathbb{N}^{\ell} \to [0,1] \tag{155}$$

such that for any Borel set, $\mathcal{A} \subset \mathcal{B}([0,1]$,

$$|\mathcal{A}| = \hat{\mu}_{\beta}(\theta^{-1}(\mathcal{A})). \tag{156}$$

Then define the overlap, γ_1, on $[0,1]$, by

$$\gamma_1(x,y) = f^{-1}\left(d\left(\theta^{-1}(x),\theta^{-1}(y)\right)\right). \tag{157}$$

Note that this overlap structure is now random, and in fact contains all the remaining randomness of the system. Then we can write $\mathcal{K}_{\beta}^{f}$ as

$$\mathcal{K}_{\beta}^{f} = \int_0^1 \mathrm{d}x\, \delta_{|\{y:\gamma_1(x,y)>\cdot\}|}. \tag{158}$$

This representation allows, in fact, to put all GREMs on a single footing. Namely, one can show that the random overlaps γ_1 can all be obtained by a deterministic time change from the *genealogical distance* of a particular continuous time branching process, the so-called *Neveu CBP*. This observation goes back to an unpublished paper of Neveu [24] and was elaborated on by Bertoin and Le Gall [3] and the present authors [8].

5. Gaussian comparison and applications

We now return to the study of the SK type models. We will emphasize here the rôle of classical comparison results for Gaussian processes. A clever use of them will allow to connect the SK models with the GREMs discussed above. We begin by recalling the basic comparison theorem.

5.1. A theorem of Slepian-Kahane

Lemma 5.1. *Let X and Y be two independent n-dimensional Gaussian vectors. Let D_1 and D_2 be subsets of $\{1,\dots,n\}\times\{1,\dots,n\}$. Assume that*

$$\begin{aligned} \mathbb{E}X_iX_j &\leq \mathbb{E}Y_iY_j, \quad \text{if} \quad (i,j)\in D_1,\\ \mathbb{E}X_iX_j &\geq \mathbb{E}Y_iY_j, \quad \text{if} \quad (i,j)\in D_2,\\ \mathbb{E}X_iX_j &= \mathbb{E}Y_iY_j, \quad \text{if} \quad (i,j)\notin D_1\cup D_2. \end{aligned} \tag{159}$$

Let f be a function on $\mathbb{R}^n$, such that its second derivatives satisfy

$$\begin{aligned} \frac{\partial^2}{\partial x_i\partial x_j}f(x) &\geq 0, \quad \text{if} \quad (i,j)\in D_1,\\ \frac{\partial^2}{\partial x_i\partial x_j}f(x) &\leq 0, \quad \text{if} \quad (i,j)\in D_2. \end{aligned} \tag{160}$$

Then

$$\mathbb{E}f(X)\leq \mathbb{E}f(Y). \tag{161}$$

Proof. The first step of the proof consists of writing

$$f(X)-f(Y)=\int_0^1 \mathrm{d}t\frac{\mathrm{d}}{\mathrm{d}t}f(X^t) \tag{162}$$

where we define the interpolating process

$$X^t\equiv\sqrt{t}\,X+\sqrt{1-t}\,Y. \tag{163}$$

Next observe that

$$\frac{\mathrm{d}}{\mathrm{d}t}f(X^t)=\frac{1}{2}\sum_{i=1}^n\frac{\partial}{\partial x_i}f(X^t)\left(t^{-1/2}X_i-(1-t)^{-1/2}Y_i\right). \tag{164}$$

Finally, we use the generalization of the standard Gaussian integration by parts formula to the multivariate setting, namely:

Lemma 5.2. *Let X_i, $i \in \{1, \dots, n\}$ be a multivariate Gaussian process, and let $g\colon \mathbb{R}^n \to \mathbb{R}$ be a differentiable function of at most polynomial growth. Then*

$$\mathbb{E}g(X)X_i = \sum_{j=1}^{n} \mathbb{E}(X_i X_j)\mathbb{E}\frac{\partial}{\partial x_j} g(X). \tag{165}$$

Applied to the mean of the left-hand side of (164) this yields

$$\mathbb{E}f(X) - \mathbb{E}f(Y) = \frac{1}{2}\sum_{i,j}\int_{0,1} \mathrm{d}t\,(\mathbb{E}X_iX_j - \mathbb{E}Y_iY_j)\,\mathbb{E}\frac{\partial^2}{\partial x_j \partial x_i} f(X^t), \tag{166}$$

from which the assertion of the theorem can be read off. □

Note that Equation (166) has the flavor of the fundamental theorem of calculus on the space of Gaussian processes.

5.2. The thermodynamic limit through comparison

Theorem 5.1 ([17]). *Assume that X_σ is a normalized Gaussian process on $\mathcal{S}_N$ with covariance*

$$\mathbb{E}X_\sigma X_\tau = \xi(R_N(\sigma,\tau)) \tag{167}$$

where $\xi\colon [-1,1] \to [0,1]$ is convex and even. Then

$$\lim_{N\uparrow\infty} \frac{-1}{\beta N}\mathbb{E}\ln \mathbb{E}_\sigma \mathrm{e}^{\beta\sqrt{N}X_\sigma} \equiv f_\beta \tag{168}$$

exists.

Proof. The proof of this fact is frightfully easy, once you think about using Theorem 5.1. Choose any $1 < M < N$. Let $\sigma = (\hat\sigma, \check\sigma)$ where $\hat\sigma = (\sigma_1, \sigma_2, \dots, \sigma_M)$, and $\check\sigma = (\sigma_{M+1}, \dots, \sigma_N)$. Define independent Gaussian processes $\hat X$ and $\check X$ on $\mathcal{S}_M$ and $\mathcal{S}_{N-M}$, respectively, such that

$$\mathbb{E}\hat X_{\hat\sigma}\hat X_{\hat\tau} = \xi(R_M(\hat\sigma,\hat\tau)) \tag{169}$$

and

$$\mathbb{E}\check X_{\check\sigma}\check X_{\check\tau} = \xi(R_{N-M}(\check\sigma,\check\tau)). \tag{170}$$

Set

$$Y_\sigma \equiv \sqrt{\tfrac{M}{N}}\hat X_{\hat\sigma} + \sqrt{\tfrac{N-M}{N}}\check X_{\check\sigma}. \tag{171}$$

Clearly,

$$\begin{aligned}\mathbb{E}Y_\sigma Y_\tau &= \tfrac{M}{N}\xi(R_M(\hat\sigma,\hat\tau)) + \tfrac{N-M}{N}\xi(R_{N-M}(\check\sigma,\check\tau)) \\ &\geq \xi\left(\tfrac{M}{N}R_M(\hat\sigma,\hat\tau) + \tfrac{N-M}{N}R_{N-M}(\check\sigma,\check\tau)\right) = \xi(R_N(\sigma,\tau)).\end{aligned} \tag{172}$$

Define real-valued functions $F_N(x) \equiv \ln \mathbb{E}_\sigma \mathrm{e}^{\beta\sqrt{N}x_\sigma}$ on $\mathbb{R}^{2^N}$. It is straightforward that

$$\mathbb{E}F_N(Y) = \mathbb{E}F_M(X) + \mathbb{E}F_{N-M}(X). \tag{173}$$

A simple computation shows that, for $\sigma \neq \tau$,

$$\frac{\partial^2}{\partial x_\sigma \partial x_\tau} F_N(x) = -\frac{2^{-2N}\beta^2 N e^{\beta\sqrt{N}(x_\sigma + x_\tau)}}{Z_{\beta,N}^2} \leq 0. \tag{174}$$

Thus, Theorem 5.1 tells us that

$$\mathbb{E}F_N(X) \geq \mathbb{E}F_N(Y) = \mathbb{E}F_M(X) + \mathbb{E}F_{N-M}(X). \tag{175}$$

This implies that the sequence $-\mathbb{E}F_N(X)$ is subadditive, and this in turn implies (see Section 1.2) that the free energy exists, provided it is bounded, which is easy to verify (see, e.g., the discussion on the correct normalization in the SK model). □

The same ideas can be used for other types of Gaussian processes, e.g., the GREM-type models discussed above [10].

Convergence of the free energy in mean implies readily almost sure convergence. This follows from a general *concentration of measure* principle for functions of Gaussian random variables.

5.3. An extended comparison principle

As we have mentioned, comparison of the free energy of SK models to simpler models does not immediately work. The idea is to use comparison on a much richer class of processes. Basically, rather than comparing one process to another, we construct an extended process on a product space and use comparison on this richer space. Let us first explain this in an abstract setting. We have a process X on a space $\mathcal{S}$ equipped with a probability measure $\mathbb{E}_\sigma$. We want to compute as usual the average of the logarithm of the partition function $F(X) = \ln \mathbb{E}_\sigma e^{\beta X_\sigma}$. Now consider a second space $\mathcal{T}$ equipped with a probability law $\mathbb{E}_\alpha$. Choose a Gaussian process, Y, independent of X, on this space, and define a further independent process, Z, on the product space $\mathcal{S} \times \mathcal{T}$. Define real-valued functions, G, H, on the space of real-valued functions on $\mathcal{T}$ and $\mathcal{S} \times \mathcal{T}$, respectively, via $G(y) \equiv \ln \mathbb{E}_\alpha e^{\beta y_\alpha}$ and $H(z) = \ln \mathbb{E}_\sigma \mathbb{E}_\alpha e^{\beta z_{\sigma,\alpha}}$. Note that $H(X+Y) = F(X) + G(Y)$. Assume that the covariances are chosen such that

$$\operatorname{cov}(X_\sigma, X_{\sigma'}) + \operatorname{cov}(Y_\alpha, Y_{\alpha'}) \geq \operatorname{cov}(Z_{\sigma,\alpha}, Z_{\sigma',\alpha'}). \tag{176}$$

Since we know that the second derivatives of H are negative, we get from Theorem 5.1 that

$$\mathbb{E}F(X) + \mathbb{E}G(Y) = \mathbb{E}H(X+Y) \leq \mathbb{E}H(Z). \tag{177}$$

This is a useful relation if we know how to compute $\mathbb{E}G(Y)$ and $\mathbb{E}H(Z)$. This idea may look a bit crazy at first sight, but we must remember that we have a lot of freedom in choosing the auxiliary spaces and processes to our convenience. Before turning to the issue whether we can find useful computable processes Y and Z, let us see why we could hope to find in this way *sharp* bounds.

5.4. The extended variational principle and thermodynamic equilibrium

To do so, we will show that, in principle, we can represent the free energy in the thermodynamic limit in the form $\mathbb{E}H(Z) - \mathbb{E}G(Y)$. To this end let $\mathcal{S} = \mathcal{S}_M$ and $\mathcal{T} = \mathcal{S}_N$, both equipped with their natural probability measure $\mathbb{E}_\sigma$. We will think of $N \gg M$, and both tending to infinity eventually. We write again $\mathcal{S} \times \mathcal{T} \ni \sigma = (\hat\sigma, \check\sigma)$. Consider the process X_σ on $\mathcal{S}_{N+M}$ with covariance $\xi(R_{N+M}(\sigma,\sigma'))$. We would like to write this as

$$X_\sigma = \hat X_{\hat\sigma} + \check X_{\check\sigma} + Z_\sigma \tag{178}$$

where all three processes are independent. Note that here and in the sequel equalities between random variables are understood to hold in distribution. Moreover, we demand that

$$\mathrm{cov}(\hat X_{\hat\sigma}, \hat X_{\hat\sigma'}) = \xi\left(\tfrac{M}{N+M} R_M(\hat\sigma,\hat\sigma')\right) \tag{179}$$

and

$$\mathrm{cov}(\check X_{\check\sigma}, \check X_{\check\sigma'}) = \xi\left(\tfrac{N}{N+M} R_N(\check\sigma,\check\sigma')\right). \tag{180}$$

Obviously, this implies that

$$\begin{aligned}\mathrm{cov}(Z_\sigma, Z_{\sigma'}) = {} & \xi\left(\tfrac{M}{N+M} R_M(\hat\sigma,\hat\sigma') + \tfrac{N}{N+M} R_N(\check\sigma,\check\sigma')\right) \\ & - \xi\left(\tfrac{M}{N+M} R_M(\hat\sigma,\hat\sigma')\right) - \xi\left(\tfrac{N}{N+M} R_N(\check\sigma,\check\sigma')\right)\end{aligned} \tag{181}$$

(we will not worry about the existence of such a decomposition; if $\xi(x) = x^p$, we can use the explicit representation in terms of p-spin interactions to construct them). Now we first note that, by super-additivity [2]

$$\lim_{M\uparrow\infty} \frac{1}{\beta M} \liminf_{N\uparrow\infty} \mathbb{E}\log\frac{Z_{\beta,N+M}}{Z_{\beta,N}} = -f_\beta. \tag{182}$$

Thus we need a suitable representation for $\frac{Z_{\beta,N+M}}{Z_{\beta,N}}$. But

$$\frac{Z_{\beta,N+M}}{Z_{\beta,N}} = \frac{\mathbb{E}_\sigma e^{\beta\sqrt{N+M}(\check X_{\check\sigma} + Z_\sigma + \hat X_{\hat\sigma})}}{\mathbb{E}_{\check\sigma} e^{\beta\sqrt{N+M}\left(\sqrt{(1-M/(N+M))}X_{\check\sigma}\right)}}. \tag{183}$$

Now we want to express the random variables in the denominator in the form

$$\sqrt{(1-M/(N+M))}X_{\check\sigma} = \check X_{\check\sigma} + Y_{\check\sigma} \tag{184}$$

where Y is independent of $\check X$. Comparing covariances, this implies that

$$\mathrm{cov}(Y_{\check\sigma}, Y_{\check\sigma'}) = (1 - M/(N+M))\xi(R_N(\check\sigma,\check\sigma')) - \xi\left(\tfrac{N}{N+M} R_N(\check\sigma,\check\sigma')\right). \tag{185}$$

As we will be interested in taking the limit $N\uparrow\infty$ before $M\uparrow\infty$, we may expand in $M/(N+M)$ to see that to leading order in $M/(N+M)$,

$$\begin{aligned}\mathrm{cov}(Y_{\check\sigma}, Y_{\check\sigma'}) \sim {} & \tfrac{M}{N+M} R_N(\check\sigma,\check\sigma')\xi'\left(\tfrac{N}{N+M} R_N(\check\sigma,\check\sigma')\right) \\ & - \tfrac{M}{N+M}\xi\left(\tfrac{N}{N+M} R_N(\check\sigma,\check\sigma')\right).\end{aligned} \tag{186}$$

Finally, we note that the random variables $\hat{X}_{\hat{\sigma}}$ are negligible in the limit $N \uparrow \infty$, since their variance is smaller than $\xi(M/(N+M))$ and hence their maximum is bounded by $\sqrt{\xi(M/(N+M))M \ln 2}$, which even after multiplication with $\sqrt{N+M}$ gives no contribution in the limit if ξ tends to zero faster than linearly at zero, which we can safely assume. Thus we see that we can indeed express the free energy as

$$f_\beta = -\lim_{M\uparrow\infty} \liminf_{N\uparrow\infty} \frac{1}{\beta M} \mathbb{E} \ln \frac{\mathbb{E}_{\hat{\sigma}} \widetilde{\mathbb{E}}_{\check{\sigma}} e^{\beta\sqrt{N+M} Z_{\hat{\sigma},\check{\sigma}}}}{\widetilde{\mathbb{E}}_{\check{\sigma}} e^{\beta\sqrt{N+M} Y_{\check{\sigma}}}} \tag{187}$$

where the measure $\widetilde{\mathbb{E}}_{\check{\sigma}}$ can be chosen as a probability measure defined by $\widetilde{\mathbb{E}}_{\check{\sigma}}(\cdot) = \mathbb{E}_{\check{\sigma}} e^{\beta\sqrt{N+M}\check{X}_{\check{\sigma}}}(\cdot)/\check{Z}_{\beta,N,M}$ where $\check{Z}_{\beta,N,M} \equiv \mathbb{E}_{\check{\sigma}} e^{\beta\sqrt{N+M}\check{X}_{\check{\sigma}}}$. Of course this representation is quite pointless, because it is certainly uncomputable, since $\widetilde{\mathbb{E}}$ is effectively the limiting Gibbs measure that we are looking for. However, at this point there occurs a certain miracle: the (asymptotic) covariances of the processes X, Y, Z satisfy

$$\xi(x) + y\xi'(y) - \xi(y) \geq x\xi'(y) \tag{188}$$

for all $x, y \in [-1, 1]$, if ξ is convex and even. This comes as a surprise, since we did not do anything to impose such a relation! But it has the remarkable consequence that asymptotically, by virtue of Lemma 5.1 it implies the bound

$$\mathbb{E} \ln \mathbb{E}_{\hat{\sigma}} e^{\beta\sqrt{M} X_{\hat{\sigma}}} \leq \mathbb{E} \ln \frac{\mathbb{E}_{\hat{\sigma}} \widehat{\mathbb{E}}_{\check{\sigma}} e^{\beta\sqrt{N+M} Z_{\hat{\sigma},\check{\sigma}}}}{\widehat{\mathbb{E}}_{\check{\sigma}} e^{\beta\sqrt{N+M} Y_{\check{\sigma}}}} \tag{189}$$

(if the processes are taken to have the asymptotic form of the covariances). Moreover, this bound will hold *even* if we replace the measure $\widehat{\mathbb{E}}$ by some other probability measure, and even if we replace the overlap R_N on the space $\mathcal{S}_N$ by some other function, e.g., the ultrametric d_N. Seen the other way around, we can conclude that a lower bound of the form (177) can actually be made as good as we want, provided we choose the right measure $\widehat{\mathbb{E}}$. This observation is due to Aizenman, Sims, and Starr [2]. They call the auxiliary structure made from a space $\mathcal{T}$, a probability measure $\mathbb{E}_\alpha$ on $\mathcal{T}$, a normalized distance q on $\mathcal{T}$, and the corresponding processes, Y and Z, a *random overlap structure*

$$\operatorname{cov}(Y_\alpha, Y_{\alpha'}) = q(\alpha,\alpha')\xi'(q(\alpha,\alpha')) - \xi(q(\alpha,\alpha')) \tag{190}$$

and the process $Z_{\sigma,\alpha}$ on $\mathcal{S}_N \times [0,1]$ with covariance

$$\operatorname{cov}(Z_{\sigma,\alpha}, Z_{\sigma',\alpha'}) \equiv R_N(\sigma,\sigma')\xi'(q(\alpha,\alpha')). \tag{191}$$

With these choices, and naturally X_σ our original process with covariance $\xi(R_N)$, the equation (176) is satisfied, and hence the inequality (177) holds, no matter what choice of q and $\mathbb{E}_\alpha$ we make. Restricting these choices to the random genealogies obtained from Neveu's process by a time change with some probability distribution function m, and $\mathbb{E}_\alpha$ the Lebesgue measure on $[0,1]$, gives the bound we want.

This bound would be quite useless if we could not compute the right-hand side. Fortunately, one can get rather explicit expressions. We need to compute two objects:

$$\mathbb{E}_\alpha \mathbb{E}_\sigma e^{\beta\sqrt{N} Z_{\sigma,\alpha}} \tag{192}$$

and

$$\mathbb{E}_\alpha e^{\beta\sqrt{N} Y_\alpha}. \tag{193}$$

In the former we use that Z has the representation

$$Z_{\sigma,\alpha} = N^{-1/2} \sum_{i=1}^{N} \sigma_i z_{\alpha,i} \tag{194}$$

where the processes $z_{\alpha,i}$ are independent for different i and have covariance

$$\operatorname{cov}(z_{\alpha,i}, z_{\alpha',i}) = \xi'(q(\alpha,\alpha')). \tag{195}$$

Thus at least the σ-average is trivial:

$$\mathbb{E}_\alpha \mathbb{E}_\sigma e^{\beta\sqrt{N} Z_{\sigma,\alpha}} = \mathbb{E}_\alpha \prod_{i=1}^{N} e^{\ln\cosh(\beta z_{\alpha,i})}. \tag{196}$$

Thus we see that, in any case, we obtain bounds that only involve objects that we introduced ourselves and that thus can be manipulated to be computable. In fact, such computations have been done in the context of the Parisi solution [23]. A useful mathematical reference is [4].

This is the form derived in Aizenman, Sims, and Starr [2].

5.5. Parisi auxiliary systems

The key idea of the Parisi solution is to choose as an auxiliary system the asymptotic model of the GREM.

That is, the space $\mathcal{T}$ in this case is chosen as an n-level infinite tree $\mathbb{N}^n$ equipped with the measure $\hat{\mu}_\beta$ defined in (146).

$\mathcal{T}$ is naturally endowed with its tree overlap, $d(\boldsymbol{i},\boldsymbol{j}) \equiv n^{-1}(\min\{\ell : i_\ell \neq j_\ell\} - 1)$. This distance will play the rôle of the distance q on $\mathcal{T}$. Finally, we define the processes $Y_{\boldsymbol{i}}$ and $Z_{\boldsymbol{i},\sigma}$ with covariances

$$\operatorname{cov}(Y_{\boldsymbol{i}}, Y_{\boldsymbol{j}}) = d(\boldsymbol{i},\boldsymbol{j})\xi'(d(\boldsymbol{i},\boldsymbol{j})) - \xi(d(\boldsymbol{i},\boldsymbol{j})) \equiv h(d(\boldsymbol{i},\boldsymbol{j})) \tag{197}$$

and the process $Z_{\sigma,\boldsymbol{i}}$ on $\mathcal{S}_N \times \mathcal{T}$ with covariance

$$\operatorname{cov}(Z_{\sigma,\boldsymbol{j}}, Z_{\sigma',\boldsymbol{j}}) \equiv R_N(\sigma,\sigma')\xi'(d(\boldsymbol{i},\boldsymbol{j})). \tag{198}$$

It is easy to see that such processes can be constructed as long as h, ξ' are increasing functions. E.g.,

$$Y_{\boldsymbol{i}} = \sum_{\ell=1}^{n} \sqrt{h(\ell/n) - h((\ell-1)/n)}\, Y^{(\ell)}_{\boldsymbol{i}_1\dots\boldsymbol{i}_\ell} \tag{199}$$

where $Y^{(\ell)}_{\boldsymbol{i}_1\dots\boldsymbol{i}_\ell}$ are independent standard normal random variables. In this way, we have constructed an explicit random overlap structure, which corresponds indeed to the one generating the Parisi solution.

Note that also the auxiliary structure depends only on the information contained in the empirical distance distribution, $\mathcal{K}_\beta$, associated with the asymptotic model. In fact we could alternatively use $\mathcal{T} = [0,1]$ equipped with the Lebesgue measure and the random overlap γ_1, as defined in (158). While this is nice conceptually, for actual computations the form above will be more useful, however.

5.6. Computing with Poisson cascades

Lemma 5.3. *Assume that $\mathcal{P}$ be a Poisson process with intensity measure $\mathrm{e}^{-x}\mathrm{d}x$. and let $Z_{i,j}$, $i \in \mathbb{N}$, $j \in \mathbb{N}$ and Y be i.i.d. standard normal random variables. Let $g_j : \mathbb{R} \to \mathbb{R}$, $j = 1, \dots, M$, be smooth functions, such that, for all $|m| \leq 2$, there exist $C < \infty$, independent of N, such that*

$$\mathbb{E}_Y \mathrm{e}^{m g_i(Y)} \equiv \mathrm{e}^{L_i(m)} < C. \tag{200}$$

Let x_i be the atoms of the Poisson process $\mathcal{P}$ with intensity measure $\mathrm{e}^{-x}\mathrm{d}x$. Then

$$\mathbb{E} \ln \frac{\sum_{i=1}^{\infty} \mathrm{e}^{\alpha x_i + \sum_{j=1}^{M} g_j(Z_{i,j})}}{\sum_{i=1}^{\infty} \mathrm{e}^{\alpha x_i}} = \sum_{j=1}^{M} \alpha L_j(1/\alpha). \tag{201}$$

Proof. Let for simplicity $M = 1$. The numerator on the left in (201) can be written as

$$\int \mathrm{e}^{\alpha z} \widetilde{\mathcal{P}}(\mathrm{d}z)$$

where $\widetilde{\mathcal{P}}$ is the point process

$$\widetilde{\mathcal{P}} \equiv \sum_j \delta_{z_j + \alpha^{-1} g(Y_j)}.$$

This follows from a general fact about Poisson point processes: if $\mathcal{N} \equiv \sum_i \delta_{x_i}$ is a Poisson point process with intensity measure λ on E, and Y_i are i.i.d. random variables with distribution ρ, then

$$\widetilde{\mathcal{N}} \equiv \sum_i \delta_{x_i + Y_i}$$

is a Poisson process with intensity measure $\lambda * \rho$ on the set $E + \operatorname{supp} \rho$. This follows from the representation of $\mathcal{N}$ as

$$\mathcal{N} = \sum_{i=1}^{N_\lambda} \delta_{X_i}$$

where N_λ is Poisson with parameter $\int_E \lambda(\mathrm{d}x) \equiv |\lambda|$ (if this is finite), and X_i i.i.d. random variables with distribution $\lambda/|\lambda|$. Clearly

$$\widetilde{\mathcal{N}} = \sum_i \delta_{x_i + Y_i} = \sum_{i=1}^{N_\lambda} \delta_{X_i + Y_i}$$

is again of the form of a PPP, and the distribution of $X_i + Y_i$ is $\lambda * \rho / |\lambda|$. Since the total intensity of the process is the parameter of N_λ, $|\lambda|$, it follows that the intensity measure of this process is the one we claimed.

Thus, in our case, $\widetilde{\mathcal{P}}$ is a PPP whose intensity measure is the convolution of the measure $\mathrm{e}^{-z}\mathrm{d}z$ and the distribution of the random variable $\alpha^{-1} g(Y)$. A simple computation shows that this is $\mathbb{E}_Y e^{g(Y)/\alpha} e^{-z} dz$, i.e., a multiple of the original intensity measure!

Finally, one makes the elementary but surprising and remarkable observation that the Poisson point process $\sum_j \delta_{z+\ln \mathbb{E}_Y e^{g(Y)/\alpha}}$ has the same intensity measure, and therefore, $\sum_j e^{\alpha z_j + g(Y_j)}$ has the same law as $\sum_j e^{\alpha z_j} [\mathbb{E}_Y e^{g(Y)/\alpha}]^\alpha$: multiplying each atom with an i.i.d. random variable leads to the same process as multiplying each atom by a suitable constant! The assertion of the lemma follows immediately. □

Remark 5.1. A slightly different proof from the one above can be found in [26].

Let us look first at (193). We can then write

$$\sum_i e^{\mathcal{H}_\gamma^n (\boldsymbol{i} + \beta\sqrt{M} Y_{\boldsymbol{i}}} = \sum_i e^{\mathcal{H}_\gamma^n(\boldsymbol{i}) + \beta\sqrt{M} Y_{\boldsymbol{i}_{n-1}} + \sqrt{h(x_n) - h(x_{n-1})} Y_{\boldsymbol{i}}^{(n)}} \tag{202}$$

$$= \sum_{\boldsymbol{i}_1 \dots \boldsymbol{i}_{n-1}} e^{\sum_{\ell=1}^{n-1} \gamma_\ell x_{\boldsymbol{i}_1 \dots \boldsymbol{i}_\ell} + \beta\sqrt{M} Y_{\boldsymbol{i}_{n-1}}} \times \sum_{\boldsymbol{i}_n} e^{\gamma_n x_{\boldsymbol{i}_{n-1}, i_n} + \beta\sqrt{M}\sqrt{h(1)-h(1-1/n)} Y^{(n)}_{\boldsymbol{i}_{n-1}, i_n}}.$$

Using Lemma 5.3, the last factor can be replaced by

$$\mathbb{E}_{\boldsymbol{i}_n} e^{\gamma_n x_{\boldsymbol{i}_{n-1}, i_n} + \beta\sqrt{M}\sqrt{h(1)-h(1-1/n)} Y^{(n)}_{\boldsymbol{i}_{n-1}, i_n}} \tag{203}$$

$$\to \left[\int \frac{dz}{\sqrt{2\pi}} e^{-\frac{z^2}{2}} e^{z m_n \beta \sqrt{M} \sqrt{h(1)-h(1-1/n)}} \right]^{1/m_n} \sum_{\boldsymbol{i}_n} e^{\gamma_n x_{\boldsymbol{i}}} \tag{204}$$

$$= e^{\frac{\beta^2 M}{2} m_n (h(1) - h(1-1/n))} \sum_{\boldsymbol{i}_n} e^{\gamma_n x_{\boldsymbol{i}}} \tag{205}$$

(we use throughout $m_n = 1/\gamma_n$). Note that the last factor is independent of the random variables $x_{i_1, \dots, i_\ell}$ with $\ell < n$. Thus

$$\mathbb{E} \ln \sum_{\boldsymbol{i}} e^{\alpha x_{\boldsymbol{i}} + \beta\sqrt{M} Y_{\boldsymbol{i}}} = \mathbb{E} \ln \sum_{i_1, \dots, i_{n-1}} e^{\sum_{\ell=1}^{n-1} \gamma_\ell x_{i_1, \dots, i_{n-1}} + \beta\sqrt{M} Y_{i_1, \dots, i_{n-1}}} \tag{206}$$

$$+ \frac{\beta^2 M}{2} m_n (h(1) - h(1-1/n)) + \mathbb{E} \ln \sum_{\boldsymbol{i}_n} e^{\gamma_n x_{\boldsymbol{i}}}.$$

The first term now has the same form as the original one with n replaced by $n-1$, and thus the procedure can obviously be iterated. As the final result, we get that

$$M^{-1} \mathbb{E} \ln \frac{\sum_{\boldsymbol{i}} e^{x_{\boldsymbol{i}} + \beta\sqrt{M} Y_{\boldsymbol{i}}}}{\sum_{\boldsymbol{i}} e^{x_{\boldsymbol{i}}}} = \sum_{\ell=1}^{n} \frac{\beta^2}{2} m_\ell (h(1-\ell/n) - h(1-(\ell-1)/n))$$

$$= \frac{\beta^2}{2} \int_0^1 m(x) x \xi''(x) dx. \tag{207}$$

The computation of the expression (192) is now very similar, but gives a more complicated result since the analogs of the expressions (203) cannot be computed explicitly. Thus, after the kth step, we end up with a new function of the remaining random variables $Y_{i_1 \dots i_{n-k}}$. The result can be expressed in the form

$$\frac{1}{M}\mathbb{E}\ln\mathbb{E}_{\boldsymbol{i}}\mathbb{E}_\sigma \mathrm{e}^{\beta\sqrt{M}Z_{\sigma,i}} = \zeta(0,h,m,\beta). \tag{208}$$

Here h is the magnetic field (which we have so far hidden in the notation) that can be taken as a parameter of the a priori distribution on the σ such that

$$\mathbb{E}_{\sigma_i}(\cdot) \equiv \frac{1}{2\cosh(\beta h)} \sum_{\sigma_i=\pm 1} \mathrm{e}^{\beta h \sigma_i}(\cdot)$$

where $\zeta(1,h) = \ln\cosh(\beta h)$, and

$$\zeta(x_{a-1},h) = \frac{1}{m_a}\ln\int \frac{\mathrm{d}z}{\sqrt{2\pi}}\mathrm{e}^{-z^2/2}\mathrm{e}^{m_a\zeta(x_a,h+z\sqrt{\xi'(x_a)-\xi'(x)})} \tag{209}$$

(we put $x_a = a/n$).

In all of the preceding discussion, the choice of the parameter n and of the numbers $m_i = 1/\gamma_1$ is still free.

We can now announce Guerra's bound in the following form:

Theorem 5.2 (**[16]**). *Let $\zeta(t,h,m,b)$ be the function defined in terms of the recursion* (209). *Then*

$$\lim_{N\uparrow\infty} N^{-1}\mathbb{E}\ln Z_{\beta,h,N} \le \inf_m \zeta(0,h,m,\beta) - \frac{\beta^2}{2}\int_0^1 m(x)x\xi''(x)\mathrm{d}x \tag{210}$$

where the infimum is over all probability distribution functions m on the unit interval.

Remark 5.2. It is also interesting to see that the recursive form of the function ζ above can also be represented in a closed form as the solution of a partial differential equation. Consider the case $\xi(x) = x^2/2$. Then ζ is the solution of the differential equation

$$\frac{\partial}{\partial t}\zeta(t,h) + \frac{1}{2}\left(\frac{\partial^2}{\partial h^2}\zeta(t,h) + m(t)\left(\frac{\partial}{\partial h}\zeta(t,h)\right)^2\right) = 0 \tag{211}$$

with final condition

$$\zeta(1,h) = \ln\cosh(\beta h). \tag{212}$$

If m is a step function, it is easy to see that a solution is obtained by setting, for $x \in [x_{a-1}, x_a)$,

$$\zeta(x,h) = \frac{1}{m_a}\ln\mathbb{E}_z \mathrm{e}^{m_a\zeta(x_a,h+z\sqrt{x_a-x})}. \tag{213}$$

For general convex ξ, analogous expressions can be obtained through changes of variables [16].

5.7. Talagrand's theorem

In both approaches, it pays to write down the expression of the difference between the free energy and the lower bound, since this takes a very suggestive form.

To do this, we just have to use formula (166) with

$$X^t_{\sigma,\alpha} \equiv \sqrt{t}(X_\sigma + Y_\alpha) + \sqrt{1-t}Z_{\sigma,\alpha} \tag{214}$$

and $f(X^t)$ replaced by $H(X^t) = \ln \mathbb{E}_\sigma \mathbb{E}_\alpha e^{\beta\sqrt{N}Z^t_{\sigma,\alpha}}$. This gives the equality

$$\begin{aligned} H(X+Y) - H(Z) = \frac{1}{2}\mathbb{E}\int_0^1 \mathrm{d}t\, \widetilde{\mu}^{\otimes 2}_{\beta,t,N}(\mathrm{d}\sigma, \mathrm{d}\alpha)\Big(\xi(R_N(\sigma,\sigma'))\\ + q(\alpha,\alpha')\xi'(q(\alpha,\alpha'))\\ - \xi(q(\alpha,\alpha')) - R_N(\sigma,\sigma')\xi'(q(\alpha,\alpha'))\Big) \end{aligned} \tag{215}$$

where the measure $\widetilde{\mu}_{\beta,t,N}$ is defined as

$$\widetilde{\mu}_{\beta,t,N}(\cdot) \equiv \frac{\mathbb{E}_\sigma \mathbb{E}_\alpha e^{\beta\sqrt{N}X^t_{\sigma,\alpha}}(\cdot)}{\mathbb{E}_\sigma \mathbb{E}_\alpha e^{\beta\sqrt{N}X^t_{\sigma,\alpha}}} \tag{216}$$

where we interpret the measure $\widetilde{\mu}_{\beta,t,N}$ as a joint distribution on $\mathcal{S}_N \times [0,1]$. Note that for convex and even ξ, the function $\xi(R_N(\sigma,\sigma')) + q(\alpha,\alpha')\xi'(q(\alpha,\alpha')) - \xi(q(\alpha,\alpha'))$ vanishes if and only if $R_N(\sigma,\sigma') = q(\alpha,\alpha')$. Thus for the left-hand side of (215) to vanish, the replicated interpolating product measure should (for almost all t), concentrate on configurations where the overlaps in the σ-variables coincide with the genealogical distances of the α-variables. Thus we see that the inequality in Theorem 5.2 will turn into an equality if it is possible to choose the parameters of the reservoir system in such a way that the overlap distribution on $\mathcal{S}_N$ aligns with the genealogical distance distribution in the reservoir once the systems are coupled by the interpolation.

This latter fact was proven very recently, and not long after the discovery of Guerra's bound, by M. Talagrand [29].

Theorem 5.3 **([29]).** *Let $\zeta(t,h,m,b)$ be the function defined in terms of* (211) *and* (212)*. Then*

$$\lim_{N\uparrow\infty} N^{-1}\mathbb{E}\ln Z_{\beta,h,N} = \inf_m \left(\zeta(0,h,m,\beta) - \frac{\beta^2}{2}\int_0^1 m(x)x\xi''(x)\mathrm{d}x\right) \tag{217}$$

where the infimum is over all probability distribution functions m on the unit interval.

We will not give the complex proof which the interested reader should study in the original paper [29], but we will make some comments on the key ideas. First, Talagrand proves more than the assertion (217). What he actually proves is the following. For any $\epsilon > 0$, there exists a positive integer $n(\epsilon) < \infty$, and a probability distribution function m_n that is a step function with n steps, such that for all $t > \epsilon$,

$$\lim_{N\uparrow\infty} \mathbb{E}\widetilde{\mu}^{\otimes 2}_{\beta,t,N}(d\sigma, d\alpha)\big(\xi(R_N(\sigma,\sigma')) + q(\alpha,\alpha')\xi'(q(\alpha,\alpha')) \\ -\xi(q(\alpha,\alpha')) - R_N(\sigma,\sigma')\xi'(q(\alpha,\alpha'))\big) = 0 \qquad (218)$$

if the measure $\widetilde{\mu}_{b,t,N}$ corresponds to the genealogical distance obtained from this function m. That is to say, if the coupling parameter t is large enough, the SK model can be aligned to a GREM with any desired number of hierarchies.

References

[1] R.J. Adler. On excursion sets, tube formulas and maxima of random fields. *Ann. Appl. Probab.*, 10(1):1–74, 2000.

[2] M. Aizenman, R. Sims, and S.L. Starr. An extended variational principle for the SK spin-glass model. *Phys. Rev. B*, 6821(21):4403–4403, 2003.

[3] J. Bertoin and J.-F. Le Gall. The Bolthausen-Sznitman coalescent and the genealogy of continuous-state branching processes. *Probab. Theory Related Fields*, 117(2):249–266, 2000.

[4] E. Bolthausen and A.-S. Sznitman. On Ruelle's probability cascades and an abstract cavity method. *Comm. Math. Phys.*, 197(2):247–276, 1998.

[5] Erwin Bolthausen and Nicola Kistler. On a nonhierarchical version of the generalized random energy model. *Ann. Appl. Probab.*, 16(1):1–14, 2006.

[6] A. Bovier and I. Kurkova. Derrida's generalised random energy models. I. Models with finitely many hierarchies. *Ann. Inst. H. Poincaré Probab. Statist.*, 40(4):439–480, 2004.

[7] A. Bovier and I. Kurkova. Derrida's generalized random energy models. II. Models with continuous hierarchies. *Ann. Inst. H. Poincaré Probab. Statist.*, 40(4):481–495, 2004.

[8] A. Bovier and I. Kurkova. Much ado about Derrida's GREM. In *Spin glasses*, volume 1900 of *Lecture Notes in Math.* Springer, Berlin, 2007.

[9] D. Capocaccia, M. Cassandro, and P. Picco. On the existence of thermodynamics for the generalized random energy model. *J. Statist. Phys.*, 46(3-4):493–505, 1987.

[10] P. Contucci, M. Degli Esposti, C. Giardinà, and S. Graffi. Thermodynamical limit for correlated Gaussian random energy models. *Comm. Math. Phys.*, 236(1):55–63, 2003.

[11] B. Derrida. A generalisation of the random energy model that includes correlations between the energies. *J. Phys. Lett.*, 46:401–407, 1985.

[12] R.S. Ellis. *Entropy, large deviations, and statistical mechanics*, volume 271 of *Grundlehren der Mathematischen Wissenschaften* [*Fundamental Principles of Mathematical Sciences*]. Springer-Verlag, New York, 1985.

[13] E. Gardner and B. Derrida. Magnetic properties and function $q(x)$ of the generalised random energy model. *J. Phys. C*, 19:5783–5798, 1986.

[14] E. Gardner and B. Derrida. Solution of the generalised random energy model. *J. Phys. C*, 19:2253–2274, 1986.

[15] E. Gardner and B. Derrida. The probability distribution of the partition function of the random energy model. *J. Phys. A*, 22(12):1975–1981, 1989.

[16] F. Guerra. Broken replica symmetry bounds in the mean field spin glass model. *Comm. Math. Phys.*, 233(1):1–12, 2003.

[17] F. Guerra and F.L. Toninelli. The thermodynamic limit in mean field spin glass models. *Comm. Math. Phys.*, 230(1):71–79, 2002.

[18] E. Ising. Beitrag zur Theorie des Ferro- und Paramagnetismus. Ph.D. thesis, Univ. Hamburg, 1924.

[19] E. Ising. *Zeitschrift. f. Physik*, 31:253, 1925.

[20] Nabin Kumar Jana. Contributions to random energy models. Thesis, 2008.

[21] M.R. Leadbetter, G. Lindgren, and H. Rootzén. *Extremes and related properties of random sequences and processes.* Springer Series in Statistics. Springer-Verlag, New York, 1983.

[22] M. Ledoux and M. Talagrand. *Probability in Banach spaces*, volume 23 of *Ergebnisse der Mathematik und ihrer Grenzgebiete* (3) [*Results in Mathematics and Related Areas* (3)]. Springer-Verlag, Berlin, 1991.

[23] M. Mézard, G. Parisi, and M.A. Virasoro. *Spin glass theory and beyond*, volume 9 of *World Scientific Lecture Notes in Physics.* World Scientific Publishing Co. Inc., Teaneck, NJ, 1987.

[24] J. Neveu. A continuous state branching process in relation with the GREM model of spin glass theory. rapport interne 267, Ecole Polytechnique Paris, 1992.

[25] D. Ruelle. A mathematical reformulation of Derrida's REM and GREM. *Comm. Math. Phys.*, 108(2):225–239, 1987.

[26] A. Ruzmaikina and M. Aizenman. Characterization of invariant measures at the leading edge for competing particle systems. *Ann. Probab.*, 33(1):82–113, 2005.

[27] D. Sherrington and S. Kirkpatrick. Solvable model of a spin glass. *Phys. Rev. Letts.*, 35:1792–1796, 1972.

[28] B. Simon. *The statistical mechanics of lattice gases. Vol. I.* Princeton Series in Physics. Princeton University Press, Princeton, NJ, 1993.

[29] Michel Talagrand. The Parisi formula. *Ann. of Math.* (2), 163(1):221–263, 2006.

Anton Bovier
Institut für Angewandte Mathematik
Rheinische Friedrich–Wilhelms–Universität Bonn
Endenicher Allee 60
D-53115 Bonn, Germany
e-mail: `bovier@uni-bonn.de`

Irina Kurkova
Laboratoire de Probabilités et Modèles Aléatoires
Université Paris 6
4, place Jussieu, B.C. 188
F-75252 Paris Cedex 05, France
e-mail: `irina.kourkova@upmc.fr`

Progress in Probability, Vol. 62, 45–84

REM Universality for Random Hamiltonians

Gérard Ben Arous and Alexey Kuptsov

Abstract. We survey in this paper a universality phenomenon which shows that some characteristics of complex random energy landscapes are model-independent, or universal. This universality, called REM-universality, was discovered by S. Mertens and H. Bauke in the context of combinatorial optimization. We survey recent advances on the extent of this REM-universality for equilibrium as well as dynamical properties of spin glasses. We also focus on the limits of REM-universality, i.e., when it ceases to be valid.

Mathematics Subject Classification (2000). 82B44, 82D30, 82C44, 60G15, 60G55.

Keywords. Spin glasses, random energy model, extreme values, Gaussian processes, statistical mechanics, disordered media

1. Introduction

We survey in this paper a universality phenomenon which shows that some characteristics of complex random energy landscapes are model-independent or universal. In a probabilistic language this means that some characteristics of random vectors in large dimensions are insensitive to correlations between the components of these vectors, for a wide class of such vectors.

This universality goes under the name of *REM-universality* and has been discovered by S. Mertens and H. Bauke in the context of combinatorial optimization (see [30, 4]). For a previous survey of this universality we refer to [23]. Here we want to concentrate on the recent progress made in [6, 12, 13, 28].

The basic phenomenon, first observed numerically by S. Mertens and H. Bauke, is that the micro-canonical distribution of energies of a very wide class of models of Statistical Mechanics of disordered media is very close to the distribution of energies for the Random Energy Model. This universality is valid in the "bulk", i.e., for micro-canonical windows of energies far enough from the ground state. Knowing how far is far enough is a difficult question that we also address here

The work of Gerard Ben Arous has been supported partly by NSF Grants DMS-0806180 and OISE-0730136.

thoroughly. We also give another version of this universality phenomenon which is more suited to dynamics of these disordered media. This new version has been introduced in the PhD thesis [28], and studied in the papers [12, 13]. It originated in a question raised by one of us and A. Bovier in view of potential applications to dynamics questions. We will show how this universality phenomenon does apply to dynamical questions in our last section.

We will start in Section 2 by giving the scope of REM-universality. We first explain in Section 2.1 some basic textbook facts about i.i.d. random variables which form the core of the phenomenon of REM-universality. We will then give in Section 2.2 a list of fundamental models of statistical mechanics where REM-universality is expected/observed/proved. We then give in Section 2.3 the definition of a more recent variant of REM-universality (the re-sampled REM-universality) and discuss briefly in Section 2.4 the implications of REM-universality for non-equlibrium questions.

In Section 3 we give the known results for local (or micro-canonical) REM-universality for examples given in Section 2.2, including results when REM universality does not hold. This section is based on the results of [1, 2, 3, 4, 13, 14, 15, 16, 19, 22, 23, 29, 28, 30].

In Section 4 we delve deeper in the new "re-sampling" REM-universality introduced in [28], and studied in [12, 13].

Finally, in Section 5 we give the REM-universality results for dynamics of mean field spin glasses as recently proved in [6].

2. The scope of REM-universality

2.1. The basic results for i.i.d. random variables

We first recall a few text-book facts about i.i.d. random variables. Even though the framework could be much more general we will restrict ourselves here to the simplest case, i.e., to the case of i.i.d. standard normal random variables $X_1, \dots, X_n$. We describe the asymptotic behavior of the point process defined by those random variables in different locations and scales. More precisely, we want to consider the values of the X_i's in a window $[a_n + \alpha b_n, a_n + \beta b_n]$, i.e., $a_n > 0$ will be the center of the window and $b_n > 0$ its width. The best way to do so is to consider a random point process on $\mathbb{R}$, defined as

$$\mathcal{P}_n = \sum_{i=1}^{n} \delta_{b_n^{-1}(X_i - a_n)}. \tag{1}$$

For $A \subset \mathbb{R}$, $\mathcal{P}_n(A)$ is a random variable equal to the number of points X_i inside the set $a_n + b_n A$. It is clear that $\mathbb{E}\mathcal{P}_n([\alpha, \beta])$, i.e., the mean number of points in the window $[a_n + \alpha b_n, a_n + \beta b_n]$ is of order $ne^{-a_n^2/2} b_n/\sqrt{2\pi}$. It is thus natural to assume that this number is finite. We will assume

$$\lim_{n\to\infty} \frac{ne^{-a_n^2/2} b_n}{\sqrt{2\pi}} = 1. \tag{2}$$

Depending on the sequence a_n the point process introduced in (1) describes the behavior of the values $X_1, X_2, \dots, X_n$ in the "bulk" of the sample or at the edge of the sample. Let us first look at the behavior in the "bulk", i.e., the case when

$$\lim_{n\to\infty} \frac{a_n e^{a_n^2/2}}{n} = 0. \tag{3}$$

Theorem 2.1 (Poissonian behavior in the bulk). *Under assumptions* (2) *and* (3) *the sequence of point processes* $\mathcal{P}_n$*, defined in* (1)*, converges in distribution to a Poisson point process on* $\mathbb{R}$ *whose intensity measure is the Lebesgue measure.*

This implies in particular that the spacings between the values X_i, which fall in the window $[a_n + \alpha b_n, a_n + \beta b_n]$ are asymptotically independent and exponentially distributed. Indeed, let us consider the order statistics

$$X^{(1)} \geq X^{(2)} \geq \dots \geq X^{(n)}, \tag{4}$$

i.e., the values $X_1, X_2, \dots, X_n$ in non-increasing order.

Corollary 2.1. *Define* r_n *so that* $X^{(r_n)} < a_n \leq X^{(r_n+1)}$. *Then under assumptions* (2) *and* (3) *for any fixed positive integer* $\ell \geq 1$

$$\frac{n e^{-a_n^2/2}}{\sqrt{2\pi}} (X^{(r_n+1)} - a_n, X^{(r_n+2)} - a_n, \dots, X^{(r_n+\ell)} - a_n) \tag{5}$$

converges in distribution as $n \to \infty$ *to* $(W_1, W_1+W_2, \dots, W_1+\dots+W_\ell)$*, where* W_i *are i.i.d. random variables which are distributed exponentially with parameter* 1.

The Poissonian behavior is valid as long as condition (3) is satisfied, i.e., as long as the window is "in the bulk". For instance, one could choose a_n to be $a_n = o(\sqrt{\log n})$ or $a_n = c\sqrt{\log n}$ with $c < \sqrt{2}$ or even $a_n = \sqrt{2\log n}(1 - \varepsilon_n)$ with $\varepsilon_n > 0$ satisfying

$$\liminf_{n\to\infty} \varepsilon_n \frac{\log n}{\log\log n} > \frac{1}{4}. \tag{6}$$

For a_n growing larger the result is different since the influence of the extreme values of the $X_i's$ is then dominant.

Theorem 2.2 (Poissonian behavior at the edge). *Assume* (2) *and*

$$\lim_{n\to\infty} \frac{\sqrt{2\pi} a_n e^{a_n^2/2}}{n} = c > 0. \tag{7}$$

Then the sequence of point processes $\mathcal{P}_n$*, defined in* (1)*, converges in distribution to a Poisson point process with intensity measure given by*

$$\mu(dt) = e^{-ct} dt. \tag{8}$$

Corollary 2.2. *Under assumptions of Theorem* 2.2 *for any positive integer* k

$$\left(\frac{X^{(1)} - a_n}{b_n}, \dots, \frac{X^{(k)} - a_n}{b_n} \right) \tag{9}$$

converges in distribution to a non-degenerate random vector as $n \to \infty$. In particular, the normalized maximum

$$\frac{X^{(1)} - a_n}{b_n} \tag{10}$$

converges in distribution to a Gumbel random variable:

$$\mathbb{P}\left(\frac{X^{(1)} - a_n}{b_n} > x\right) = \mathbb{P}(\mathcal{P}_n((x,\infty)) = 0) \to \exp(-\mathrm{e}^{-cx}/c). \tag{11}$$

2.2. Important models for local REM-universality

After this brief survey of the results for the i.i.d. case, we now want to come to our universality question and examine the influence of correlations on these Poissonian results. Consider the case of a Gaussian process H_N, defined on a large finite space S_N, when $|S_N|$ tends to infinity as $N \to \infty$. An element $\sigma \in S_N$ will be called a *configuration*. For given $\sigma \in S_N$ we call the corresponding value of the function $H_N(\sigma)$ the energy of the configuration σ. We assume that the process H_N is centered

$$\forall \sigma \in S_N \quad \mathbb{E}(H_N(\sigma)) = 0 \tag{12}$$

and that its covariance is given by

$$\forall \sigma, \sigma' \in S_N \quad \mathbb{E}(H_N(\sigma)H_N(\sigma')) = K_N(\sigma,\sigma') \text{ and } K_N(\sigma,\sigma) = 1. \tag{13}$$

If the covariance function is

$$K_N(\sigma,\sigma') = \mathbf{1}_{\sigma=\sigma'}, \tag{14}$$

then the values $(H_N(\sigma))_{\sigma \in S_N}$ are i.i.d. Gaussian random variables and then Theorems 2.1 and 2.2 are valid with $n = |S_N|$.

The local or micro-canonical REM-universality result we want to explain first is that, under general conditions on the covariance (13), results analogous to Theorem 2.1 are valid, i.e., we can observe Poissonian statistics in the bulk of the sample.

Let us consider an important and specific example, which will occupy most of this survey. Let S_N be the N-dimensional hypercube $\{-1,1\}^N$ and consider a centered Gaussian process $(H_N(\sigma))_{\sigma \in S_N}$ indexed by S_N with covariance

$$\mathrm{Cov}(H_N(\sigma), H_N(\sigma')) = \nu(R(\sigma,\sigma')), \tag{15}$$

where $R(\sigma,\sigma') = N^{-1}\sum_{i=1}^{N} \sigma_i\sigma_i'$ is the so-called *overlap* and ν is a smooth function with $\nu(1) = 1$. We will always assume that

$$\nu(0) = 0, \tag{16}$$

which will be our key hypothesis. Note that the covariance is a function of the Hamming distance $d(\sigma,\sigma') = \#\{i \le N : \sigma_i \ne \sigma_i'\}$:

$$\mathrm{Cov}(H_N(\sigma), H_N(\sigma')) = \nu\left(1 - \frac{2}{N}d(\sigma,\sigma')\right). \tag{17}$$

Thus our hypothesis (16) means that the covariance vanishes when the configurations σ, σ' are at the typical distance $N/2$ on the hypercube.

Example 2.1 (Random Energy Model). Consider the case when

$$\nu(r) = \mathbf{1}_{r=1}. \tag{18}$$

Then the random variables $H_N(\sigma), \sigma \in S_N$, are i.i.d. standard normal random variables and they are the energy levels of the simplest spin-glass model, the Random Energy Model (REM), introduced by B. Derrida in [25]. In this case Theorems 2.1 and 2.2 are clearly valid with n substituted by 2^N.

Example 2.2 (Number Partitioning Problem). Consider the case when

$$\nu(r) = r. \tag{19}$$

Then the random variables $H_N(\sigma), \sigma \in S_N$, are the energy levels of the Gaussian version of the number partitioning problem (NPP), which is usually described in the following equivalent way:

$$H_N(\sigma) = \frac{1}{\sqrt{N}} \sum_{i=1}^{N} g_i \sigma_i, \tag{20}$$

where $(g_i)_{1 \leq i \leq N}$ are i.i.d. standard normal rv's.

It is not at all clear that a Poissonian behavior analogous to the one given in Theorem 2.1 could be valid here because of the strong correlations between the energy levels. Nevertheless, Bauke and Mertens (see [3]) observed a Poissonian behavior in the bulk (initially when a_N was finite). The Poissonian behavior has then been proven to be correct if and only if $a_N = o(N^{1/4})$, (see [22, 14, 15]). Notice that the maximum value of the Hamiltonian is of order $N^{1/2}$. Thus there is a very large gap between the scales in the bulk where REM-universality holds, i.e., $o(N^{1/4})$, and the edge of the sample of order $N^{1/2}$. This will be detailed in Section 3.1.1.

Example 2.3 (Sherrington–Kirkpatrick mean-field spin glass). Consider the case when

$$\nu(r) = r^2. \tag{21}$$

This is the celebrated Sherrington–Kirkpatrick (SK) mean-field model of a spin glass. It is predicted that the equilibrium picture in this model is rather involved exhibiting a full continuous replica symmetry breaking at low temperatures. The subject of our interest though are the energy levels in the bulk which clearly do not influence the Gibbs distribution and thus the equilibrium properties of the model. The energy levels of the SK could be equivalently written as

$$H_N(\sigma) = \frac{1}{N} \sum_{1 \leq i_1, i_2 \leq N} g_{i_1,i_2} \sigma_{i_1} \sigma_{i_2}, \tag{22}$$

where $(g_{i_1,i_2})_{1 \leq i_1, i_2 \leq N}$ are i.i.d. standard normal random variables. The Poissonian behavior has been shown if and only if the energy scales a_N satisfy $a_N = o(N^{1/2})$, (see [22, 15]). Thus the REM-universality is valid up to scales of order $o(N^{1/2})$ and the edge of the sample is of order $N^{1/2}$. The gap is much smaller than in Example 2.2 but still exists. For details see Section 3.1.2.

Example 2.4 (p-spin models). Consider the case when

$$\nu(r) = r^p, \quad p \geq 3. \tag{23}$$

This model is well known in statistical mechanics of spin glass systems as the p-spin interaction model. Both, NPP and SK are versions of the p-spin model with $p = 1$ and $p = 2$ respectively. Similarly, its energy levels could be rewritten in the following equivalent form:

$$H_N(\sigma) = \frac{1}{\sqrt{N^p}} \sum_{1 \leq i_1, \ldots, i_p \leq N} g_{i_1,\ldots,i_p} \sigma_{i_1} \ldots \sigma_{i_p}, \tag{24}$$

where $(g_{i_1,\ldots,i_p})_{1 \leq i_1,\ldots,i_p \leq N}$ are i.i.d. standard normal random variables. It is predicted that the behavior in equilibrium of p-spin models with $p \geq 3$ is remarkably different from the behavior of the SK model. Physicists predict the onset of a so-called one step replica symmetry breaking picture.

Here the REM-universality persists for energy scales $a_N < c_p N^{1/2}$ with $c_p > 0$ which is remarkably of the same order as the edge. Thus the gap between the region where we observe a Poissonian behavior of energy levels and the edge of the sample is rather small. For details we refer to Section 3.1.3.

We will give in Section 3.1 results on the micro-canonical REM-universality for general covariance of the form (15) on the hypercube. This will show that the behavior given by Examples 2.2, 2.3 and 2.4 is typical.

We want to emphasize now that REM-universality is not restricted to the cases described above where S_N is the N-dimensional hypercube and to a covariance structure of the form (15). To support this we give Examples 2.5 and 2.6 for which REM-universality has been established rigorously, while in Examples 2.7 and 2.8 it was conjectured but the conjecture is still open (see for instance H. Bauke's PhD thesis [1]).

Example 2.5 (Edwards–Anderson short range spin glass). Let Λ_N be the d-dimensional torus of length N, i.e., $\Lambda_N = \{0, 1, \ldots, N-1\}^d$ and $|\Lambda_N| = N^d$. We consider the configuration space $S_N = \{-1, 1\}^{\Lambda_N}$ and the Hamiltonian corresponding to $\sigma \in S_N$ of the form

$$H_N(\sigma) = \frac{1}{\sqrt{N^d}} \sum_{x \sim y} g_{x,y} \sigma_x \sigma_y, \tag{25}$$

where $g_{x,y}$, $x, y \in \Lambda_N$ are i.i.d. standard normal random variables and $x \sim y$ means that the sites x and y are neighbors on Λ_N. Then

$$\mathrm{Cov}(H_N(\sigma), H_N(\sigma')) = \frac{1}{N^d} \sum_{x \sim y} \sigma_x \sigma_y \sigma'_x \sigma'_y. \tag{26}$$

Up to the multiplicative factor $1/\sqrt{N^d}$ the model with Hamiltonian (25) is the Gaussian version of the famous Edwards–Anderson (EA) model. The debate about the equilibrium properties of this model is still open. We will see that, as long as the REM-universality is concerned, it is most likely that the EA model behaves as the NPP in Example 2.2. More precisely, it is established that the micro-canonical REM-universality holds if $a_N = o(N^{d/4}/\log N)$ but we believe that, in fact, REM-universality holds if and only if $a_N = o(N^{d/4}) = o(|\Lambda_N|^{1/4})$. In other words, we believe that the threshold on the validity of the REM-like behavior diverges with the size of the system with critical exponent 1/4 while the scale of the ground state is $N^{d/2} = |\Lambda_N|^{1/2}$. Bovier and Kurkova proved in [22] that the Poissonian behavior holds for $a_N = cN^\gamma, c \in \mathbb{R}, \gamma \in [0, d/4)$ and by adapting the proof of Section 4.2 one can prove that it persists up to $a_N = o(N^{d/4}/\log N)$. For details we refer to Sections 3.2 and 4.2 where we also explain what combinatorial step is missing to get $a_N = o(N^{d/4})$.

Example 2.6 (Directed Polymers). Let Ω be the path space of a simple random walk on the d-dimensional lattice $\mathbb{Z}^d$, i.e., $\Omega = \{(w_n)_{n\geq 0}; w_n \in \mathbb{Z}^d, w_0 = 0\}$. We define the configuration space S_N as the space of paths of length N, i.e., the set of graphs $\{(w_i)\}_{i=1}^N$ in $\mathbb{Z}^d$. A configuration $\sigma \in S_N$ will be called a *polymer chain.* Let $\{\eta(n,x), n \in \mathbb{N}, x \in \mathbb{Z}^d\}$ be a sequence of independent identically distributed normal random variables of zero mean and variance 1. They describe the random environment, i.e., the impurities at site $x \in \mathbb{Z}^d$ at time n. We define the energy of the chain $\sigma = (w_i)_{i=0}^N$ to be

$$H_N(\sigma) = \frac{1}{\sqrt{N}} \sum_{i=1}^N \eta(i, w_i). \tag{27}$$

Then for a given pair $\sigma, \sigma' \in S_N$

$$\mathrm{Cov}(H_N(\sigma), H_N(\sigma')) = \frac{1}{N} \sum_{i=1}^N 1_{w_i(\sigma)=w_i(\sigma')}, \tag{28}$$

and thus the covariance is a function of the number of sites common to both chains σ and σ'. The REM-universality for directed polymers is proved in [29] in dimension $d = 1$ for energy scales $a_N = cN^\gamma, c \in \mathbb{R}, \gamma \in [0, 1/4)$ and in dimension $d \geq 2$ for energy scales $a_N = cN^\gamma, c \in \mathbb{R}, \gamma \in [0, 1/2)$. It is an open question whether 1/4 and 1/2 are the true critical exponents.

Example 2.7 (Minimum Spanning Tree Problem). Assume $G_N = (V_N, E_N)$ is a sequence of complete graphs, where V_N is the set of vertices and $E_N = V_N \times V_N$ is the set of edges of G_N, we fix $|V_N| = N$. Define S_N to be the set of spanning trees of G_N with Hamiltonian (or weight) corresponding to a configuration $\sigma \in S_N$ given by

$$H_N(\sigma) = \frac{1}{\sqrt{N-1}} \sum_{e\in\sigma} X(e), \tag{29}$$

where $(X(e))_{e\in E_N}$ is a set of i.i.d. standard normal random variables. Then clearly

$$\mathrm{Cov}(H_N(\sigma), H_N(\sigma')) = \frac{1}{N-1} \sum_{e\in E_N} \mathbf{1}_{e\in\sigma, e\in\sigma'}, \tag{30}$$

i.e., the covariance is a function of the number of common edges of configurations σ and σ'. It is a Gaussian version of the minimal spanning tree problem (MSTP).

The REM-universality has been conjectured by Mertens and Bauke but, to our knowledge, there are no rigorous results, even for constant energy scales.

Example 2.8 (Travelling Salesman Problem). As in Example 2.7 we consider a sequence of complete graphs $G_N = (V_N, E_N)$ on N vertices. Define S_N to be the set of closed paths of length N on the graph G_N. The Hamiltonian corresponding to the configuration $\sigma \in S_N$ (the weight corresponding to the path σ) is defined as

$$H_N(\sigma) = \frac{1}{\sqrt{N}} \sum_{e \in \sigma} X(e), \tag{31}$$

where $(X(e))_{e \in E_N}$ is a set of i.i.d. standard normal rv's. Clearly, the covariance

$$\mathrm{Cov}(H_N(\sigma), H_N(\sigma')) = \frac{1}{N} \sum_{e \in E_N} \mathbf{1}_{e \in \sigma, e \in \sigma'} \tag{32}$$

is a function of the number of common edges of paths σ and σ'. Again, as in Example 2.7 it is conjectured by Bauke and Mertens that REM-universality holds for TSP, but no rigorous results are known to us.

Let us now explain very vaguely the reason why correlations can be ignored safely for the bulk behavior in Examples 2.2, 2.3, 2.4 and thus the rough mechanism for the proof of REM-universality in these examples. The "rough" mechanism is the following: the finite set of configurations σ corresponding to values $H_N(\sigma)$ in a window $[a_N + \alpha b_N, a_N + \beta b_N]$ are "sparse" in the configuration space and they are "typical", i.e., they are close to being distributed uniformly on S_N. So that the overlap of two such configurations σ and σ' is close to zero and thus by the critical assumption that $\nu(0) = 0$ the correlation between the values $H_N(\sigma)$ and $H_N(\sigma')$ is also close to zero, which makes the distribution of the values of the Hamiltonian in the window $[a_N + \alpha b_N, a_N + \beta b_N]$ close to that of Example 2.2, i.e., to the Random Energy Model.

2.3. Re-sampling REM-universality

For many random Hamiltonians Theorem 2.1 is valid for a wide region in the bulk. But nothing like Theorem 2.2 about the edge bevavior can be proven or even expected, since in order to feel the intensity $\mathrm{e}^{-ct}\mathrm{d}t$ one must get near the edge where the correlations seem too strong to be controlled. We will now present a different form of REM-universality, a different way to "kill" correlations, which will enable us to extend fully Theorems 2.1 and 2.2 to a wide class of random Hamiltonians. The natural way to do that, different from the micro-canonical point of view given above is the following. Instead of looking at the values of the Hamiltonian restricted to a small window, one could re-sample those values, i.e., draw at random a subset of all possible values, and hope that the same mechanism as above will also kill the influence of correlations in this context. Hence the name of this new form of REM-universality: *re-sampling REM-universality.*

Let us be more specific about this new variant of REM-universality for models on the hypercube. We again assume that the configuration space is $S_N = \{-1, 1\}^N$

and we consider a random Hamiltonian $H_N(\sigma)$ on S_N, which is a Gaussian process satisfying conditions (12) and (13). Now consider a sequence of positive numbers $M(N) \leq N$ and for fixed N define a set of i.i.d. Bernoulli random variables $(\mathcal{X}_\sigma)_{\sigma \in S_N}$ with

$$\mathbb{P}(\mathcal{X}_\sigma = 1) = 1 - \mathbb{P}(\mathcal{X}_\sigma = 0) = \frac{2^M}{2^N}. \tag{33}$$

Let us define a sequence of random subsets $\mathcal{X}_N \subset S_N$ as $\mathcal{X}_N = \{\sigma \in S_N : \mathcal{X}_\sigma = 1\}$ and for a given realization of random variables $(\mathcal{X}_\sigma)_{\sigma \in S_N}$ examine the statistics of the energy levels $H_N(\sigma)$ restricted to $\mathcal{X}_N$. The main object of our study will be the sequence of point processes $\mathcal{P}_N$ on $\mathbb{R}$, defined as

$$\mathcal{P}_N = \sum_{\sigma \in \mathcal{X}_N} \delta_{b_N^{-1}(H_N(\sigma) - a_N)}, \tag{34}$$

where the sequences a_N and b_N will be chosen later. As before, we assume that for many disordered systems the asymptotic behavior of the sequence $\mathcal{P}_N$ is universal.

In order to realize what this universality behavior could be let us examine the case when $\nu(r) = \mathbf{1}_{r=1}$, i.e., the case of the Random Energy Model. For the REM the random variables $H_N(\sigma), \sigma \in \mathcal{X}_N$, are independent and if $M(N)$ is not too small then $\mathbb{P}$-almost surely there are $2^M(1+o(1))$ configurations in the random subset $\mathcal{X}_N$. Thus results analogous to Theorems 2.1 and 2.2 are valid with $n = 2^M$, i.e., under normalization satisfying

$$\lim_{N\to\infty} \frac{2^M \mathrm{e}^{-a_N^2/2} b_N}{\sqrt{2\pi}} = 1 \tag{35}$$

$\mathbb{P}$-almost surely the sequence of point processes $\mathcal{P}_N$ converges in distribution to the Poisson point process on $\mathbb{R}$ with intensity measure equal to the Lebesgue measure if

$$\lim_{N\to\infty} \frac{a_N \mathrm{e}^{a_N^2/2}}{2^M} = 0 \tag{36}$$

and to the Poisson point process on $\mathbb{R}$ with intensity measure given by

$$\mu(\mathrm{d}t) = \mathrm{e}^{-ct}\mathrm{d}t \tag{37}$$

if

$$\lim_{N\to\infty} \frac{\sqrt{2\pi} a_N \mathrm{e}^{a_N^2/2}}{2^M} = c > 0. \tag{38}$$

Therefore the re-sampling REM-universality is that for a wide class of random Hamiltonians one finds the Poissonian statistics not only in the “bulk” of the random sample $\mathcal{X}_N$ but also at the edge of the sample.

If $M = N$, i.e., the random sample $\mathcal{X}_N$ coincides with S_N, then we do not expect that the analog of Theorem 2.2 is valid unless it is the Random Energy Model. Instead, we expect that there is a gap between the largest sequence $M = M(N)$ such that the re-sampling REM-universality still holds and N. As we will see in the following the size of the gap will depend on the behavior of the function ν at zero.

Let us examine whether Examples 2.2–2.8 exhibit the re-sampling REM-universality and if so what is the threshold on the validity of this REM-universality.

Example 2.2. For the Gaussian version of the number partitioning problem it is proved in [12] that the re-sampling REM-universality holds if and only if $M = o(\sqrt{N})$. More precisely, $\mathbb{P}$-almost surely the sequence of point processes $\mathcal{P}_N$, defined in (34) with normalization given by (35) and (38), converges weakly to the Poisson point process with intensity defined by (37) if and only if $M = o(\sqrt{N})$. Thus there is a very large gap in the case of the NPP. For details see Section 4.1.1.

Example 2.3. For the SK model from Example 2.3 it is proved in [12] that the re-sampling REM-universality is valid as long as $M = o(N)$. Thus, just like in the micro-canonical picture, the gap for the SK model still exists but is much smaller than for the NPP. This will be detailed in Section 4.1.2.

Example 2.4. It is proved in [12] that for the p-spin models, $p \geq 3$, the re-sampling REM-universality is valid as long as $M = o(N)$. In [13] it is further shown that the true threshold is actually of the form cN with some $c > 0$ and thus the gap is rather small. For details we refer to Section 4.1.3.

Even though by now we considered the re-sampling REM-universality only on the hypercube S_N it goes beyond this class of models as we will see in the following example.

Example 2.5. The configuration space of the Edwards–Anderson short range spin glass is $S_N = \{-1, 1\}^{\Lambda_N}$, where Λ_N is the d-dimensional torus of length N. To state the re-sampling REM-universality for the EA spin glass we consider a sequence of positive numbers $M(N) \leq N$ and for each N consider a set of i.i.d. Bernoulli random variables $(\mathcal{X}_\sigma)_{\sigma \in S_N}$ with

$$\mathbb{P}(\mathcal{X}_\sigma = 1) = 1 - \mathbb{P}(\mathcal{X}_\sigma = 0) = \frac{2^{M^d}}{2^{N^d}}. \tag{39}$$

Then the average size of the random set $\mathcal{X}_N = \{\sigma \in S_N : \mathcal{X}_\sigma = 1\}$ is 2^{M^d}. The re-sampling REM-universality reduces to the following. The sequence of point processes $\mathcal{P}_N$, defined in (34), under normalization (2) with $n = 2^{M^d}$, i.e.,

$$\lim_{N \to \infty} \frac{2^{M^d} e^{-a_N^2/2} b_N}{\sqrt{2\pi}} = 1, \tag{40}$$

and (38) with $M = M^d$, i.e.,

$$\lim_{N \to \infty} \frac{\sqrt{2\pi} a_N e^{a_N^2/2}}{2^{M^d}} = c > 0, \tag{41}$$

converges weakly to the Poisson point process with intensity measure given by (37). In Section 4.2 we establish that the re-sampling REM-universality holds if $M = o(N^{1/2}/\log N)$. In fact, we believe that the re-sampling REM-universality holds for the EA spin glass if and only if $M = o(N^{1/2})$ and thus, just like in the micro-canonical framework, the EA model has the same critical exponent as the NPP.

In Examples 2.6, 2.7 and 2.8 no rigorous work is known to us but it is likely that the re-sampling REM-universality holds for small enough random subsets.

2.4. Dynamics and REM-universality

We will show in Section 5 how the equilibrium REM-universality described above translates for dynamics questions. We will restrict ourselves to the case of the mean-field spin glasses, mainly the REM itself (Example 2.1) and the p-spin models, $p \geq 3$, (Example 2.4). We report in Section 5 on works done jointly (by one of us) with A. Bovier, J. Černý and V. Gayrard. This line of research explains how the aging results of the REM are universal for p-spin models ($p \geq 3$) in a wide range of time scales e^{cN}, for $0 < c < c(p)$. One should note that these time scales are exponentially long but still "transient," i.e., shorter than the mixing time, or time to reach equilibrium. One cannot hope that this dynamic REM- universality is valid up to the mixing time since the last phase of approach to equilibrium must be model-dependent, as the equilibrium is.

3. The local or micro-canonical REM-universality

In this section we study in more depth the local or micro-canonical REM-universality introduced in Section 1. More precisely, we consider a Gaussian process H_N, defined on a large finite space $S_N, |S_N| \to \infty$, and study the point process, defined by values $(H_N(\sigma))_{\sigma\in S_N}$, in the bulk of the sample. To do so we study the asymptotic behavior of the sequence of point processes $\mathcal{P}_N$, defined as

$$\mathcal{P}_N = \sum_{\sigma\in S_N} \delta_{b_N^{-1}(H_N(\sigma)-a_N)}, \tag{42}$$

where a_N, b_N satisfy

$$\lim_{N\to\infty} \frac{a_N e^{a_N^2/2}}{|S_N|} = 0 \text{ and } \lim_{N\to\infty} \frac{|S_N| e^{-a_N^2/2} b_N}{\sqrt{2\pi}} = 1. \tag{43}$$

Micro-canonical REM-universality says that for a wide class of random Hamiltonians H_N and, for a wide range of scales a_N, the sequence $\mathcal{P}_N$ converges weakly to a Poisson point process whose intensity measure is the Lebesgue measure.

We give rigorous results on the local REM-universality for the mean-field models on the hypercube in Section 3.1 and on the Edwards–Anderson short range spin glass in Section 3.2.

3.1. Mean-field models on the hypercube

Let us consider the case of a Gaussian Hamiltonian H_N indexed by the N-dimensional hypercube. We assume that H_N is centered, i.e.,

$$\forall\sigma \in S_N \qquad \mathbb{E}(H_N(\sigma)) = 0, \tag{44}$$

and its covariance is a function of the overlap, $R(\sigma,\sigma') = N^{-1}\sum_{i=1}^{N} \sigma_i\sigma'_i$, i.e., it is given by

$$\forall\sigma,\sigma' \in S_N \qquad \mathbb{E}(H_N(\sigma)H_N(\sigma')) = \nu(R(\sigma,\sigma')). \tag{45}$$

Moreover, we always assume that the covariance ν is a smooth function satisfying the key hypothesis

$$\nu(0) = 0. \tag{46}$$

We will show here that the types of behavior observed in Examples 2.2, 2.3 and 2.4 are typical and actually are the only possible.

Definition 3.1. We will say that Hamiltonian H_N is of:

- the NPP-type if $\nu'(0) \neq 0$;
- the SK-type if $\nu'(0) = 0$ and $\nu''(0) \neq 0$;
- the p-spin type if $\nu'(0) = \nu''(0) = 0$.

In the rest of the section we will consider separately each of these three types and show that the gap between the region where the micro-canonical REM-universality holds and the edge of the sample $(H_N(\sigma))_{\sigma \in S_N}$ depends only on the type of the Hamiltonian H_N.

3.1.1. The NPP-type models. For the number partitioning problem that was introduced in Section 1 (see Example 2.2) the micro-canonical REM-universality holds if and only if $a_N = o(N^{1/4})$. In this section we give results for the whole class of the NPP-type models. We also briefly describe the approach used in the proof, the so-called *moment method.*

Theorem 3.1. *Assume the Hamiltonian H_N is of the NPP-type.*

(a) *If $a_N = o(N^{1/4})$ then the sequence of point processes $\mathcal{P}_N$, defined as*

$$\mathcal{P}_N = \sum_{\sigma \in S_N} \delta_{b_N^{-1}(H_N(\sigma) - a_N)}, \tag{47}$$

converges weakly to a Poisson point process whose intensity measure is the Lebesgue measure, where b_N satisfies (2), *i.e.,*

$$\lim_{N \to \infty} \frac{2^N e^{-a_N^2/2} b_N}{\sqrt{2\pi}} = 1. \tag{48}$$

(b) *If the sequence a_N is such that*

$$\limsup_{N \to \infty} \frac{a_N}{N^{1/4}} \in (0, \infty), \tag{49}$$

then the sequence of point processes $\mathcal{P}_N$ does not converge to a Poisson point process.

Let us indicate briefly how one can prove Theorem 3.1. The proof is based on the following

Proposition 3.1. *Let $\xi_n = \sum_{i=1}^{k_n} \delta_{X_{i,n}}$ be a sequence of point processes on $\mathbb{R}$ defined on a common probability space $(\Omega, \mathcal{F}, \mathbb{P})$ and ξ be a Poisson point process on $\mathbb{R}$ with intensity measure μ. Assume that for all bounded Borel sets $A \subset \mathbb{R}$ and for all positive integers $\ell \geq 1$*

$$\lim_{N \to \infty} \mathbb{E}(\xi_N(A))_\ell = (\mu(A))^\ell, \tag{50}$$

where for a random variable Z we define $\mathbb{E}(Z)_\ell \equiv \mathbb{E}Z(Z-1)\dots(Z-\ell+1)$ and call it the ℓth factorial moment. Then the sequence of point processes ξ_N converges weakly to a Poisson point process on $\mathbb{R}$ with intensity measure μ.

Proof. The proof of the proposition is based on the following classical theorem which is easily derived from Theorem 4.7 in [27].

Theorem 3.2. *With the notation of Proposition 3.1 the sequence ξ_n converges weakly to ξ if the following conditions are satisfied:*

(1) *For all intervals $A = [a, b)$*

$$\lim_{n\to\infty} \mathbb{E}(\xi_n(A)) = \mu(A). \tag{51}$$

(2) *For all finite unions of intervals $A = \bigcup_{i=1}^{k} [a_i, b_i)$*

$$\lim_{n\to\infty} \mathbb{P}(\xi_n(A) = 0) - \mathrm{e}^{-\mu(A)}. \tag{52}$$

Taking $\ell = 1$ in (50) we obtain that condition (1) is satisfied. Thus we are left with the proof of the condition (2).

For a Poisson random variable with parameter $\lambda < \infty$, which we will denote $\mathrm{Po}(\lambda)$, its moments unambiguously determine its distribution. Therefore given a sequence of random variables $(Z_i)_{i\geq 1}$, defined on a common probability space $(\Omega, \mathcal{F}, \mathbb{P})$, moment convergence

$$\lim_{i\to\infty} \mathbb{E}Z_i^\ell \to \mathbb{E}(\mathrm{Po}(\lambda))^\ell \text{ for all } \ell \geq 1 \tag{53}$$

implies weak convergence $Z_i \xrightarrow{w} \mathrm{Po}(\lambda)$ and thus

$$\lim_{i\to\infty} \mathbb{P}(Z_i = 0) \to \mathbb{P}(\mathrm{Po}(\lambda) = 0) = \mathrm{e}^{-\lambda}. \tag{54}$$

It is clear from the definition that convergence of the factorial moments

$$\lim_{i\to\infty} \mathbb{E}(Z_i)_\ell \to \mathbb{E}(\mathrm{Po}(\lambda))_\ell = \lambda^\ell \text{ for all } \ell \geq 1 \tag{55}$$

is equivalent to convergence of the ordinary moments (53) and thus it implies (54). Thus (50) implies condition (2) of Theorem 3.2 and the sequence ξ_N converges weakly to a Poisson point process with intensity measure μ. □

One then establishes REM-universality in part (a) of Theorem 3.1 using Proposition 3.1 and the following result which we give without a proof.

Theorem 3.3. *If the Hamiltonian H_N is of the NPP-type and $a_N = o(N^{1/4})$ then for every integer $\ell \geq 1$ and every bounded Borel set A*

$$\lim_{N\to\infty} \mathbb{E}(\mathcal{P}_N(A))_\ell = |A|^\ell, \tag{56}$$

where $|A|$ denotes the Lebesgue measure of A.

Theorem 3.3 is the core of the proof of universality. For a proof we refer the reader to [22] where it is proved for $a_N = cN^\gamma, c \in \mathbb{R}, \gamma \in [0, 1/4)$ but it could be easily adjusted to work for $a_N = o(N^{1/4})$.

We now want to concentrate on the limits of the REM-universality, i.e., we want to prove that for large a_N the sequence of point processes $\mathcal{P}_N$ does not converge. For that we use the following

Theorem 3.4. *With the notation of Proposition* 3.1, *assume that for some Borel set A the following conditions hold:*

(1) $\lim_{N\to\infty} (\xi_N(A))_1$ *exists,*

(2) $\lim_{N\to\infty} (\xi_N(A))_2$ *does not equal to* $\left(\lim_{N\to\infty} (\xi_N(A))_1\right)^2$,

(3) $\lim_{N\to\infty} (\xi_N(A))_3 < \infty$.

Then the sequence of point processes ξ_N *does not converge weakly to a Poisson point process.*

This theorem is a consequence of the following simple consideration. Given a sequence of random variables $(Z_i)_{i\geq 1}$ assume that $Z_i \xrightarrow{w} \text{Po}(\lambda)$ and for some integer $\ell_0 > 1$

$$\limsup_{i\to\infty} \mathbb{E}(Z_i)_{\ell_0} < \infty. \tag{57}$$

Then necessarily $\mathbb{E}(Z_i)_\ell \to \lambda^\ell$ for all $\ell < \ell_0$. Therefore, if $Z_i \xrightarrow{w} \text{Po}(\lambda)$ and for some $\ell_0 > 1$ (57) holds and also for some $\ell < \ell_0$ sequence $\mathbb{E}(Z_i)_\ell$ does not converge to λ^ℓ then sequence Z_i does not converge weakly to $\text{Po}(\lambda)$. Taking $\ell_0 = 3$ and $\ell = 2$ implies Theorem 3.4.

Thus Theorem 3.4 together with the following theorem establishes part (b) of Theorem 3.1.

Theorem 3.5. *Assume the Hamiltonian* H_N *is of the NPP-type. For every bounded Borel set A,*

$$\lim_{N\to\infty} \mathbb{E}(\mathcal{P}_N(A))_1 = |A|. \tag{58}$$

Moreover, if $\limsup_{N\to\infty} \frac{a_N}{N^{1/4}} = \varepsilon < \infty$, *then*

(i) $\limsup_{N\to\infty} \mathbb{E}(\mathcal{P}_N(A))_2 = \mathrm{e}^{\varepsilon^2(\nu'(0))^2/2}|A|^2$,

(ii) $\limsup_{N\to\infty} \mathbb{E}(\mathcal{P}_N(A))_3 < \infty$.

The proof of Theorem 3.5 could be found in [14] (not just for the Gaussian version of the number partitioning problem but for quite general distribution of the random variables $(g_i)_{i\leq N}$ in (20)).

Let us conclude the part on the local REM-universality for the number partitioning problem with an open problem.

Problem 3.1. What type of statistics can one expect when observing the energy levels of a Gaussian version of the number partitioning problem inside a micro-canonical window $[a_N + \alpha b_N, a_N + \beta b_N]$ with $a_N = cN^{1/4}, c > 0$? It was shown in [15] that in that case the factorial moments are

$$\lim_{N\to\infty} \mathbb{E}(\mathcal{P}_N(A))_\ell = |A|^\ell e^{c^4 \ell(\ell-1)/4}, \tag{59}$$

but the moments growing so fast with ℓ do not determine the asymptotic distribution of $\mathcal{P}_N(A)$. Actually, it is not even clear that the point process with factorial moments defined by (59) exists, so the first problem would be to build such a process.

3.1.2. The SK-type models. For the Sherrington–Kirkpatrick model introduced in Section 1 (see Example 2.3) the micro-canonical REM-universality holds if and only if $a_N = o(N^{1/2})$. In this section we give the same REM-universality results for all SK-type models.

Theorem 3.6. *Let the Hamiltonian H_N be of the SK-type.*

(a) *If $a_N = o(N^{1/2})$ then the sequence of point processes $\mathcal{P}_N$, defined in* (47), *converges weakly to a Poisson point process whose intensity measure is the Lebesgue measure.*

(b) *If the sequence a_N satisfies*

$$\limsup_{N\to\infty} \frac{a_N}{N^{1/2}} \in \Big(0, \frac{1}{\sqrt{8\nu''(0)}}\Big) \tag{60}$$

then the sequence of point processes $\mathcal{P}_N$, defined in (47), *does not converge to a Poisson point process.*

To prove Theorem 3.6 one can use the same method as for Theorem 3.1 – the moment method. More precisely, Theorem 3.6 is a direct consequence of the following two theorems.

Theorem 3.7. *If H_N is of the SK-type and $a_N = o(N^{1/2})$ then for every $\ell \in \mathbb{N}$ and every bounded Borel set A,*

$$\lim_{N\to\infty} \mathbb{E}(\mathcal{P}_N(A))_\ell = |A|^\ell. \tag{61}$$

Theorem 3.8. *Assume that H_N is of the SK-type. Then for every bounded Borel set A,*

$$\lim_{N\to\infty} \mathbb{E}(\mathcal{P}_N(A))_1 = |A|. \tag{62}$$

Moreover, if the sequence a_N is such that

$$\limsup_{N\to\infty} \frac{a_N}{N^{1/2}} = \varepsilon < \frac{1}{\sqrt{8\nu''(0)}}, \tag{63}$$

then

(i) $\displaystyle\limsup_{N\to\infty} \mathbb{E}(\mathcal{P}_N(A))_2 = \frac{|A|^2}{\sqrt{1-2\nu''(0)\varepsilon^2}}$, (ii) $\displaystyle\limsup_{N\to\infty} \mathbb{E}(\mathcal{P}_N(A))_3 < \infty$.

We omit the proofs but refer the reader to [22] for the proof of Theorem 3.7 when $a_N = cN^\gamma, c \in \mathbb{R}, \gamma \in [0, 1/2)$. This proof extends easily to the case $a_N = o(N^{1/2})$. For the proof of a result equivalent to Theorem 3.8 we refer to [15].

Remark 3.1. The somewhat cumbersome restriction (63) is due to the fact that for large ε the third moment becomes infinite and thus Theorem 3.4 is not applicable. Condition (63) is not optimal and could be improved, the reason for such a choice is that for ε satisfying (63) the third moment estimate is quite simple.

Remark 3.2. In fact, our statement of Theorem 3.6 is not really accurate if ν is an even function. Then it is clear that $H_N(\sigma) = H_N(-\sigma)$ for every configuration $\sigma \in S_N$. Thus each value of the point process $\mathcal{P}_N$ is repeated twice which excludes a Poissonian limit. To circumvent this trivial symmetry effect we must redefine $\mathcal{P}_N$ to be

$$\mathcal{P}_N = \sum_{\sigma \in \Sigma_N} \delta_{b_N^{-1}(H_N(\sigma) - a_N)}, \tag{64}$$

where Σ_N is the set of residual classes modulo the group of automorphisms, G, of S_N, that leave $H_N(\sigma)$ invariant and a_N, b_N satisfy

$$\lim_{N\to\infty} \frac{a_N e^{a_N^2/2}}{|\Sigma_N|} = 0 \text{ and } \lim_{N\to\infty} \frac{|\Sigma_N| e^{-a_N^2/2} b_N}{\sqrt{2\pi}} = 1. \tag{65}$$

We will not mention this symmetry problem in the following.

3.1.3. The p-spin type models. For the p-spin models from Section 1, see Example 2.4, the micro-canonical REM-universality holds if $a_N < c_p N^{1/2}$. Here we give a proof of this fact, more precisely, we show that this behavior is common to all the p-spin type models (see Definition 3.1). We also show that the usual approach, the moment method, does not work for p-spin type models because the explosion of the factorial moments (see Theorem 3.11).

To formulate the main theorem of this section we define the function

$$\mathcal{J}(x) = \begin{cases} \frac{1}{2}(1-x)\log(1-x) + \frac{1}{2}(1+x)\log(1+x) & \text{if } x \in [-1,1], \\ +\infty & \text{otherwise.} \end{cases} \tag{66}$$

Theorem 3.9. *Assume the Hamiltonian H_N be of the p-spin type.*

(a) *If the sequence $a_N \geq 0$ satisfies*

$$\limsup_{N\to\infty} \frac{a_N}{N^{1/2}} < c_\nu, \tag{67}$$

where $c_\nu > 0$ is defined by

$$c_\nu^2 = \inf_{x\in(0,1)} \frac{\mathcal{J}(x)(1+\nu(x))}{\nu(x)}, \tag{68}$$

then the sequence of point processes $\mathcal{P}_N$, defined in (47), converges weakly to a Poisson point process whose intensity measure is the Lebesgue measure.

(b) *If the sequence a_N satisfies*

$$\limsup_{N\to\infty} \frac{a_N}{N^{1/2}} > c'_\nu, \tag{69}$$

where $c'_\nu > 0$ is defined by

$$(c'_\nu)^2 = \inf_{x\in(0,1)} \frac{2\mathcal{J}(x)}{\nu^2(x)}, \tag{70}$$

then the sequence of point processes $\mathcal{P}_N$, defined in (47), does not converge weakly to a Poisson point process.

Remark 3.3. In the case of pure p-spin models, i.e., when $\nu(r) = r^p$, $p \geq 3$ then c_ν coincides with the well-known estimate on the critical value of inverse temperature, see [31]. The reason is that the second-order calculations are used in both cases.

It is also established in [19] for the pure p-spin models, i.e., $\nu(r) = r^p$, that the constants c_ν and c'_ν satisfy for large p

$$\sqrt{2\log 2}\,(1 - 2^{-p}) \leq c_\nu < c'_\nu \leq \sqrt{2\log 2}\,(1 - 2^{-5p}). \tag{71}$$

Roughly speaking, it means that in the limit $p \to \infty$ the micro-canonical REM-universality persists even when we examine the energy levels in a small window very close to the maximum energy. This is not surprising since in the limit $p \to \infty$ p-spin models formally "converge" to the Random Energy Model for which the statistics of the extremes are Poissonian.

We draw the reader's attention to the fact that for the NPP-type and for the SK-type models we know the exact threshold on the validity of the REM-universality while for the p-spin type models there is a region for a_N,

$$c_\nu \leq \limsup_{N\to\infty} \frac{a_N}{N^{1/2}} \leq c'_\nu, \tag{72}$$

for which we can neither prove nor reject that the statistics of energy levels inside the micro-canonical window $[a_N + \alpha b_N, a_N + \beta b_N]$ are Poissonian.

Let us address the proof of Theorem 3.9. Part (b) is trivial. Indeed, for c'_ν, defined in (70), one can show by modifying the proof given in [19] that $\mathbb{P}$-almost surely

$$\limsup_{N\to\infty} \frac{\max_{\sigma\in S_N} H_N(\sigma)}{\sqrt{N}} < c'_\nu. \tag{73}$$

Thus for a sequence a_N satisfying (69) the observed window $[a_N+\alpha b_N, a_N+\beta b_N]$ is above the edge of the sample and clearly, we cannot have the Poissonian statistics because there are just no energy levels there.

To prove part (a) of Theorem 3.9 in case $a_N = o(N^{1/2})$ one can use the moment method, more precisely it follows from Proposition 3.1 and the following theorem.

Theorem 3.10. *If the Hamiltonian H_N is of the p-spin type and $a_N = o(N^{1/2})$ then for every $\ell \in \mathbb{N}$ and every bounded Borel set A*

$$\lim_{N\to\infty} \mathbb{E}(\mathcal{P}_N(A))_\ell = |A|^\ell. \tag{74}$$

Theorem 3.10 was established in [22] for $a_N = cN^\gamma, c \in \mathbb{R}, \gamma \in [0, 1/2)$ and $\nu(r) = r^p$ but the proof given there could be extended for $a_N = o(N^{1/2})$ and for a general ν.

It was actually conjectured in a follow-up paper, see [23], that the moment method would work to prove the REM-universality for $a_N = O(N^{1/2})$. However it is not the case and the moment method fails to establish the REM-universality in part (a) of Theorem 3.9 when

$$\limsup_{N\to\infty} \frac{a_N}{N^{1/2}} > 0. \tag{75}$$

The reason is that for models with $\nu'(0) = 0$, high enough factorial moments of the random variable $\mathcal{P}_N(A)$ blow up as $N \to \infty$.

Theorem 3.11. *Assume $\nu'(0) = 0$. Then for every sequence $(a_N)_{N\geq 1}$ satisfying*

$$\limsup_{N\to\infty} \frac{a_N}{N^{1/2}} > 0, \tag{76}$$

there exists ℓ_0 such that for all $\ell \geq \ell_0$ and for all Borel sets $A \subset \mathbb{R}$ with nonzero Lebesque measure

$$\limsup_{N\to\infty} \mathbb{E}(\mathcal{P}_N(A))_\ell = \infty, \tag{77}$$

where the sequence of point processes $\mathcal{P}_N$ is defined in (47).

We give the proof of Theorem 3.11 in Section 3.1.5, it was first proved in [13].

Remark 3.4. In [23], using the moment method, REM-universality was proved up to $a_N = cN^{1/2}$ for an explicit constant c. The conjecture was that the constant c is positive. It is a direct consequence of the proof of Theorem 3.11 that, in fact, this constant c is 0.

The fact that factorial moments explode does not imply the absence of REM-universality since moment convergence is only a sufficient condition for weak convergence. It turns out that the explosion of the factorial moments is caused by the "highly" correlated energy levels that are inside the window $[a_N + \alpha b_N, a_N + \beta b_N]$. Therefore, the idea is to exclude them from our consideration. We show that for

$$\limsup_{N\to\infty} \frac{a_N}{N^{1/2}} < c_\nu, \tag{78}$$

where $c_\nu > 0$ is defined in (68), the probability to have two energy levels that are inside the window and are highly correlated (i.e., levels corresponding to configurations with a large overlap or, equivalently, to configurations close enough on the hypercube) is negligible in the limit $N \to \infty$ and an event with probability approaching zero has no influence on the weak convergence. More quantitatively, we fix sequences $R_N \to 0$ and $\gamma_N \to \infty$, define the event

$$\mathcal{A}_N = \Big\{\exists\, \sigma^1, \sigma^2 \in S_N : |H_N(\sigma^1) - a_N| \leq b_N \gamma_N, \\ |H_N(\sigma^2) - a_N| \leq b_N \gamma_N, |R(\sigma^1, \sigma^2)| > R_N\Big\}, \tag{79}$$

and show that under condition (78) the probability of $\mathcal{A}_N$ tends to zero as $N \to \infty$ and that on $\mathcal{A}_N^c$ the moment method "works", i.e., for all bounded Borel sets A and all integers $\ell \geq 1$

$$\lim_{N\to\infty} \mathbb{E}((\mathcal{P}_N(A))_\ell \mathbf{1}_{\mathcal{A}_N^c}) = (\mu(A))^\ell. \tag{80}$$

The details are covered in the next part.

3.1.4. Proof of REM-universality for the p-spin type models. As we already mentioned, the idea of the proof is to exclude the possibility to have two energy levels inside the considered micro-canonical window such that the corresponding configurations have "large" overlap. To define event $\mathcal{A}_N$ in (79) we specify $\gamma_N = \log N$ and $R_N = \log N/\sqrt{N}$.

Proposition 3.2. *If the sequence a_N satisfies* (78) *then* $\mathbb{P}(\mathcal{A}_N) \to 0$.

Proof. By the definition of the event $\mathcal{A}_N$

$$\mathbb{P}(\mathcal{A}_N) = \mathbb{P}\Big(\bigcup_{\sigma^1,\sigma^2} \big\{H'_N(\sigma^1) \leq \gamma_N, H'_N(\sigma^2) \leq \gamma_N\big\}\Big) \\ \leq \sum_{\sigma^1,\sigma^2} \mathbb{P}\big(H'_N(\sigma^1) \leq \gamma_N, H'_N(\sigma^2) \leq \gamma_N\big), \tag{81}$$

where the union and the sum are over all pairs of configurations σ^1, σ^2 with $|R(\sigma^1, \sigma^2)| > R_N$. In (81) we group together terms with fixed overlap, i.e.,

$$\mathbb{P}(\mathcal{A}_N) \leq \sum_{R_{12}\in\Delta_N} \sum_{\substack{\sigma^1,\sigma^2: \\ R(\sigma^1,\sigma^2)=R_{12}}} \mathbb{P}\big(H'_N(\sigma^1) \leq \gamma_N, H'_N(\sigma^2) \leq \gamma_N\big), \tag{82}$$

where

$$\Delta_N \equiv \Big\{1 - \frac{2k}{N}, k = \overline{0, N}\Big\} \cap \Big\{(-1, -R_N) \cup (R_N, 1)\Big\}. \tag{83}$$

An individual term in (82) is

$$\mathbb{P}\big(H'_N(\sigma^1) \leq \gamma_N, H'_N(\sigma^2) \leq \gamma_N\big) \\ = \frac{b_N^2}{2\pi\sqrt{\det B}} \int_{-\gamma_N}^{\gamma_N} \int_{-\gamma_N}^{\gamma_N} \mathrm{e}^{-\frac{1}{2}(\vec{x}, B^{-1}\vec{x}) b_N^2 - a_N b_N (\vec{x}, B^{-1}\vec{1}) - \frac{1}{2}(\vec{1}, B^{-1}\vec{1}) a_N^2} \, \mathrm{d}\vec{x}, \tag{84}$$

where $B = (\mathrm{Cov}(H_N(\sigma^i), H_N(\sigma^j))_{1\le i,j\le 2}$ is the covariance matrix, $\vec{1} = (1,1)^t$ and $\vec{x} = (x_1, x_2)^t$. From (48) it follows that b_N is exponentially small in N and after a little algebra, up to multiplication by $1 + o(1)$,

$$\mathbb{P}\big(H'_N(\sigma^1) \le \gamma_N, H'_N(\sigma^2) \le \gamma_N\big) = \frac{\gamma_N^2}{2^{2N}} \frac{\mathrm{e}^{a_N^2 \nu(R_{12})/(1+\nu(R_{12}))}}{\sqrt{1-\nu^2(R_{12})}}. \tag{85}$$

For every fixed configuration $\sigma \in S_N$ the number of its neighbors at the distance d is given by the binomial coefficient, $\binom{N}{d}$. Therefore, the number of terms in the inner sum in (82), i.e., the number of pairs $\sigma^1, \sigma^2 \in S_N$ with overlap fixed to R_{12}, is

$$\begin{aligned}\#\{(\sigma^1,\sigma^2) \in S_N^2 : R(\sigma^1,\sigma^2) = R_{12}\} &= 2^N \binom{N}{\frac{N}{2}(1-R_{12})} \\ &= \sqrt{\frac{2}{\pi}} \frac{2^{2N}\mathrm{e}^{-N\mathcal{J}(R_{12})}}{\sqrt{N(1-R_{12}^2)}} \Big(1 + O\Big(\frac{1}{N(1-R_{12})}\Big)\Big),\end{aligned} \tag{86}$$

where the second equality follows from the Stirling's formula and where $\mathcal{J}$ is defined in (66). Thus for some $C > 0$

$$\#\{(\sigma^1,\sigma^2) \in S_N^2 : R(\sigma^1,\sigma^2) = R_{12}\} \le C\, 2^{2N} \mathrm{e}^{-N\mathcal{J}(R_{12})}. \tag{87}$$

Together with (85) it implies that the inner sum in (82) is bounded by

$$\sqrt{N}\gamma_N^2 \mathrm{e}^{-N\mathcal{J}(R_{12}) + a_N^2 \nu(R_{12})/(1+\nu(R_{12}))}. \tag{88}$$

From (82) and (88) we further get that

$$\mathbb{P}(\mathcal{A}_N) \le \sqrt{N}\gamma_N^2 \sum_{R_{12}\in\Delta_N} \mathrm{e}^{-N\mathcal{J}(R_{12}) + a_N^2 \nu(R_{12})/(1+\nu(R_{12}))}. \tag{89}$$

From assumption (78) and from the definition of c_ν in (68) there exists $\delta > 0$ such that

$$\min_{x\in\Delta_N} \Big\{N\mathcal{J}(x) - a_N^2 \frac{\nu(x)}{1+\nu(x)}\Big\} > \delta N \mathcal{J}(R_N). \tag{90}$$

Therefore

$$\sum_{R_{12}\in\Delta_N} \mathrm{e}^{-N\mathcal{J}(R_{12}) + a_N^2 \nu(R_{12})/(1+\nu(R_{12}))} \le N \mathrm{e}^{-\delta N \mathcal{J}(R_N)} \tag{91}$$

decays faster than any polynomial. This finishes the proof of Proposition 3.2. □

Let us turn to the proof of part (a) of Theorem 3.9. By Theorem 3.2 it is sufficient to prove that for every bounded Borel set A

(1) $\lim_{N\to\infty} \mathbb{E}\mathcal{P}_N(A) = |A|$,

(2) $\lim_{N\to\infty} \mathbb{P}(\mathcal{P}_N(A) = 0) = \mathrm{e}^{-|A|}$.

Condition (1) is a consequence of our normalization, more precisely, of (48). We, therefore, concentrate on the proof of condition (2). Since $\mathbb{P}(\mathcal{A}_N) = o(1)$

$$\begin{aligned}\mathbb{P}(\mathcal{P}_N(A) = 0) &= \mathbb{P}(\{\mathcal{P}_N(A) = 0\} \cap \mathcal{A}_N) + \mathbb{P}(\{\mathcal{P}_N(A) = 0\} \cap \mathcal{A}_N^c)\\ &= \mathbb{P}(\{\mathcal{P}_N(A) = 0\} \cap \mathcal{A}_N^c) + o(1).\end{aligned} \tag{92}$$

Using the inclusion-exclusion principle we write

$$\begin{aligned}&\mathbb{P}(\{\mathcal{P}_N(A) = 0\} \cap \mathcal{A}_N^c) \\ &\quad= \sum_{\ell=0}^{\infty} \frac{(-1)^\ell}{\ell!} \times \sum_{\sigma^1,\dots,\sigma^\ell} \mathbb{P}(\{H'_N(\sigma^1) \in A, \dots, H'_N(\sigma^\ell) \in A\} \cap \mathcal{A}_N^c),\end{aligned} \tag{93}$$

where the inner sum is over ordered collections of distinct configurations $\sigma^1, \dots, \sigma^\ell$ belonging to S_N. Let A be bounded. Thus $A \subset [0, \gamma_N]$ when N is large enough. Therefore from the definition of the event $\mathcal{A}_N$

$$\begin{aligned}&\mathbf{1}_{H'_N(\sigma^1)\in A} \cdots \mathbf{1}_{H'_N(\sigma^\ell)\in A} \mathbf{1}_{\mathcal{A}_N^c}\\ &\quad= \mathbf{1}_{R_{\max}(\sigma^1,\dots,\sigma^\ell)\le R_N} \mathbf{1}_{H'_N(\sigma^1)\in A} \cdots \mathbf{1}_{H'_N(\sigma^\ell)\in A} \mathbf{1}_{\mathcal{A}_N^c},\end{aligned} \tag{94}$$

where we define the maximal overlap as

$$R_{\max}(\sigma^1, \dots, \sigma^\ell) = \max_{1\le i<j\le \ell} |R(\sigma^i, \sigma^j)|. \tag{95}$$

Taking expectation

$$\begin{aligned}&\mathbb{P}(\{H'_N(\sigma^1) \in A, \dots, H'_N(\sigma^\ell) \in A\} \cap \mathcal{A}_N^c)\\ &\quad= \mathbb{E}\left(\mathbf{1}_{H'_N(\sigma^1)\in A} \cdots \mathbf{1}_{H'_N(\sigma^\ell)\in A} \mathbf{1}_{\mathcal{A}_N^c}\right)\\ &\quad= \mathbf{1}_{R_{\max}(\sigma^1,\dots,\sigma^\ell)\le R_N} \mathbb{E}\left(\mathbf{1}_{H'_N(\sigma^1)\in A} \cdots \mathbf{1}_{H'_N(\sigma^\ell)\in A} \mathbf{1}_{\mathcal{A}_N^c}\right),\end{aligned} \tag{96}$$

and summing over all ordered sequences of distinct configurations $\sigma^1, \dots, \sigma^\ell$ belonging to S_N we obtain

$$\begin{aligned}&\sum_{\sigma^1,\dots,\sigma^\ell} \mathbb{P}(\{H'_N(\sigma^1) \in A, \dots, H'_N(\sigma^\ell) \in A\} \cap \mathcal{A}_N^c)\\ &\quad= \sum_{\sigma^1,\dots,\sigma^\ell} \mathbf{1}_{R_{\max}(\sigma^1,\dots,\sigma^\ell)\le R_N} \mathbb{P}(\{H'_N(\sigma^1) \in A, \dots, H'_N(\sigma^\ell) \in A\} \cap \mathcal{A}_N^c).\end{aligned} \tag{97}$$

Lemma 3.1. *For every bounded Borel set* $A \subset \mathbb{R}_+$

$$\sum_{\sigma^1,\dots,\sigma^\ell} \mathbf{1}_{R_{\max}(\sigma^1,\dots,\sigma^\ell)\le R_N} \mathbb{P}(H'_N(\sigma^1) \in A, \dots, H'_N(\sigma^\ell) \in A) \to |A|^\ell, \tag{98}$$

where the sum is over ordered sequences of distinct configurations $\sigma^1, \dots, \sigma^\ell$ *belonging to* S_N.

Proof. For a collection $(\sigma^1,\dots,\sigma^\ell)\in S_N^\ell$ satisfying $R_{\max}(\sigma^1,\dots,\sigma^\ell)\le R_N$ the covariance matrix $B=B(\sigma^1,\dots,\sigma^\ell)$ of the vector $(H_N(\sigma^1),\dots,H_N(\sigma^\ell))$ is invertible and therefore

$$\begin{aligned}&\mathbb{P}(\{H'_N(\sigma^1)\in A,\dots H'_N(\sigma^\ell)\in A\})\\&=\frac{b_N^\ell}{(2\pi)^{\ell/2}}\frac{\mathrm{e}^{-\frac{1}{2}a_N^2(\vec{1},B^{-1}\vec{1})}}{\sqrt{\det B}}\int_A\cdots\int_A \mathrm{e}^{-(\vec{x},B^{-1}\vec{x})b_N^2/2-a_Nb_N(\vec{x},B^{-1}\vec{1})}d\vec{x}.\end{aligned}\tag{99}$$

Since b_N is exponentially small in N, the dominated convergence implies that the integral in (99) is $|A|^\ell(1+o(1))$. Substituting b_N we further get, up to a multiplicative term of order $1+o(1)$,

$$\mathbb{P}\big(H'_N(\sigma^1)\in A,\dots,H'_N(\sigma^\ell)\in A\big)=\frac{\mathrm{e}^{\frac{1}{2}a_N^2(\ell-(\vec{1},B^{-1}\vec{1}))}}{2^{N\ell}}|A|^\ell.\tag{100}$$

Since $\ell-(\vec{1},B^{-1}\vec{1})=O\left(\nu(R_{\max}(\sigma^1,\dots,\sigma^\ell))\right)$ and

$$N\nu(R_N)=O(NR_N^3)=O(\log^3 N/\sqrt{N})$$

we obtain

$$\mathrm{e}^{\frac{1}{2}a_N^2(\ell-(\vec{1},B^{-1}\vec{1}))}=\mathrm{e}^{O(N\nu(R_N))}=1+o(1),\tag{101}$$

and thus

$$\mathbb{P}\big(H'_N(\sigma^1)\in A,\dots,H'_N(\sigma^\ell)\in A\big)=\frac{|A|^\ell}{2^{N\ell}}.\tag{102}$$

By Lemma 2.2 in [12] there are $2^{N\ell}(1+O(\mathrm{e}^{-\frac{1}{8}NR_N^2}))$ ordered collections of distinct configurations $(\sigma^1,\dots,\sigma^\ell)\in S_N^\ell$ satisfying $R_{\max}(\sigma^1,\dots,\sigma^\ell)\le R_N$. Since $NR_N^2=\log^2 N$ Lemma 3.1 follows. □

Lemma 3.1 together with the dominated convergence theorem imply that

$$\sum_{\sigma^1,\dots,\sigma^\ell}\mathbf{1}_{R_{\max}(\sigma^1,\dots,\sigma^\ell)\le R_N}\mathbb{P}\big(\{H'_N(\sigma^1)\in A,\dots,H'_N(\sigma^\ell)\in A\}\cap\mathcal{A}_N\big)\to 0,\tag{103}$$

and thus we obtain from (97) that

$$\sum_{\sigma^1,\dots,\sigma^\ell}\mathbb{P}\big(\{H'_N(\sigma^1)\in A,\dots,H'_N(\sigma^\ell)\in A\}\cap\mathcal{A}_N^c\big)\to|A|^\ell.\tag{104}$$

The partial sums of the right-hand side of (93) provide the upper and lower bounds for $\mathbb{P}(\{\mathcal{P}_N(A)=0\}\cap\mathcal{A}_N^c)$, i.e., for any $\ell_0\in\mathbb{N}$

$$\begin{aligned}&\sum_{\ell=0}^{2\ell_0+1}\frac{(-1)^\ell}{\ell!}\sum_{\sigma^1,\dots,\sigma^\ell}\mathbb{P}\big(\{H'_N(\sigma^1)\in A,\dots,H'_N(\sigma^\ell)\in A\}\cap\mathcal{A}_N^c\big)\\&\quad\le\mathbb{P}(\{\mathcal{P}_N(A)=0\}\cap\mathcal{A}_N^c)\\&\quad\le\sum_{\ell=0}^{2\ell_0}\frac{(-1)^\ell}{\ell!}\sum_{\sigma^1,\dots,\sigma^\ell}\mathbb{P}\big(\{H'_N(\sigma^1)\in A,\dots,H'_N(\sigma^\ell)\in A\}\cap\mathcal{A}_N^c\big).\end{aligned}\tag{105}$$

Letting $N \to \infty$ in (105) and applying (104) we see that for all $\ell_0 \geq 0$

$$\sum_{\ell=0}^{2\ell_0+1} \frac{(-1)^\ell}{\ell!}|A|^\ell \leq \lim_{N\to\infty} \mathbb{P}(\{\mathcal{P}_N(A)=0\} \cap \mathcal{A}_N^c) \leq \sum_{\ell=0}^{2\ell_0} \frac{(-1)^\ell}{\ell!}|A|^\ell. \tag{106}$$

This easily implies that $\mathbb{P}(\mathcal{P}_N(A) = 0) \to \mathrm{e}^{-|A|}$ and thus the micro-canonical REM-universality in part (a) of Theorem 3.1 is proved.

3.1.5. Explosion of moments. To prove Theorem 3.11 we estimate the number of equidistant subsets on the hypercube, i.e., the number of sequences $(\sigma^1, \dots, \sigma^\ell) \in S_N^\ell$ with all the overlaps between nonequal configurations equal to some $a \in [0, 1)$. To be exact, we are interested in the asymptotic result as $N \to \infty$, i.e., we estimate the number of configurations where all the non-diagonal overlaps are $a + o(1)$. In terms of the Hamming distance, the problem is equivalent to estimating the number of the subsets of the hypercube with ℓ elements such that the distance between any two of them is $\frac{N}{2}(1-a)$. In the following discussion let us use C to denote an arbitrary positive constant, not necessarily the same from case to case.

Proposition 3.3. *For $a \in [0, 1)$ there are at least*

$$2^{N\ell(1-a)} \frac{C}{N^{(2^{\ell-1}-1)/2}}(1+o(1)) \tag{107}$$

ordered sequences $(\sigma^1, \dots, \sigma^\ell) \in S_N^\ell$ with $R(\sigma^i, \sigma^j) = a$ for all $1 \leq i < j \leq \ell$.

Remark 3.5. In the statement of the theorem and in the following proof we ignore the divisibility questions since one can easily resolve them by introducing an error of order $o(1)$ to the off-diagonal overlaps.

Proof. In general, the structure of the set consisting of all ordered sequences of configurations $(\sigma^1, \dots, \sigma^\ell) \in S_N^\ell$ with $R(\sigma^i, \sigma^j) = a$ for all $1 \leq i < j \leq \ell$ can be quite complicated. Therefore we explicitly construct a subset W whose size is easy to estimate. We build W inductively.

(1) Let $I = \{1, 2, \dots, N\}$ be the index set. To construct σ^1 we fix set $B_0 \subset I$ of cardinality aN and let the spins of σ^1 with indices belonging to the set B_0 be fixed to 1. The values of all other spins, i.e., of the spins belonging to the set $B_1 = I \backslash B_0$, are chosen in an arbitrary manner. If we denote by W_1 the set of the configurations constructed in this way then its size is given by the number of ways to choose the signs of spins with indices in B_1, i.e.,

$$|W_1| = 2^{N(1-a)}. \tag{108}$$

Below we fix $\sigma^1 \in W_1$ and construct a sequence $\sigma^2, \dots, \sigma^\ell$ such that $R(\sigma^i, \sigma^j) = a$ for all $1 \leq i < j \leq \ell$.

(2) Let us construct a configuration σ^2 with $R(\sigma^1, \sigma^2) = a$. We first split the set B_1 into two nonintersecting subsets B_2 and B_3, both of cardinality $\frac{N(1-a)}{2}$.

Then σ^2 is obtained by putting $\sigma_i^2 = \sigma_i^1$ for all $i \in B_2 = B_1 \backslash B_3$ and $\sigma_i^2 = -\sigma_i^1$ for $i \in B_3$. By construction

$$R(\sigma^1, \sigma^2) = \frac{|B_0| + |B_2| - |B_3|}{N} = a. \tag{109}$$

We denote by $W_2 = W_2(\sigma^1)$ the set of all configurations σ^2 built in this way. It is easy to see that the size of W_2 is given by the number of ways to split B_1 into B_2, B_3, i.e.,

$$|W_2| = \binom{N(1-a)}{\frac{N(1-a)}{2}} = 2^{N(1-a)} \frac{C}{\sqrt{N}}(1 + o(1)). \tag{110}$$

(3) To construct a configuration σ^3 with $R(\sigma^1, \sigma^3) = a$ and $R(\sigma^2, \sigma^3) = a$ we split B_2 and B_3 into nonintersecting subsets B_4, B_5 and B_6, B_7 respectively, each of them having cardinality $\frac{N(1-a)}{4}$. We put $\sigma_i^3 = \sigma_i^2$ for $i \in B_4 \cup B_6 \cup B_0$ and $\sigma_i^3 = -\sigma_i^2$ for $i \in B_5 \cup B_7$. Then it is easy to see that σ^3 satisfies

$$R(\sigma^1, \sigma^3) = \frac{|B_4| - |B_5| + |B_6| - |B_7| + |B_0|}{N} = a \tag{111}$$

and

$$R(\sigma^2, \sigma^3) = \frac{|B_4| - |B_5| - |B_6| + |B_7| + |B_0|}{N} = a. \tag{112}$$

The set of configurations σ^3 built in this way is denoted by $W_3 = W_3(\sigma^1, \sigma^2)$ and its size is determined by the number of ways to split B_2 and B_3 into sets B_4, B_5 and B_6, B_7 respectively, i.e.,

$$\begin{aligned} |W_3| &= \binom{\frac{N(1-a)}{2}}{\frac{N(1-a)}{4}} \binom{\frac{N(1-a)}{2}}{\frac{N(1-a)}{4}} \\ &= \left(\frac{C}{\sqrt{N}} 2^{\frac{N(1-a)}{2}} (1 + o(1)) \right)^2 = 2^{N(1-a)} \frac{C}{N}(1 + o(1)). \end{aligned} \tag{113}$$

(4) At this point it is clear how to construct σ^4 and, more generally, how to construct σ^k, having constructed $\sigma^1, \ldots, \sigma^{k-1}$. In particular, for every $j = 2^{k-2}, \ldots, 2^{k-1} - 1$ we split B_j into nonintersecting sets B_{2j}, B_{2j+1} and then we put $\sigma_i^k = \sigma_i^{k-1}$ for $i \in B_0 \cup \left(\bigcup_j B_{2j} \right)$ and $\sigma_i^k = -\sigma_i^{k-1}$ for $i \in \bigcup_j B_{2j+1}$. It is easy to show that $R(\sigma^j, \sigma^k) = a$ for all $j \leq k - 1$. The set of configurations σ^k built in this way is denoted by $W_k = W_k(\sigma^1, \ldots, \sigma^{k-1})$ and its cardinality is given by the number of ways to split the B_j's, i.e.,

$$|W_k| = 2^{N(1-a)} \frac{C}{N^{2^{k-3}}} (1 + o(1)). \tag{114}$$

To conclude the argument it remains to notice that the constructed collections $\sigma^1, \ldots, \sigma^\ell$ are all distinct which is a direct consequence of the fact that we are interested in ordered sequences $(\sigma^1, \ldots, \sigma^\ell)$. Let us now define

$$W = \{(\sigma^1, \ldots, \sigma^\ell) : \sigma^1 \in W_1, \sigma^2 \in W_2(\sigma^1), \ldots, \sigma^\ell \in W_\ell(\sigma^1, \ldots, \sigma^{\ell-1})\}, \tag{115}$$

which clearly has cardinality

$$|W| = 2^{N\ell(1-a)} \frac{C}{N^{(2^{\ell-1}-1)/2}}(1+o(1)). \tag{116}$$

Thus Proposition 3.3 is proved. □

Proof. We can now proceed with the proof of Theorem 3.11.

Let us begin by fixing notations. For a vector

$$\vec{R} = (R_{ij})_{1\le i<j\le \ell} \in [-1,1]^{\ell(\ell-1)/2}$$

we define $B_{\ell,\nu} = B_{\ell,\nu}(\vec{R})$ to be an $\ell \times \ell$ symmetric matrix with entries

$$(B_{\ell,\nu})_{ij} = \begin{cases} 1 & \text{if } i = j, \\ \nu(R_{ij}) & \text{if } i < j. \end{cases} \tag{117}$$

We also denote $S_{N,\vec{R}}$ the set of all ordered sequences $(\sigma^1,\dots,\sigma^\ell) \in S_N^\ell$ compatible with a given set of overlaps $\vec{R} = (R_{ij})_{1\le i<j\le \ell}$, i.e.,

$$S_{N,\vec{R}} = \left\{(\sigma^1,\dots,\sigma^\ell) \in S_N^\ell : \forall\, 1 \le i < j \le \ell \quad R(\sigma^i,\sigma^j) = R_{ij}\right\}. \tag{118}$$

In particular, setting $\vec{a} = (a,\dots,a) \in \mathbb{R}^\ell$,

$$S_{N,\vec{a}} = \left\{(\sigma^1,\dots,\sigma^\ell) \in S_N^\ell : \forall 1 \le i < j \le \ell \;\; R(\sigma^i,\sigma^j) = a\right\}. \tag{119}$$

It is an easy exercise to show that the ℓth factorial moment of $\mathcal{P}_N(A)$ is

$$\mathbb{E}(\mathcal{P}_N(A))_\ell = \sum_{\sigma^1,\dots,\sigma^\ell} \mathbb{P}\big(H_N(\sigma^1) \in A,\dots,H_N(\sigma^\ell) \in A\big), \tag{120}$$

where the summation is over all ordered sequences of different configurations $\sigma^1,\dots,\sigma^\ell \in S_N$. Using (120) we can obviously bound

$$\mathbb{E}(\mathcal{P}_N(A))_\ell \ge \sum_{(\sigma^1,\dots,\sigma^\ell)\in S_{N,\vec{a}}} \mathbb{P}\big(H'_N(\sigma^1) \in A,\dots,H'_N(\sigma^\ell) \in A\big). \tag{121}$$

We will show that with appropriately chosen a the sum (121) grows exponentially fast as $N \to \infty$. To do this first note that by Proposition 3.3

$$|S_{N,\vec{a}}| \ge 2^{N\ell(1-a)} \frac{C}{N^{2^\ell}}(1+o(1)). \tag{122}$$

Let us now assume that $(\sigma^1,\dots,\sigma^\ell) \in S_{N,\vec{a}}$ and estimate the general term in (121) that is given by formula (99). We rewrite

$$B_{\ell,\nu}(\vec{a}) = (1-\nu(a))\,\mathbb{I} + \nu(a)\,\mathbb{J}, \tag{123}$$

where $\mathbb{J}$ denotes the $\ell \times \ell$ matrix with all entries equal to 1, i.e., $\mathbb{J}_{ij} = 1$, and $\mathbb{I}$ denotes the $\ell \times \ell$ identity matrix. Inverting $B_{\ell,\nu}(\vec{a})$ we get

$$\begin{aligned} B^{-1}_{\ell,\nu}(\vec{a}) &= ((1-\nu(a))\,\mathbb{I} + \nu(a)\,\mathbb{J})^{-1} \\ &= \frac{1}{1-\nu(a)}\,\mathbb{I} - \frac{\nu(a)}{(1-\nu(a))(1+(\ell-1)\nu(a))}\,\mathbb{J}. \end{aligned} \tag{124}$$

Let us choose $a_{\ell,\nu}$ in such a way that $\nu(a_{\ell,\nu}) = 1/\ell$. Then

$$B_{\ell,\nu}^{-1}(\vec{a}_{\ell,\nu}) = \frac{1}{1-1/\ell}\mathbb{I} - \frac{1/\ell}{(1-1/\ell)(1+(\ell-1)/\ell)}\mathbb{J}. \tag{125}$$

and

$$\det B_{\ell,\nu}(\vec{a}_{\ell,\nu}) = (1-1/\ell)^{\ell-1}(1+(\ell-1)/\ell). \tag{126}$$

We thus conclude that

$$\mathbb{P}(H'_N(\sigma^1) \in A, \ldots, H'_N(\sigma^\ell) \in A) \geq \frac{C}{2^{N\ell}} \mathrm{e}^{\frac{1}{2}a_N^2(\ell-(\vec{1},B_{\ell,\nu}^{-1}\vec{1}))}. \tag{127}$$

Using (127) we can further bound (121) from below by

$$\frac{C}{2^{N\ell}}\,\mathrm{e}^{\frac{1}{2}a_N^2(\ell-(\vec{1},B_{\ell,\nu}^{-1}\vec{1}))}\,|S_{N,\vec{a}}|$$
$$\geq C\exp\Big\{\frac{1}{2}a_{\ell,\nu}N\ell\log 2\Big(\frac{a_N^2}{N}\frac{\ell-(\vec{1},B_{\ell,\nu}^{-1}\vec{1})}{a_{\ell,\nu}\ell\log 2} - 2 + O\Big(\frac{\log N}{N}\Big)\Big)\Big\}. \tag{128}$$

By (124) on $B_{\ell,\nu}^{-1}$ we obtain that $(\vec{1}, B_{\ell,\nu}^{-1}\vec{1}) = \frac{\ell}{1+(\ell-1)/\ell}$ and hence

$$\frac{\ell-(\vec{1},B_{\ell,\nu}^{-1}\vec{1})}{a_{\ell,\nu}\ell\log 2} = \frac{\ell - \frac{\ell}{1+(\ell-1)/\ell}}{a_{\ell,\nu}\ell\log 2} = \frac{\ell-1}{a_{\ell,\nu}\ell(1+(\ell-1)/\ell)\log 2}. \tag{129}$$

Since $a_{\ell,\nu}$ is the solution of the equation $\nu(a_{\ell,\nu}) = 1/\ell$ and a_N satisfies (76) we conclude from (129) that for large enough ℓ

$$\limsup \frac{a_N^2}{N}\frac{\ell-(\vec{1},B_{\ell,\nu}^{-1}\vec{1})}{a_{\ell,\nu}\ell\log 2} > 2, \tag{130}$$

which finishes the proof of Theorem 3.11. □

3.2. Edwards–Anderson short-range spin glass

Consider the Hamiltonian of Example 2.5, more precisely let us normalize it so that its variance is 1, i.e., we define

$$H_N(\sigma) = \frac{1}{\sqrt{dN^d}}\sum_{x\sim y} g_{x,y}\sigma_x\sigma_y. \tag{131}$$

Theorem 3.12. *If $a_N = o(N^{d/4}/\log N)$ then the sequence of point processes $\mathcal{P}_N$, defined in* (47)*, converges weakly to the Poisson point process on $\mathbb{R}$ whose intensity measure is the Lebesque measure.*

Theorem 3.12 was proved in [22] for $a_N = cN^\gamma$, where $c \in \mathbb{R}, \gamma \in [0, 1/4)$. The proof in [22] could be easily extended to work for $a_N = o(N^{d/4}/\log N)$.

We believe that in reality the micro-canonical REM-universality for the EA spin glass is valid as long as $a_N = o(N^{d/4})$. It, actually, can be confirmed rigorously conditionally on the fact that some combinatorial lemma is correct. We postpone to Section 4.2 the formulation of the lemma as well as the proof of Theorem 3.12, since the proof of the micro-canonical REM-universality is quite similar to the proof of the re-sampling REM-universality.

3.3. References

REM-universality was first observed by Mertens in [30] for the number partitioning problem in the case when the micro-canonical window was at zero, i.e., in the case $a_N = 0$, and when the random variables g_i in the definition of the Hamiltonian (see (20)) were uniformly distributed on $[0, 1]$. This numerical observation was rigourously proved by Borgs, Chayes and Pittel in [16]. Then Bauke, Franz and Mertens in [2] and also Bauke and Mertens in [3] generalized the observation of Mertens to the case of non-zero energy scales as well as for many other disordered systems. Motivated by this new numerical evidence Bovier and Kurkova in [22] gave an axiomatic approach to the proof of the micro-canonical REM-universality and applied it to many spin glass models. In particular, Bovier and Kurkova showed that the universality persists in the Gaussian version of the number partitioning problem for $a_N = cN^\gamma, c \in \mathbb{R}, \gamma \in [0, 1/4)$ and in p-spin models with $p \geq 2$ it persists for $a_N = cN^\gamma, c \in \mathbb{R}, \gamma \in [0, 1/2)$. They also proved the micro-canonical REM-universality for the Edwards–Anderson short-range spin glass on d-dimensional lattice when $a_N = cN^\gamma, c \in \mathbb{R}, \gamma \in [0, d/4)$ and showed the validity of the local REM-universality for a wide class of non-Gaussian short-range spin glasses. The fact that $N^{1/4}$ and $N^{1/2}$ are indeed the true thresholds on the validity of the REM-universality in the NPP-type and the SK-type models was proved by Borgs, Chayes, Mertens and Nair in [15]. Moreover, in [14] and [15] they proved that the micro-canonical REM-universality holds for the Hamiltonian from (20) for quite general assumptions on the distribution of the g_i's. Finally, we proved in [13] that for the p-spin type models REM-universality holds for $a_N < c_\nu N^{1/2}$.

4. Re-sampling REM-universality

In this section we give results on the re-sampling REM-universality introduced in Section 1. More precisely, we consider a Gaussian process H_N on a large finite set S_N, sample a random subset $\mathcal{X}_N \subset S_N$ and study the asymptotic behavior of the sequence of point processes $\mathcal{P}_N$, defined by

$$\mathcal{P}_N = \sum_{\sigma \in \mathcal{X}_N} \delta_{b_N^{-1}(H_N(\sigma) - a_N)}, \tag{132}$$

where the normalizations a_N, b_N are determined by

$$\lim_{N\to\infty} \frac{\sqrt{2\pi} a_N \mathrm{e}^{a_N^2/2}}{|\mathcal{X}_N|} = c \quad \text{and} \quad \lim_{N\to\infty} \frac{|\mathcal{X}_N| \mathrm{e}^{-a_N^2/2} b_N}{\sqrt{2\pi}} = 1. \tag{133}$$

Re-sampling REM-universality means that for sparse subsets $\mathcal{X}_N$ the sequence $\mathcal{P}_N$ converges weakly to a Poisson point process whose intensity measure is given by $\mu(\mathrm{d}t) = \mathrm{e}^{-ct}\mathrm{d}t$.

We will only study here the simplest possible random subset, i.e., a site percolation cluster $\mathcal{X}_N = \{\sigma \in S_N : \mathcal{X}_\sigma = 1\}$, where the random variables $(\mathcal{X}_\sigma)_{\sigma \in S_N}$ are i.i.d. Bernoulli random variables.

In Section 4.1 we give results on the mean-field spin glass models on the hypercube and in Section 4.2 we study in detail the re-sampling REM-universality for the Edwards–Anderson short range spin glass.

4.1. Mean-field models on the hypercube

We consider a Gaussian Hamiltonian H_N indexed by the N-dimensional hypercube. We assume that H_N is centered, i.e.,

$$\forall \sigma \in S_N \qquad \mathbb{E}(H_N(\sigma)) = 0, \tag{134}$$

and its covariance is a function of the overlap, $R(\sigma, \sigma') = N^{-1} \sum_{i=1}^{N} \sigma_i \sigma'_i$, i.e., it is given by

$$\forall \sigma, \sigma' \in S_N \qquad \mathbb{E}(H_N(\sigma) H_N(\sigma')) = \nu(R(\sigma, \sigma')). \tag{135}$$

Moreover, we always assume that the covariance ν is a smooth function satisfying the key hypothesis

$$\nu(0) = 0. \tag{136}$$

To specify the way we choose the random subset $\mathcal{X}_N$ we consider a sequence $M = M(N), 0 \leq M \leq N$, and put

$$\mathbb{P}(\mathcal{X}_\sigma = 1) = 1 - \mathbb{P}(\mathcal{X}_\sigma = 0) = \frac{2^M}{2^N}. \tag{137}$$

Then the mean size of the sampled set $\mathcal{X}_N$ is 2^M. In order to obtain results almost surely with respect to the realization of $\mathcal{X}_N$ we will assume that $\mathcal{X}_N$ is not too small, e.g., it is enough to assume that $\log N = o(2^M)$. On the other side, if $\mathcal{X}_N$ is too large, e.g., in case $M = N$, then we do not expect the re-sampling REM-universality to hold. Thus there is a gap between the largest scales of M for which REM-universality still holds and N. We will see that just like in Section 3.1 the size of the gap depends on the type of the Hamiltonian, see Definition 3.1.

4.1.1. The NPP-type models. The number partitioning problem was introduced in Section 1 (see Example 2.2). The following theorem asserts that for all models of the NPP-type the threshold on the validity of the re-sampling REM-universality is $M = o(N^{1/2})$.

Theorem 4.1. *Let the Hamiltonian H_N be of the NPP-type.*

(a) *If $M = o(N^{1/2})$ then $\mathbb{P}$-almost surely the sequence of point processes $\mathcal{P}_N$ converges weakly to a Poisson point process whose intensity measure is given by $\mu(\mathrm{d}t) = \mathrm{e}^{-ct}\mathrm{d}t$.*

(b) *If the sequence $M(N)$ is such that*

$$\limsup_{N \to \infty} \frac{M}{N^{1/2}} \in (0, \infty) \tag{138}$$

then $\mathbb{P}$-almost surely $\mathcal{P}_N$ does not converge to a Poisson point process.

Let us just mention that one can prove Theorem 4.1 using the moment method (see Theorems 3.2 and 3.4). We refer the reader to [12] for complete proofs.

4.1.2. The SK-type models. In the following theorem we show that all models of the SK-type share the same threshold on the validity of the re-sampling REM-universality, which is $M = o(N)$.

Theorem 4.2. *Let the Hamiltonian H_N be of the SK-type.*

(a) *If $M = o(N)$ then $\mathbb{P}$-almost surely the sequence of point processes $\mathcal{P}_N$ converges weakly to a Poisson point process whose intensity measure is given by $\mu(\mathrm{d}t) = \mathrm{e}^{-ct}\mathrm{d}t$.*

(b) *If the sequence $M(N)$ is such that*

$$\limsup_{N\to\infty} \frac{M}{N} \in \Big(0, \frac{1}{4\nu''(0)\log 2}\Big) \tag{139}$$

then $\mathbb{P}$-almost surely $\mathcal{P}_N$ does not converge to a Poisson point process.

We omit the proofs here and refer the reader to [12].

4.1.3. The p-spin models. Finally, we examine the case of the p-spin type models, i.e., models with $\nu'(0) = \nu''(0) = 0$. We show that the re-sampling REM-universality holds for $M < c_\nu N$.

Theorem 4.3. *Let the Hamiltonian H_N be of the p-spin type.*

(a) *If the sequence $M(N)$ satisfies*

$$\limsup_{N\to\infty} \frac{M}{N} < c_\nu, \tag{140}$$

where $c_\nu > 0$ is defined by

$$c_\nu = \inf_{x\in(0,1)} \frac{\mathcal{J}(x)(1+\nu(x))}{2\nu(x)\log 2}, \tag{141}$$

then the sequence of point processes $\mathcal{P}_N$ converges weakly to the Poisson point process whose intensity measure is given by $\mu(\mathrm{d}t) = \mathrm{e}^{-ct}\mathrm{d}t$.

(b) *If the sequence $M(N)$ satisfies*

$$\limsup_{N\to\infty} \frac{M}{N} > c'_\nu, \tag{142}$$

where $c'_\nu > 0$ is defined by

$$c'_\nu = \inf_{x\in(0,1)} \frac{\mathcal{J}(x)}{\nu^2(x)\log 2}, \tag{143}$$

then $\mathbb{P}$-almost surely $\mathcal{P}_N$ does not converge weakly to a Poisson point process.

Just like in Theorem 3.9, part (b) of Theorem 4.3 is trivial. Indeed, one can show that for $M > c'_\nu N$, where c'_ν is defined in (143),

$$\limsup_{N\to\infty} \frac{\max_{\sigma\in\mathcal{X}_N} H_N(\sigma)}{\sqrt{M}} < \sqrt{2\log 2}, \quad \mathbb{P}\times\mathbb{P}\text{-a.s.} \tag{144}$$

which is incompatible with the Poissonian convergence at scales defined by (133). For a proof of part (a) we refer to [13].

4.2. Edwards–Anderson short range spin glass

Here we give a proof of the re-sampling REM-universality for the EA short range spin glass. Let us redefine the Hamiltonian from Example 2.5 as follows

$$H_N(\sigma) = \frac{1}{\sqrt{dN^d}} \sum_{x\sim y} g_{x,y}\sigma_x\sigma_y, \tag{145}$$

then H_N are standard normal random variables indexed by $S_N = \{-1,1\}^{\Lambda_N}$.

To specify the way we choose the random subset $\mathcal{X}_N$ we consider a sequence $M = M(N), 0 \le M \le N$, and define a sequence of i.i.d. Bernoulli random variables $(\mathcal{X}_\sigma)_{\sigma\in S_N}$ as

$$\mathbb{P}(\mathcal{X}_\sigma = 1) = 1 - \mathbb{P}(\mathcal{X}_\sigma = 0) = \frac{2^{M^d}}{2^{N^d}}. \tag{146}$$

We prove that the statistics of energy levels restricted to the set $\mathcal{X}_N = \{\sigma \in S_N : \mathcal{X}_\sigma = 1\}$ is Poissonian at scales (133) if $M = o(N^{1/2}/\log N)$.

Theorem 4.4. *If $M = o(N^{1/2}/\log N)$ then $\mathbb{P}$-almost surely the sequence of point processes $\mathcal{P}_N$ converges weakly to a Poisson point process with intensity measure given by $\mu(\mathrm{d}t) = \mathrm{e}^{-ct}\mathrm{d}t$.*

The natural question is whether the sparseness assumption of Theorem 4.4 is optimal. We believe that the re-sampling REM-universality holds for the EA spin glass if and only if $M = o(N^{1/2})$. Actually, we can prove it if the following conjecture is correct.

Conjecture 4.1. For all sequences $\theta_N \to 0$, $N\theta_N \in \mathbb{Z}$,

$$\frac{|U_{N,2}(\theta_N)|}{2^{2N^d}} = \sqrt{\frac{2r}{\pi N^d}}\mathrm{e}^{-N^d\mathcal{J}^\Psi(\theta_N)}(1+o(1)), \tag{147}$$

where set $U_{N,2}$ is defined in (149) and $\mathcal{J}^\Psi$ is defined in Lemma 4.1.

The rest of the section is concentrated on the proof of Theorem 4.4. First we need to give a few auxiliary results.

4.2.1. Combinatorial estimates. For a given pair of configurations $\sigma,\rho \in S_{\Lambda_N}$ we define the overlap

$$R^{sr}(\sigma,\rho) = \frac{1}{dN^d}\sum_{x\sim y}\sigma_x\sigma_y\rho_x\rho_y. \tag{148}$$

Then for a sequence of configurations $\sigma^1,\sigma^2,\ldots,\sigma^\ell$ belonging to S_N and a Borel set $A \subset \mathbb{R}_+$ we define

$$U_{N,\ell}(A) = \left\{(\sigma^1,\ldots,\sigma^\ell) \in S_\Lambda^\ell : \; R^{sr}_{\max} \in A\right\}, \tag{149}$$

where $R^{sr}_{\max}(\sigma^1,\ldots,\sigma^\ell) = \max_{1\le i<j\le \ell}|R^{sr}(\sigma^i,\sigma^j)|$. We will write $U_{N,\ell}(\theta_1,\theta_2)$ and $U_{N,\ell}(\theta)$ instead of $U_{N,\ell}(\{\theta_1,\theta_2\})$ and $U_{N,\ell}(\{\theta\})$ respectively.

Lemma 4.1. *For all Borel subsets $A \subset \mathbb{R}$*

$$\lim_{N\to\infty} \frac{1}{N^d} \log \frac{|U_{N,2}(A)|}{2^{2N^d}} = -\inf_{\theta\in A} \mathcal{J}^{\Psi}(\theta), \tag{150}$$

where $\mathcal{J}^{\Psi}(\theta) = \sup_{t\in\mathbb{R}}(\theta t - \mathcal{P}(-t\Psi))$ and $\mathcal{P}$ is the pressure (or specific free Gibbs energy) of the Ising nearest-neighbor ferromagnet at inverse temperature $-t/d$.

Proof. The lemma is a direct application of Corollary 2.3 in [26]. □

We will need a similar result for the N-dependent sets A. Therefore we use a result of Bovier and Kurkova (see [22, Proposition 3.8]):

Proposition 4.1. *There exists a positive constant K such that for every $R \in [0,1]$ for large enough N*

$$|U_{N,2}(R,1)| \leq 2^{2N^d} \mathrm{e}^{-KR^2N^d}. \tag{151}$$

Corollary 4.1. *For every $\ell \geq 2$ and $R \in [0,1]$, and for large enough N,*

$$|U_{N,\ell}(R,1)| \leq 2^{\ell N^d} \mathrm{e}^{-KR^2N^d}. \tag{152}$$

Corollary 4.2. $\mathbb{P}$*-a.s.* $\max_{\sigma,\rho\in X} |R^{sr}(\sigma,\rho)| < \theta_N \equiv \sqrt{\frac{2M^d \log 2 + 2\log N}{KN^d}}$.

For a set $Y \subset S_N^\ell$ we denote by Y^X its intersection with $\mathcal{X}_N^\ell$.

Corollary 4.3. *Assume R_N satisfies $\log N = o(N^d R_N^2)$. Then for some $\alpha \in (0,1)$*

$$|U_{N,\ell}^X(R_N,1)| < 2^{\ell M^d} \mathrm{e}^{-(1-\alpha)KN^dR_N^2}, \quad \mathbb{P}\text{-a.s.} \tag{153}$$

Proof. By Markov's inequality, for any α,

$$\mathbb{P}\Big(|U_{N,\ell}^X(R_N,1)| \geq \mathrm{e}^{\alpha KN^dR^2} \mathbb{E}|U_{N,\ell}^X(R_N,1)|\Big) \leq \mathrm{e}^{-\alpha KN^dR^2}. \tag{154}$$

By the assumption $\log N = o(N^d R_N^2)$ sum $\sum_{i=1}^{\infty} \mathrm{e}^{-\alpha N^d R_N^2}$ is finite for $\alpha > 0$ and thus by the Borel-Cantelli Lemma we obtain that $\mathbb{P}$-a.s.

$$|U_{N,\ell}^X(R_N,1)| < \mathrm{e}^{\alpha KN^dR^2} \mathbb{E}|U_{N,\ell}^X(R_N,1)|. \tag{155}$$

We further obtain from Corollary 4.1 that $\mathbb{P}$-a.s.

$$|U_{N,\ell}^X(R_N,1)| < 2^{\ell M^d} \mathrm{e}^{-(1-\alpha)KN^dR_N^2}. \tag{156}$$

The corollary is proven. □

4.2.2. Proof of Theorem 4.4. The proof follows the standard outline: for $M = o(N^{1/2}/\log N)$ we show that $\mathbb{P}$-almost surely $\mathbb{E}(\mathcal{P}_N(A))_\ell \to (\mu(A))^\ell$ and thus Proposition 3.1 applies.

First, by (120) we write the ℓth factorial moment as

$$\mathbb{E}(\mathcal{P}_N(A))_\ell = \sum_{\sigma^1,\dots,\sigma^\ell} \mathbb{P}(H'_N(\sigma^1) \in A, \dots, H'_N(\sigma^\ell) \in A), \tag{157}$$

where the summation is over all ordered sequences $\sigma^1, \dots, \sigma^\ell$ belonging to $\mathcal{X}_N$.

By Corollary 4.2 $\mathbb{P}$-a.s. $R^{sr}_{\max} < \theta_N = o(1)$ and thus $B = \mathbb{I}_\ell(1 + O(R^{sr}_{\max}))$. Therefore for all collections $(\sigma^1, \dots, \sigma^\ell) \in \mathcal{X}^\ell_N$ covariance matrix $B(\sigma^1, \dots, \sigma^\ell)$ is invertible and $B^{-1} = \mathbb{I}_\ell(1 + O(R^{sr}_{\max}))$. It implies that

$$\mathbb{P}(H'_N(\sigma^1) \in A, \dots, H'_N(\sigma^\ell) \in A) = \frac{\mathrm{e}^{(\ell-(\vec{1},B^{-1}\vec{1}))(M^d \log 2 - 1/2 \log 2M^d)}}{2^{\ell M^d}\sqrt{\det B}} \times \int_A \cdots \int_A \mathrm{e}^{-(\vec{x},B^{-1}\vec{x})b_N^2 - a_N b_N(\vec{x},B^{-1}\vec{1})} d\vec{x}. \tag{158}$$

To simplify the above formula we further notice that

1. $\ell - (\vec{1}, B^{-1}\vec{1}) = O(R^{sr}_{\max})$ uniformly in N.
2. By the dominated convergence theorem

$$\int_A \cdots \int_A \mathrm{e}^{-(\vec{x},B^{-1}\vec{x})b_N^2 - a_N b_N(\vec{x},B^{-1}\vec{1})} d\vec{x} = (\mu(A))^\ell(1 + o(1)). \tag{159}$$

3. $\det B = 1 + O(R^{sr}_{\max})$ uniformly in N.

It implies that up to a multiplication by $1 + o(1)$

$$\mathbb{P}(H'_N(\sigma^1) \in A, \dots, H'_N(\sigma^\ell) \in A) = \mathrm{e}^{O(R^{sr}_{\max} M^d)} \frac{(\mu(A))^\ell}{2^{\ell M^d}}. \tag{160}$$

Next we split the summation in (157) in two parts: the first one over the set $U^X_{N,\ell}(0, R_N)$ and the other one over $U^X_{N,\ell}(R_N, \theta_N)$, where we put

$$R_N = \frac{c_N}{N^{d/2}} \text{ and } c_N = \frac{\log N}{\log \log N.} \tag{161}$$

By Corollary 4.3 $|U^X_{N,\ell}(0, R_N)| = 2^{\ell M^d}(1+o(1))$ and thus applying (160) we obtain that the contribution from the set $U^X_{N,\ell}(0, R_N)$ is $(\mu(A))^\ell(1 + o(1))$.

To calculate the asymptotic contribution from the set $U^X_{N,\ell}(R_N, \theta_N)$ we re-write it as

$$\sum_{\substack{R_N < R < \theta_N \\ NR \in \mathbb{Z}}} \sum_{(\sigma^1,\dots,\sigma^\ell) \in U^X_{N,\ell}(R)} \mathbb{P}(H'_N(\sigma^1) \in A, \dots, H'_N(\sigma^\ell) \in A). \tag{162}$$

From (160) we conclude that for some N-independent constant C_1

$$\mathbb{P}(H'_N(\sigma^1) \in A, \dots, H'_N(\sigma^\ell) \in A) \le \mathrm{e}^{C_1 M^d R} \frac{(\mu(A))^\ell}{2^{\ell M^d}}. \tag{163}$$

If $R \in (R_N, \theta_N)$ then $N^d R^2 > c_N^2$. Thus by Corollary 4.3 for some $\alpha \in (0,1)$ $\mathbb{P}$-a.s.

$$|U_{N,\ell}^X(R)| \le |U_{N,\ell}^X(R,1)| \le 2^{\ell M^d} e^{-(1-\alpha)N^d R^2}. \tag{164}$$

Combining it with (163) we obtain that the sum (162) is bounded by

$$\sum_{\substack{R_N<R<\theta_N \\ NR\in\mathbb{Z}}} e^{C_1 M^d R-(1-\alpha)N^d R^2}. \tag{165}$$

If $R \in (R_N, \theta_N)$ then $M^d R = o(N^d R^2)$ and thus for some constant $C_2 > 0$ the sum (162) is bounded by

$$\sum_{\substack{R_N<R<\theta_N \\ NR\in\mathbb{Z}}} e^{-C_2 N^d R^2}. \tag{166}$$

There are at most N^d terms in the last sum and every term is bounded by $e^{-C_2 c_N^2}$, which decays faster than any polynomial in N. Thus the contribution from the set $U_{N,\ell}^X(R_N, \theta_N)$ is $o(1)$ and it finishes the proof of Theorem 4.4.

5. REM-universality for dynamics of spin glasses

We explain here how the REM-universality translates for non-equilibrium questions. In particular we show the universal character of the aging mechanism proved for the REM.

These aging results are proven only for a very simple instance of Glauber dynamics, i.e., the trap model dynamics (introduced by Bouchaud in [17], for a review of the trap models see [10]). It is a major open question to extend them to more general Glauber dynamics, including for instance Metropolis dynamics.

These dynamical universality results explain the domain of validity for the aging mechanism of p-spin models proposed by J.-P. Bouchaud. Its failure for $p = 2$ also points to a very different behavior for the SK model.

We will start by describing very briefly the trap model on a general graph in Section 5.1, then explain the large time behavior of this model on the simplest graph, i.e., the complete graph, see Section 5.2. We then give in Section 5.3 the aging behavior of dynamics for the REM on a very broad range of time scales (proven in [8, 9, 7] for the longest possible time scales before equilibrium and in [11] for shorter exponential time scales). This is a short survey of Section 6 in [10]. We then give in Section 5.4 the full REM-universality results for p-spin models, proven very recently in [6].

The best way to summarize these results is the following: the trap dynamics for a mean-field spin glass consists in a simple random walk on the hypercube time-changed so that it waits at each configuration a time exponentially distributed with mean proportional to the Gibbs weight. The dynamics thus samples the random landscape in a different but not unrelated way than in the re-sampling question of Section 4. Here the random sampling of the energy landscape is through the sampling of the path of a random walk. Thus the question of universality is harder than in Section 4, since close points and high "local energy correlations" are inevitable on paths of random walks.

Nevertheless the picture emerging is the following: the energy landscape as seen from the path of long random walk on the hypercube is universal, in the sense that its statistics are insensitive to correlations, including the local correlations mentioned above. In order to state this universality result the best tool is the internal clock of trapped dynamics. The result is that the internal clock converges to an α-stable subordinator in all cases mentioned above (large complete graphs, REM and p-spin models with $p \geq 3$ for appropriate exponential time scales).

5.1. The trap model on a finite graph

In [17] J.-P. Bouchaud introduced the following phenomenological model for the dynamics of mean-field spin glasses.

Let us consider a finite graph $G = (\mathcal{V}, \mathcal{E})$, non-oriented and connected, and collection of "depths" $\tau(x)$, $x \in \mathcal{V}$ such that $\tau(x) \in (0, \infty)$. The collection $(\tau(x), x \in \mathcal{V})$, is called a *trapping landscape* or *trapping environment.* We define a continuous time Markov chain $X(t)$ on $\mathcal{V}$ by its jump rates

$$c(x,y) = \begin{cases} \nu(\tau(x))^{-(1-a)}(\tau(y))^a, & \text{if } x \sim y, \\ 0, & \text{otherwise}, \end{cases} \tag{167}$$

where $\nu > 0$ fixes the time unit and $a \in [0,1]$ is a symmetry parameter. Since the measure $\boldsymbol{\tau} \equiv \sum_{x \in \mathcal{V}} \tau(x)\delta_x$ satisfies the detailed balance condition:

$$\tau(x)c(x,y) = \tau(y)c(y,x), \tag{168}$$

it is an invariant and reversible measure for the Markov chain $X(t)$.

As we already mentioned our main object of study will be the clock process S of the chain $X(t)$ where $S(k)$, $k = 1, 2, \ldots$ is the time of the kth jump of $X(t)$ and $S(t) = S(\lfloor t \rfloor)$. The embedded discrete time random walk $Y(k)$ is defined as

$$Y(k) = X(t) \quad \text{for} \quad S(k) \leq t < S(k+1). \tag{169}$$

In the case $a = 0$ the process $X(t)$ is particularly simple. Its jumping rates $c(x,y)$ do not depend on the depth of the target vertex y. Therefore, the process X waits at the vertex x exponentially distributed time with mean $\tau(x)(\nu d(x))^{-1}$, where $d(x)$ is the degree of the vertex x in G. After this time it jumps to one of the neighbors of x chosen uniformly at random. Hence, the name of this dynamics - Random Hopping Times dynamics. The embedded Markov chain $Y(k)$ is a standard random walk and $X(t)$ is its time change.

In the case $a \neq 0$, the Markov chain $Y(k)$ is no longer a standard random walk, in fact, it is a random walk in a random environment.

5.2. Aging for the complete graph

Trap dynamics on the complete graph were first introduced in [17] and [18] as a paradigm for dynamics of mean field spin glasses. This section gives a short version of Section 5.1 of [10].

Let $G_N = (\mathcal{V}_N, \mathcal{E}_N)$ be a complete graph on N vertices $\mathcal{V}_N = \{1, 2, \dots, N\}$ and $\mathcal{E}_N = \mathcal{V}_N \times \mathcal{V}_N$. We assume that the "depths" of the traps are random, in particular, we let $\tau(x)$ be i.i.d. with $\mathbb{E}\tau(x) = \infty$. In fact, we choose

$$\mathbb{P}(\tau \geq u) = \frac{1}{u^\alpha}, \qquad 0 < \alpha < 1 \tag{170}$$

and $L(u)$ is a slowly varying function.

We will consider only the case $a = 0$, i.e., the case where the embedded random walk $Y(k)$ is a standard random walk and the continuous time Markov chain X_N is a time change of Y. If $\tau(Y(i))$ is the time spent by the random walk Y at site $Y(i)$ then the clock process is

$$S_N(t) = S_N(\lfloor t \rfloor) = \sum_{i=1}^{\lfloor t \rfloor} \tau(Y(i)) e_i, \tag{171}$$

where $\{e_i\}$ is an i.i.d. sequence of exponential random variables with mean one.

If the random variables $\tau(Y(i))$, $i = 1, \dots, \lfloor t \rfloor$ are independent then $S_N(t)$ should be close to an α-stable subordinator, when properly normalized. For the random variables $\tau(Y(i))$ to be independent, we must see that the probability of self-intersections of the trajectory $Y(\cdot)$ is negligible. This is true for short enough time scales.

Theorem 5.1. *Assume $0 < \kappa < 1/\alpha$ and choose the time scale $t(N) = N^\kappa$. Then for almost every realization of the environment $\boldsymbol{\tau}$ the rescaled clock process converges to a stable subordinator V_α with the Lévy measure $\alpha\Gamma(1+\alpha)u^{-\alpha-1}\mathrm{d}u$:*

$$\frac{S_N(sN^\kappa)}{N^{\kappa/\alpha}} \to V_\alpha(s). \tag{172}$$

This theorem suggests a natural aging result. Let us introduce the following two-point function

$$\Pi_N(t, t+s; \boldsymbol{\tau}) = \mathbb{P}(X(t+u) = X(t) \ \ \forall u \in [0, s] | \boldsymbol{\tau}), \tag{173}$$

which is the probability that, conditionally on $\boldsymbol{\tau}$, the system does not jump during the specified time interval $[t, t+s]$.

Theorem 5.2. *For a.e. realization of the random environment $\boldsymbol{\tau}$*

$$\lim_{N \to \infty} \Pi_N(tN^\kappa, (1+\theta)tN^\kappa; \boldsymbol{\tau}) = F_\alpha\Big(\frac{1}{1+\theta}\Big), \tag{174}$$

where

$$F_\alpha(u) = \frac{\sin(\alpha\pi)}{\pi} \int_0^u u^{\alpha-1}(1-u)^{-\alpha} du, \quad u \in (0,1). \tag{175}$$

Indeed, the dimension of the range of an α-stable subordinator is $\alpha < 1$ and the probability that a given interval (a, b) does not intersect the range of V_α is $F_\alpha(a/b)$, where F_α is defined in (174). This fact together with Theorem 5.1 suggests Theorem 5.2 which gives aging for our model.

A similar result, proved in [18] and [21], holds for the shortest possible time scale $\kappa = 0$. Then

$$\lim_{t\to\infty}\lim_{N\to\infty}\Pi_N(t,(1+\theta)t;\boldsymbol{\tau}) = F_\alpha\left(\frac{1}{1+\theta}\right). \tag{176}$$

For the largest possible time scales, i.e., for $\kappa = 1/\alpha$, a double limiting procedure is also necessary. One can show (see [10]) that $\Pi_N(tN^{1/\alpha},(1+\theta)tN^{1/\alpha};\boldsymbol{\tau})$ converges to $F_\alpha(1/1+\theta)$ in probability, i.e., for any $\varepsilon > 0$

$$\lim_{t\to 0}\lim_{N\to\infty}\mathbb{P}\left(\left|\Pi_N(tN^{1/\alpha},(1+\theta)tN^{1/\alpha};\boldsymbol{\tau}) - F_\alpha(1/1+\theta)\right| > \varepsilon\right) = 0. \tag{177}$$

5.3. Trap dynamics for the REM

The first model for which the paradigm given in the preceding section applies is the REM. We summarize here the results of [8, 9, 7, 11, 24].

Let $G_N = (\mathcal{V}_N,\mathcal{E}_N)$, where $\mathcal{V}_N$ is the N-dimensional hypercube $\{-1,1\}^N$ and $\mathcal{E}_N$ is the set of pairs of vertices at Hamming distance 1. We choose here the depths to be given by the un-normalized Gibbs measure of the REM, i.e.,

$$\tau(\sigma) = \mathrm{e}^{-\beta\sqrt{N}H_N(\sigma)}, \quad \sigma\in\mathcal{V}_N, \tag{178}$$

and $(H_N(\sigma))_{\sigma\in\mathcal{V}_N}$ are i.i.d. $\mathcal{N}(0,1)$ random variables. We also choose $a = 0$, i.e., the jump rates of the trap dynamics are

$$c(\sigma,\sigma') = \frac{1}{N}\tau(\sigma)^{-1}. \tag{179}$$

Again here, the embedded random walk Y is a standard random walk on the hypercube.

If we consider the clock process (171) then, once properly rescaled, it converges in distribution as $N\to\infty$ to an α-stable subordinator in a proper range of time scales e^{cN} and α is a function of c.

Theorem 5.3. *Let the parameters $\alpha\in(0,1)$ and $\beta > 0$ satisfy*

$$0 < \alpha^2\beta^2/2\log 2 < 1. \tag{180}$$

We choose the time scale to be $t(N) = \mathrm{e}^{\alpha\beta^2 N}$. Then for almost every realization of the random environment $\boldsymbol{\tau}$

$$N^{\alpha/2}\frac{S_N(t\mathrm{e}^{\alpha^2\beta^2N/2})}{\mathrm{e}^{\alpha\beta^2N}} \to V_\alpha(t). \tag{181}$$

Here V_α is an α-stable subordinator as in Theorem 5.1. Theorem 5.3 implies then an aging result similar to that of Theorem 5.2.

Theorem 5.4. *Let the parameters $\alpha\in(0,1)$ and $\beta > 0$ satisfy* (180). *Then for a.e. $\boldsymbol{\tau}$*

$$\lim_{N\to\infty}\Pi_N(\mathrm{e}^{\alpha\beta^2N},(1+\theta)\mathrm{e}^{\alpha\beta^2N}) = F_\alpha(1/1+\theta). \tag{182}$$

The first difference between the REM and the complete graph dynamics of Sections 5.2 and 5.3 is the potential theory for the standard random walk on the underlying graph (complete graph vs. hypercube). The second important difference is that here the depths $\tau(x)$ are i.i.d. but not heavy-tailed. The following computation explain why, on a proper time scale, the random walk can be "fooled" to believe that they are. Let $\alpha \in (0,1)$ and $u > 0$:

$$\begin{aligned}
&\mathbb{P}\big(\tau(\sigma) \geq u e^{\alpha\beta^2 N} \mid \tau(\sigma) \geq e^{\alpha\beta^2 N}\big) \\
&\quad = \mathbb{P}\big(\beta\sqrt{N} H_N(\sigma) \geq \log u + \alpha\beta^2 N \mid \beta\sqrt{N} H_N(\sigma) \geq \alpha\beta^2 N\big) \\
&\quad \sim e^{-1/2\left(\frac{\log u + \alpha\beta^2 N}{\beta\sqrt{N}}\right)^2} e^{1/2\left(\frac{\alpha\beta^2 N}{\beta\sqrt{N}}\right)^2} \sim e^{-(\log u)\alpha} = \frac{1}{u^\alpha}.
\end{aligned} \tag{183}$$

Theorem 5.3 has an interesting interpretation which we take verbatim from [5]: at a time scale $e^{\alpha\beta^2 N} N^{-\alpha/2}$ the process succeeds to make $e^{\alpha^2\beta^2 N/2} = 2^{\varrho N}$, $\varrho \equiv \alpha^2\beta^2/2\log 2$, steps, that is, it explores a subset of the configuration space that corresponds to a "little REM" in volume $n = \varrho N$. At this time scale, the process feels an effective inverse temperature $\beta_\varrho \equiv \beta/\sqrt{\varrho}$. If the effective temperature is below the critical temperature $\beta_c \equiv \sqrt{2\log 2}$ for the standard REM, then the system shows aging, otherwise it does not. It may seem somewhat counterintuitive that the system is effectively "warming up" as time goes by. The heuristics of this result is the following. When the random walk has made $2^{\varrho N}$ steps, with $\varrho < 1$, it has only explored a small fraction of the total configuration space. In particular, it has not had time to find the absolute minima of H_N, hence it is still out of equilibrium. Moreover, the random walk will essentially not have visited any configuration twice. Therefore, the minimum of H_N along those configurations that were visited is the minimum of $2^{\varrho N}$ independent Gaussian random variables of mean zero and variance 1. It is well known that this is of order $\sqrt{2\varrho N \log 2}$. Then the mean waiting time in this extreme trap is of order $e^{\beta N\sqrt{2\varrho\log 2}} = e^{\alpha\beta^2 N}$. Now the condition $\beta_\varrho > \beta_c$ implies that this time is of the same order as the total time the process has accumulated in all the other sites along its way, and, more precisely, the process will have spent all but a negligible fraction of its time in the "few" "deepest traps". Again, standard results of extreme value theory imply that the precise statistics of the times spent in the deepest traps are asymptotically governed by a Poisson process, and that the sum of these random times, after rescaling, converges to a stable subordinator.

5.4. Trap dynamics for p-spin models, $p \geq 3$

We point out that previously discussed results, for instance aging in the REM, made crucial use of the independence of energies of different spin configurations. In this section we survey a recent work [6] which studies the dynamics of p-spin models, see Example 2.4. To our knowledge, it is the first rigorous aging result for a model with correlated energies and it affirms that the aging picture for this class of spin glasses is essentially the same as for the REM, in a wide range of exponential time scales.

Theorem 5.5. *Let $\mathcal{Y}$ be the σ-algebra generated by the standard random walk random variables $Y_N(k)$, $k \in \mathbb{N}$. There exists a function $\zeta(p)$ such that for all $p \geq 3$*

and γ satisfying

$$0 < \gamma < \min(\beta^2, \zeta(p)\beta), \tag{184}$$

under the conditional distribution $\mathbb{P}(\cdot\,|\mathcal{Y})$ the law of the stochastic process

$$e^{-\gamma N} S_N(\lfloor tN^{1/2} e^{N\gamma^2/2\beta^2} \rfloor), \quad t \geq 0, \tag{185}$$

defined on the space of càdlàg functions equipped with the Skorokhod M_1-topology, converges, $\mathcal{Y}$-a.s. to the law of the γ/β^2-stable subordinator $V_{\gamma/\beta^2}(Kt)$, $t \geq 0$, where K is a positive constant depending on γ, β and p.

Moreover, the function $\zeta(p)$ is increasing and it satisfies

$$\zeta(3) \simeq 1.0291 \quad \text{and} \quad \lim_{p\to\infty} \zeta(p) = \sqrt{2\log 2}. \tag{186}$$

The difference with similar results of Section 5.2 and 5.3 is that the traps here will not consist of a single site, but consist of a deep valley (along the trajectory) whose bottom has approximately the same energy as in the i.i.d. case and whose shape and width can be described quite precisely. Moreover, the number of sites contributing to the time spent in the valley is essentially finite, and different valleys are statistically independent (see [6] or [5]).

Since the valleys contain more than one configuration we re-define the two-point function Π_N as

$$\Pi_N^\varepsilon(t, t+s) = \mathbb{P}\big(R(\sigma_N(te^{\gamma N}), \sigma_N((t+s)e^{\gamma N})) \geq 1 - \varepsilon\big), \tag{187}$$

that is Π_N^ε is the probability that the overlap at two far-distant time instants is exceptionally large.

Theorem 5.6. *Under the hypothesis of Theorem 5.5, for all $\varepsilon \in (0,1)$, $t > 0$ and $s > 0$*

$$\lim_{N\to\infty} \Pi^\varepsilon(t, t+s) = F_\alpha(t/t+s), \tag{188}$$

where F_α is defined in (174).

Note the role of the condition (184) in Theorem 5.5, which is equivalent to $0 < \gamma < \beta^2$ and $0 < \gamma < \zeta(p)\beta$. The first condition is again the statement that the effective temperature at the time scale considered is below the critical one. The second condition is related to the correlation of the energies. It implies that the REM-like behavior holds only up to time scales where the explored region is so small that the process does not feel the correlations; essentially it ensures that the process does not have enough time to get close to a point it visited before so that it is able to feel correlations.

Finally, let us recall the static REM-universality of p-spin models discussed in Sections 3.1.3 and 4.1.3. Conditions (67) and (140) of Theorem 4.3 and Theorem 3.9 are very similar to the condition $\gamma < \zeta(p)\beta$ and could be interpreted as the conditions ensuring that the energies in the micro-canonical window and in the sampled random cloud respectively are so sparse on the configuration space that they do not yet feel correlations. Moreover, the constant c_ν in (67) for the case $\nu(r) = r^p$ coincides with the constant $\zeta(p)$.

Acknowledgement

We thank our collaborators A. Bovier, J. Černý, V. Gayrard for their help with this work. We also would like to thank Heiko Bauke for illuminating discussions and private communication of his PhD thesis.

This paper is based on a series of lectures given at the Summer School "Spin Glasses" at Institut Henri Poincaré in June 2007. We thank the organizers Anne Boutet de Monvel, Anton Bovier, and Michel Talagrand for their invitation.

References

[1] Bauke, H.: Zur Universalität des Random-Energy-Modells. Ph.D. thesis, University of Magdeburg (2006).

[2] Bauke, H., Franz, S., Mertens, S.: Number partitioning as a random energy model. J. Stat. Mech. Theory Exp. page P04003 (2004).

[3] Bauke, H., Mertens, S.: Universality in the level statistics of disordered systems. Phys. Rev. E (3) **70**, 025,102(R) (2004).

[4] Bauke, H., Mertens, S.: Ubiquity of the Random-Energy Model. Poster presented at the "International Summer School: Fundamental Problems in Statistical Physics XI" in Leuven (Belgium) (2005).

[5] Ben Arous, G., Bovier, A., Černý, J.: Universality of random energy model-like ageing in mean field spin glasses. J. Stat. Mech. Theory Exp. (4), L04,003, 8 (2008).

[6] Ben Arous, G., Bovier, A., Černý, J.: Universality of the REM for dynamics of mean-field spin glasses. Comm. Math. Phys. **282**(3), 663–695 (2008).

[7] Ben Arous, G., Bovier, A., Gayrard, V.: Aging in the random energy model. Phys. Rev. Lett. **88**(8), 87,201–1–87,201–4 (2002).

[8] Ben Arous, G., Bovier, A., Gayrard, V.: Glauber dynamics of the random energy model. I. Metastable motion on the extreme states. Comm. Math. Phys. **235**(3), 379–425 (2003).

[9] Ben Arous, G., Bovier, A., Gayrard, V.: Glauber dynamics of the random energy model. II. Aging below the critical temperature. Comm. Math. Phys. **236**(1), 1–54 (2003).

[10] Ben Arous, G., Černý, J.: Dynamics of Trap Models. In: A. Bovier, F. Dunlop, A. Van Enter, F. Den Hollander (eds.) Mathematical Statistical Physics, École d'Été de Physique des Houches : Session LXXXIII : 4-29 July, 2005, pp. 331–394. Elsevier (2006).

[11] Ben Arous, G., Černý, J.: The arcsine law as a universal aging scheme for trap models. Comm. Pure Appl. Math. **61**(3), 289–329 (2008).

[12] Ben Arous, G., Gayrard, V., Kuptsov, A.: A new REM conjecture. In: V. Sidoravicius, M.E. Vares (eds.) In and Out of Equilibrium 2, *Progress in Probability*, vol. 60, pp. 59–96. Birkhäuser (2008).

[13] Ben Arous, G., Kuptsov, A.: The limits of REM Universality, in preparation.

[14] Borgs, C., Chayes, J., Mertens, S., Nair, C.: Proof of the local REM conjecture for number partitioning. I: Constant energy scales. Random Struct. Algorithms **34**(2), 217–240 (2009).

[15] Borgs, C., Chayes, J., Mertens, S., Nair, C.: Proof of the local REM conjecture for number partitioning. II. Growing energy scales. Random Struct. Algorithms **34**(2), 241–284 (2009).

[16] Borgs, C., Chayes, J., Pittel, B.: Phase transition and finite-size scaling for the integer partitioning problem. Random Structures Algorithms **19**(3-4), 247–288 (2001). Analysis of algorithms (Krynica Morska, 2000).

[17] Bouchaud, J.P.: Weak ergodicity breaking and aging in disordered systems. J. Phys. I (France) **2**, 1705–1713 (1992).

[18] Bouchaud, J.P., Dean, D.S.: Aging on Parisi's tree. J. Phys. I (France) **5**, 265 (1995).

[19] Bovier, A.: Statistical mechanics of disordered systems, *MaPhySto Lecture Notes*, vol. 10. Aarhus University (2001).

[20] Bovier, A.: Statistical mechanics of disordered systems, *Cambridge Series in Statistical and Probabilistic Mathematics*, vol. 18. Cambridge University Press, Cambridge (2006). A mathematical perspective.

[21] Bovier, A., Faggionato, A.: Spectral characterization of aging: the REM-like trap model. Ann. Appl. Probab. **15**(3), 1997–2037 (2005).

[22] Bovier, A., Kurkova, I.: Local energy statistics in disordered systems: a proof of the local REM conjecture. Comm. Math. Phys. **263**(2), 513–533 (2006).

[23] Bovier, A., Kurkova, I.: Local energy statistics in spin glasses. J. Stat. Phys. **126**(4-5), 933–949 (2007).

[24] Černý, J., Gayrard, V.: Hitting time of large subsets of the hypercube. Random Struct. Algorithms **33**(2), 252–267 (2008).

[25] Derrida, B.: Random-energy model: an exactly solvable model of disordered systems. Phys. Rev. B (3) **24**(5), 2613–2626 (1981).

[26] Georgii, H.O.: Large deviations and maximum entropy principle for interacting random fields on $\mathbf{Z}^d$. Ann. Probab. **21**(4), 1845–1875 (1993).

[27] Kallenberg, O.: Random measures, third edn. Akademie-Verlag, Berlin (1983).

[28] Kuptsov, A.: Universality of random Hamiltonians. Ph.D. thesis, New York University (2008).

[29] Kurkova, I.: Local energy statistics in directed polymers. Electron. J. Probab. **13**, no. 2, 5–25 (2008).

[30] Mertens, S.: Random costs in combinatorial optimization. Phys. Rev. Lett. **84**(6), 1347–1350 (2000).

[31] Talagrand, M.: Spin glasses: a challenge for mathematicians, *Ergebnisse der Mathematik und ihrer Grenzgebiete. 3. Folge*, vol. 46. Springer-Verlag, Berlin (2003). Cavity and mean field models.

Gérard Ben Arous and Alexey Kuptsov
Courant Institute of Mathematical Sciences
New York University
251 Mercer Street
New York, NY 10012, USA
e-mail: `benarous@cims.nyu.edu`
`kuptsov@cims.nyu.edu`

Progress in Probability, Vol. 62, 85–101

Another View on Aging in the REM

Jiří Černý

Abstract. We give a new proof of aging for a version of a Glauber dynamics in the Random Energy Model. The proof uses ideas that were developed in [2] for studying the dynamics of a p-spin Sherrington–Kirkpatrick spin glass.

Mathematics Subject Classification (2000). 82D30, 60G15, 60K37.

Keywords. Dynamics of disordered systems, spin glasses, random energy model, aging, extremal theory, Gaussian processes.

1. Introduction

Aging was proved recently for a simple Glauber-type dynamics of p-spin spin glasses [2]. In order to overcome the difficulties stemming from the correlations of the Hamiltonian of these spin glasses, several new ideas were introduced. They allowed to show that the aging behaviour of the p-spin spin glass is essentially the same as the one of the Random Energy Model (REM), at least at some time scales.

In the present paper, we take a step back and apply these ideas to the REM. A new proof of aging in the REM, even if it be shorter than the older proofs, is, however, not the principal objective of this paper. Rather, it is written as a different presentation of ideas of [2], uncluttered from rather heavy computations that were necessary for the p-spin spin glass.

Let us start with a brief summary of the efforts that lead to [2]. Aging in spin glasses was for the first time observed experimentally in the beginning of the 1980's. In order to explain the observations, trap models were introduced by Bouchaud [9, 10] in the physics literature. Trap models are effective models which can be solved analytically using simple renewal arguments, and which nevertheless reproduce the characteristic power-law behaviour observed experimentally. While trap models are heuristically motivated to capture the behaviour of the dynamics of spin glass models, there is no clear theoretical, let alone mathematical, derivation of them from an underlying spin-glass dynamics.

The first steps to provide such a derivation were taken in [3, 4] for a version of a Glauber dynamics in the REM. Technically very elaborate renewal arguments were used to prove aging and the relevance of the trap model ansatz in this case. The core of the argument is an analysis of visits of the dynamics to a finite set of extremes of the Hamiltonian. In order to allow the dynamics to discover these extremes, the time scales studied in these papers should be carefully fixed to be only slightly shorter than the equilibration scale. In particular, it means that the considered time scales should increase exponentially with the size of the system.

Another proof of aging in the REM, for a dynamics close to this of [3], was given in [5]. It was inspired by methods developed for trap models on $\mathbb{Z}^d$ [6]. The main idea of the approach of [5] is to study the extremes of the Hamiltonian along the trajectory of the dynamics instead of concentrating on visits of dynamics to the extremes of the Hamiltonian. Apart from having some slight technical advantages, the new point of view gave more freedom in choosing the time scales. In [5] time scales much shorter than the equilibration scale, but still increasing exponentially with the size of the system, were studied. The techniques can be however easily extended to apply also to the scales considered in [3].

The main technical tool of [5] is the so-called *clock process*, which is, roughly speaking, a process that records the time needed for a given number of jumps of the dynamics (see (4) for the exact definition). It was argued that this process converges, after a proper rescaling, to a stable subordinator in many situations. From the convergence to a subordinator, aging can be deduced using the classical arc-sine law. Even if techniques and time scales are different, the obtained aging results are essentially the same as those predicted by the trap models.

Both above mentioned studies of the REM used substantially the fact that the Hamiltonian is particularly simple: it is a collection of i.i.d. random variables. This ceases to be true for 'more realistic' mean-field spin glasses, like Sherrington–Kirkpatrick model or p-spin spin glass. For statics of the spin glass the correlation between the energies impose that (at low temperature) the main contribution to the Gibbs measure does not come from a finite number of distant configurations as in the REM, but from a finite number of distant 'valleys', which, however, contain many configurations.

Interestingly, in [2] it was showed that the global behaviour of the clock process is not influenced by the correlations in the dynamics of p-spin spin glass with $p \geq 3$. The rescaled clock process hardly feels the correlations and large valleys and it converges to a stable subordinator, confirming the universality of this behaviour. However, this convergence can be proved only at the expense of restricting the range of time scales to the lower part of the range of [5], remaining far from the equilibration time scale (see (14) for details).

Even if the asymptotic behaviour of the clock process remains unchanged, the non-i.i.d. character of the Hamiltonian disallows a direct application of methods of [5]. New techniques used in [2] are based on ideas from extremal theory and exploit strongly the Gaussian character of the Hamiltonian. It should be remarked that this property of the Hamiltonian had been hardly used in the studies of the REM dynamics. E.g., the only Gaussian ingredient in [5] is the standard asymptotic expression for the tail probability. The results of [5] can thus be extended to a much larger class of distributions of the Hamiltonian.

In this paper, we take full advantage of the Gaussian distribution of the Hamiltonian and we use the methods of [2] to provide new and relatively short proof of aging in the REM on time scales that are much shorter than the equilibration scale. Aside from this proof, we use the occasion to explicitly write out the modifications which are necessary to obtain a result on the longest possible time scales of [4].

To close the introduction, let us remark that the time scales of this paper (and also of [3, 5]) are 'much longer' than those used in the studies of the Langevin dynamics of soft-spin models [11, 8, 7], where one considers the infinite volume limit at fixed time t, and then analyzes the ensuing dynamics as t tends to infinity. In this paper, the time depends exponentially on the size of the system and both tend to infinity together.

2. Model and results

We will study the same dynamics of the REM as in [5]. It is defined as follows. Let $\mathcal{S}_N \equiv \{-1,1\}^N$ be an N-dimensional hypercube equipped with the distance

$$\operatorname{dist}(\sigma,\tau) = \frac{1}{2}\sum_{i=1}^{N} |\sigma_i - \tau_i|, \qquad \sigma,\tau \in \mathcal{S}_N. \tag{1}$$

The Hamiltonian of the REM is defined as $\sqrt{N}H_N$, where $H_N : \mathcal{S}_N \to \mathbb{R}$ is a centred i.i.d. Gaussian process on $\mathcal{S}_N$ with variance $\mathbb{E}[H_N(\sigma)^2] = 1$. We will use $\mathcal{H}$ to denote the σ-algebra generated by $\{H_N(\sigma), \sigma \in \mathcal{S}_N, N \in \mathbb{N}\}$. The corresponding Gibbs measure is given by

$$\mu_{\beta,N}(\sigma) \equiv Z_{\beta,N}^{-1} \mathrm{e}^{\beta\sqrt{N}H_N(\sigma)}. \tag{2}$$

We consider a nearest-neighbour continuous-time Markov dynamics $\sigma_N = (\sigma_N(t), t \geq 0)$ on $\mathcal{S}_N$ which is given by its transition rates

$$w_N(\sigma,\tau) = \begin{cases} N^{-1}\mathrm{e}^{-\beta\sqrt{N}H_N(\sigma)}, & \text{if } \operatorname{dist}(\sigma,\tau) = 1,\\ 0, & \text{otherwise.} \end{cases} \tag{3}$$

Obviously, σ_N is reversible with respect to the Gibbs measure $\mu_{\beta,N}$.

It is an important property, that this dynamics can be constructed as a time change of a simple random walk on $\mathcal{S}_N$: Let $(Y_N(k), k \in \mathbb{N})$ be the simple discrete-time random walk (SRW) on $\mathcal{S}_N$ started at some fixed point of $\mathcal{S}_N$, say at $\mathbf{1} = \{1,\dots,1\}$. For $\beta > 0$ and $k \in \mathbb{N}$ we define the *clock-process* by

$$S_N(k) = \sum_{i=0}^{k-1} e_i \exp\bigl\{\beta\sqrt{N}H_N\bigl(Y_N(i)\bigr)\bigr\}, \tag{4}$$

where $(e_i, i \in \mathbb{N})$ is a sequence of mean-one i.i.d. exponential random variables. The process σ_N can then be written as

$$\sigma_N(t) \equiv Y_N(S_N^{-1}(t)). \tag{5}$$

We consider all random processes to be defined on an abstract probability space $(\Omega, \mathcal{F}, \mathbb{P})$. We denote by $\mathcal{Y}$ the σ-algebra generated by $\{Y_N(k),\ k \in \mathbb{N},\ N \in \mathbb{N}\}$. The σ-algebra generated by $\{e_i,\ i \in \mathbb{N}\}$ will be denoted by $\mathcal{E}$. Note that the three σ-algebras $\mathcal{H}$, $\mathcal{Y}$, and $\mathcal{E}$ are all independent under $\mathbb{P}$.

For $\gamma > 0$ we define

$$\begin{aligned} r_N &= r_N(\gamma) = \mathrm{e}^{N\gamma^2/2\beta^2}, \\ t_N &= t_N(\gamma) = N^{-\frac{1}{2\alpha}}\mathrm{e}^{\gamma N}, \end{aligned} \tag{6}$$

and the rescaled clock process $\bar{S}_N^\gamma$ by

$$\bar{S}_N^\gamma(s) = t_N^{-1} S_N(\lfloor s r_N \rfloor), \qquad s \geq 0. \tag{7}$$

The function r_N is the *number-of-jumps* scale: We observe σ_N after making $\mathrm{O}(r_N)$ jumps. The function t_N, the *time scale*, then gives the time that σ_N typically needs to make this number of steps. We view $\bar{S}_N^\gamma$ as an element of the space D of càdlàg functions from $[0,\infty)$ to $\mathbb{R}$ equipped with the standard Skorokhod J_1-topology.

Let $V_\alpha(t)$ be the α-stable subordinator with the Laplace transform given by

$$\mathbb{E}[\mathrm{e}^{-\lambda V_\alpha(t)}] = \exp(-t\lambda^\alpha). \tag{8}$$

We will use $\beta_c = \sqrt{2\log 2}$ to denote the critical temperature of the REM.

The main result of this paper is the following theorem that provides the asymptotic behaviour of the clock.

Theorem 2.1. *For any fixed γ such that*

$$0 < \gamma < \min\left(\beta^2, \beta_c\beta\right), \tag{9}$$

under the conditional distribution $\mathbb{P}[\,\cdot\,|\mathcal{Y}]$, $\mathcal{Y}$-a.s., the law of the stochastic process $\bar{S}_N^\gamma$ converges to the law of α-stable subordinator $V_\alpha(\mathcal{K}\cdot)$, where $\alpha \equiv \gamma/\beta^2$, and $\mathcal{K}$ is a constant which will be computed explicitly in Lemma 3.1.

This theorem is very close to the results of [5]. However, there is one important difference. In [5], the convergence to the subordinator is proved under the law $\mathbb{P}[\,\cdot\,|\mathcal{H}]$, $\mathcal{H}$-a.s. The slightly non-physical conditioning on $\mathcal{Y}$ that appears in our theorem, and of course also in [2] whose methods we use, is the price to pay for having at hand Gaussian tools. As we have already remarked, the Gaussian character of the Hamiltonian was almost not exploited in the previous studies of the REM. Hence, the conditioning on $\mathcal{H}$ did not pose any problem. This conditioning, that is fixing the Gaussian disorder, is ruled out if we want to employ more advanced Gaussian techniques now.

It is an interesting open question whether it is possible to deduce the results of [5] from Theorem 2.1 without using too much the properties of the REM. This could allow to prove the convergence under $\mathbb{P}[\,\cdot\,|\mathcal{H}]$ also for the p-spin spin glass.

The next theorem will be used to prove aging on the longest possible time scales, that means on time scales of [4].

Theorem 2.2. *Let $\beta > \beta_c$ and $\gamma = \beta\beta_c$. Then for all $n \geq 1$ finite, for all $0 \leq s_1 < \cdots < s_n$ and $A_1, \ldots, A_n \in \mathcal{B}(\mathbb{R})$*

$$\lim_{\xi \to 0} \lim_{N \to \infty} \mathbb{P}\Big[\bigcap_{i=1}^{n} \xi^{-1/\alpha} \bar{S}_N^{\gamma}(\xi s_i) \in A_i \Big| \mathcal{Y}\Big] = \mathbb{P}\Big[\bigcap_{i=1}^{n} V_\alpha(\mathcal{K}s_i) \in A_i\Big] \tag{10}$$

in probability.

As a consequence of Theorems 2.1 and 2.2 we get the following aging result

Theorem 2.3.

(a) *Under the hypotheses of Theorem 2.1, for all $\theta > 1$, $\mathcal{Y}$-a.s.,*

$$\lim_{N \to \infty} \mathbb{P}[\sigma_N(t_N) = \sigma_N(\theta t_N) | \mathcal{Y}] = \mathsf{Asl}_\alpha(\theta), \tag{11}$$

where $\mathsf{Asl}_\alpha(\theta) = \frac{\sin(\alpha\pi)}{\pi} \int_0^{1/\theta} u^{\alpha-1}(1-u)^{-\alpha} du.$

(b) *Under the hypotheses of Theorem 2.2, for all $\theta > 1$, in probability,*

$$\lim_{\xi \to 0} \lim_{N \to \infty} \mathbb{P}[\sigma_N(\xi t_N) = \sigma_N(\xi\theta t_N) | \mathcal{Y}] = \mathsf{Asl}_\alpha(\theta). \tag{12}$$

Claim (b) of the last theorem relates to the aging result obtained in [4] just as claim (a) relates to [5]: the role of σ-algebras $\mathcal{H}$ and $\mathcal{Y}$ is inverted.

Finally, let us compare our result with the result of [2] for the p-spin spin glass. The definition of the dynamics considered there is the same as in this paper. The only change is, of course, the Hamiltonian, which is given by a centred Gaussian process on $\mathcal{S}_N$ with covariance

$$\mathrm{Cov}(H_N(\sigma), H_N(\tau)) = \Big(\frac{1}{N} \sum_{i=1}^{N} \sigma_i \tau_i\Big)^p. \tag{13}$$

With two replacements, Theorem 2.1 holds in this case: First, condition (9) should be replaced by $p \geq 3$ and

$$\gamma < \min\left(\beta^2, \zeta(p)\beta\right), \tag{14}$$

where $\zeta(p)$ is an increasing function strictly smaller than β_c which converges to β_c as $p \to \infty$. Second, the space D should be equipped with a weaker topology.

Condition (14) implies that the longest time scales we are able to treat in the p-spin model are shorter than for the REM. We do not know what happens in the p-spin model on longer scales. As $p \to \infty$ (14) approaches (9), which is not surprising, since the REM can be considered as p-spin model with $p = \infty$.

Let us close this section by a rough description of the techniques which are used to prove Theorem 2.1. The behaviour of the rescaled clock process $\bar{S}_N^{\gamma}$ is determined by the energies of spin configurations that are visited during the first $\mathrm{O}(r_N)$ steps. The energies of visited configurations form a Gaussian process $X_N^0(k) \equiv H_N(Y_N(k))$ which has random covariance $\mathrm{Cov}(X_N^0(k), X_N^0(j)) = \mathbf{1}\{Y_N(k) = Y_N(j)\}$. We are interested mainly in the visited configurations whose energy is very large, because these configurations contribute a lot to the clock process. We want thus to know how extremes of a Gaussian process X_N^0 with random correlation structure behave.

The standard method how to study such extremes is to replace the complicated Gaussian process with a simpler process, the behaviour of whose extremes can be determined more easily and whose correlation structure locally approximates well the correlation structure of the original process. A study of the behaviour of extremes of the original process then breaks into two parts. First, the behaviour of extremes of the simple process should be determined. Second, it should be proved that the approximation by the simple process is reasonable.

In the case of the REM, that is of Gaussian process X_N^0, the approximating process will be particularly simple. We define $(X^1(k), k \in \mathbb{N}, N \in \mathbb{N})$ as an i.i.d. sequence of standard Gaussian random variables. It is clearly a natural choice, since the simple random walk on the hypercube has a very small probability to return to an already visited configuration. More precisely, for any j fixed, $\mathbb{P}[\exists k : r_N > k > j, Y_N(k) = Y_N(j)] \sim 1/N$ (at least if $\gamma < \beta\beta_c$).

In the first step of the proof we will thus analyse extremes of an i.i.d. sequence, which is, of course, quite simple. This is done in Section 3 using a method that allows control the clock process immediately. The second step of the proof, that is the verification if the approximation is justified, is done in Section 5. In Section 4 we collect several estimates on the simple random walk on the hypercube. Finally, all theorems are proved in Section 6.

3. Sum of i.i.d. exponentials

As explained in the last section, we will compare the clock process with the sum of i.i.d. random variables with the same distribution, that is with the sum

$$\tilde{S}_N^\gamma(s) = t_N^{-1} \sum_{i=0}^{\lfloor s r_N \rfloor} e^{\beta\sqrt{N} X_N^1(i)}, \tag{15}$$

Sums of this type were exhaustively studied in [1]. The results of this paper imply directly that $\tilde{S}_N^\gamma$ converges to an γ/β^2 stable subordinator. For the sake of completeness we will provide here a simple proof of this claim. We start by an easy lemma.

Lemma 3.1. *For all β, γ satisfying hypotheses of Theorems* 2.1 *or* 2.2 *there exists* $\mathcal{K} = \mathcal{K}_{\beta,\gamma} > 0$ *such that, for r_N and t_N as in* (6) *and* $u > 0$,

$$\lim_{N\to\infty} r_N \Big(1 - \mathbb{E}\Big[\exp\Big\{-\frac{u}{t_N} e_i e^{\beta\sqrt{N}X_N^1(i)}\Big\}\Big]\Big) = \mathcal{K}u^\alpha. \tag{16}$$

This lemma should be viewed as a 'large deviation' statement, since values of $X_N^1(i)$ that give the largest contribution to the Laplace transform in (16) differ significantly from the typical ones. From large deviation point of view, the proof below is simply a tilting of the Gaussian distribution. This tilting makes the most contributing values typical.

Proof. Recall that $\alpha = \gamma/\beta^2$. To save the notation we set $X_i = X_N^1(i)$ and define $g(x) = \ln(1+x)$. Performing the expectation over e_i we get

$$1-\mathbb{E}\Big[\exp\Big\{-\frac{ue_i}{t_N}\mathrm{e}^{\beta\sqrt{N}X_i}\Big\}\Big] = \int_{-\infty}^{\infty}\frac{\mathrm{d}x}{\sqrt{2\pi}}\mathrm{e}^{-\frac{x^2}{2}}\Big(1-\exp\Big\{-g\Big(\frac{u}{t_N}\mathrm{e}^{\beta\sqrt{N}x}\Big)\Big\}\Big). \quad (17)$$

We now tilt the measure. Setting $x = (\beta z + \log t_N - \log u)/(\beta\sqrt{N})$ we find that (17) equals

$$u^\alpha r_N^{-1}\int_{-\infty}^{\infty}\frac{\mathrm{d}z}{\sqrt{2\pi}}\mathrm{e}^{-z^2/2N}\frac{\mathrm{e}^{-\gamma z/\beta}\mathrm{e}^{\beta z}}{1+\mathrm{e}^{\beta z}}\mathrm{e}^{zf_N(\beta,\gamma,u)}\,(1+o(1)), \quad (18)$$

where $f_N(\beta,\gamma,u) = (\beta\log N)/(2\gamma N)+\log u/(\beta N) \to 0$ as $N\to\infty$. The integrand converges to $\mathrm{e}^{(1-\alpha)\beta z}(1+\mathrm{e}^{\beta z})^{-1}$ which decays exponentially as z tends both to ∞ and $-\infty$, since $\beta/\gamma^2 < 1$. An application of the dominated convergence theorem then yields the convergence of the integral to a positive constant $\mathcal{K}$ independent of u. □

By consequence, we get the analogue of Theorem 2.1 for the process X_N^1:

Proposition 3.1. *For all β, γ satisfying hypotheses of Theorem 2.1 or 2.2, the sequence of processes $\tilde{S}_N^\gamma(s)$ converges to the stable subordinator $V_\alpha(\mathcal{K}s)$ weakly in the Skorokhod J_1-topology.*

Proof. To check the convergence of finite-dimensional marginals from Lemma 3.1 is trivial. E.g.,

$$\mathbb{E}[\mathrm{e}^{-u\tilde{S}_N^\gamma(s)}] = \mathbb{E}\Big[\exp\Big\{-\frac{u}{t_N}e_i\mathrm{e}^{\beta\sqrt{N}X_i}\Big\}\Big]^{\lfloor sr_N\rfloor} \xrightarrow{N\to\infty} \mathrm{e}^{-s\mathcal{K}u^\alpha}, \quad (19)$$

which is the Laplace transform of $V_\alpha(\mathcal{K}s)$. The proof of the tightness is an easy modification of the tightness proof for Theorem 2.1. We omit it therefore here. □

By simply changing the domains of integration in the last proof, the next lemma can be verified. We will need it later to check the tightness of $\bar{S}_N^\gamma$.

Lemma 3.2. *Let B_N^ε be such that*

$$t_N^{-1}\mathrm{e}^{\beta\sqrt{N}B_N^\varepsilon} = \varepsilon \quad (20)$$

and let

$$f_N(\varepsilon) = r_N\Big(1-\mathbb{E}\Big[\exp\Big\{-t_N^{-1}e_i\mathrm{e}^{\beta\sqrt{N}X_N^1(i)}\Big\}\mathbf{1}\{X_N^1(i)\le B_N^\varepsilon\}\Big]\Big). \quad (21)$$

Then $\lim_{\varepsilon\to 0}\limsup_{N\to\infty} f_N(\varepsilon) = 0$.

4. Random walk properties

To compare the clock processes $\tilde{S}_N^\gamma$ and $\bar{S}_N^\gamma$ we need to know to what extent the covariances of X^1 and X^0 differ. Since the non-zero covariances of X^0 rise from self-intersections of the simple random walk Y_N we should control their number.

Lemma 4.1. *Under the assumptions of Theorem* 2.1 *there exists* $C = C(\beta, \gamma)$ *such that* $\mathcal{Y}$*-a.s. for all but finitely many* N,

$$\sum_{i \neq j = 1}^{sr_N} \mathbf{1}\{Y_N(i) = Y_N(j)\} \le CN^{-1} sr_N. \tag{22}$$

Proof. Note that the assumption (9) implies that $r_N \ll 2^N$. Let $p_k^N(x,y) = \mathbb{P}[Y_N(k) = y | Y_N(0) = x]$. To bound the sum (22) for i, j that are far from each other we use the fact that the random walk on the hypercube reaches the equilibrium very quickly. The next lemma can be proved by using the coupling argument of [15]. Detailed proof is given in [2] and we will not repeat it here.

Lemma 4.2. *There exists* K *large enough such that for all* $k \ge KN^2 \log N =: \mathfrak{K}(N)$ *and* $x, y \in \mathcal{S}_N$

$$\left| \frac{p_k^N(x,y) + p_{k+1}^N(x,y)}{2} - 2^{-N} \right| \le 2^{-8N}. \tag{23}$$

Let $A_N = \{(i,j) : 0 \le i, j \le sr_N, |i-j| > \mathfrak{K}(N)\}$. From Lemma 4.2 it follows that

$$\mathbb{E}\Big[\sum_{(i,j) \in A_N} \mathbf{1}\{Y_N(i) = Y_N(j)\} \Big] \le C(sr_N)^2 2^{-N}. \tag{24}$$

Hence,

$$\mathbb{P}\Big[\sum_{(i,j) \in A_N} \mathbf{1}\{Y_N(i) = Y_N(j)\} \ge CN^{-1} sr_N \Big] \le CNsr_N 2^{-N}. \tag{25}$$

The Borel-Cantelli lemma then implies an a.s. bound for the sum over A_N.

To bound the contribution of pairs i, j with $|i-j| \le \mathfrak{K}(N)$ we need another lemma.

Lemma 4.3. *There exist* $c_2 > c_1 > 0$ *such that, for all* N,

$$\frac{c_1}{N} \le \mathbb{E}\Big[\sum_{i=1}^{\mathfrak{K}(N)} \mathbf{1}\{Y_N(i) = Y_N(0)\} \Big] \le \frac{c_2}{N}. \tag{26}$$

Proof. There are many ways how to prove this lemma. Let us sketch one of them. Let, for $A \subset \mathcal{S}_N$, $\tau_A = \min\{k \ge 1 : Y_N(k) \in A\}$, and let $B_4 = \{z : d(z, \mathbf{1}) = 4\}$. Using the fact that $d(\mathbf{1}, Y_N(k))$ is the Ehrenfest's-Urn Markov chain, one can check that $c'/N \le \mathbb{P}[\tau_1 < \tau_{B_4} | Y_N(0) = \mathbf{1}] \le c/N$. A similar argument gives also $\mathbb{P}[\tau_1 < \tau_{B_4} | Y_N(0) \in B_4] < c/N^{-4}$. Therefore to get from B_4 to 1 we need in average N^4 tries, but we have at most $\mathfrak{K}(N) = KN^2 \log N$ of them. Hence, the probability of returning to $\mathbf{1}$ before $\mathfrak{K}(N)$ is smaller than c/N, which yields the claim of the lemma. □

Let $Z_i = \mathfrak{K}(N)^{-1} \sum_{j=i+1}^{i+\mathfrak{K}(N)} \mathbf{1}\{Y_N(j) = Y_N(i)\}$. Then $Z_i \in [0,1]$, and by the last lemma $c_1(N\mathfrak{K}(N))^{-1} \le \mathbb{E}[Z_i] \le c_2(N\mathfrak{K}(N))^{-1}$. Obviously,

$$\sum_{\substack{i,j=1\\|i-j|\le\mathfrak{K}(N)}}^{sr_N} \mathbf{1}\{Y_N(i) = Y_N(j)\} \le 2\mathfrak{K}(N) \sum_{k=1}^{\mathfrak{K}(N)} \sum_{j=1}^{m} Z_{j\mathfrak{K}(N)+i}, \tag{27}$$

where $m = \lceil sr_N/\mathfrak{K}(N)\rceil$. The inner sum in the last expression is an i.i.d. sum. Hoeffding's inequality [13] applied to the sequence $\{Z_i\}$ gives for any $u > 0$,

$$\mathbb{P}\Big[\sum_{j=1}^{m} Z_{j\mathfrak{K}(N)+i} - m\mathbb{E}[Z_i] \ge um\Big] \le \exp\{-2mu^2\}. \tag{28}$$

Setting $u = \mathbb{E}[Z_i]$ and observing that the right-hand side of the last expression is summable even after a multiplication by $\mathfrak{K}(N)$, the Borel-Cantelli lemma and (27) imply that $\mathcal{Y}$-a.s., for all but finitely many N,

$$\sum_{\substack{i,j=1\\|i-j|\le\mathfrak{K}(N)}}^{sr_N} \mathbf{1}\{Y_N(i) = Y_N(j)\} \le CN^{-1}sr_N. \tag{29}$$

This completes the proof of Lemma 4.1. □

Lemma 4.4. *Let $s > 0$ and let for any $\omega \ge 1$*

$$I_\omega = I_\omega(N,s) = \{\sigma \in \mathcal{S}_N : \exists\, i_1 < \cdots < i_\omega \le r_N s,\ Y_N(i_1) = \cdots = Y_N(i_\omega) = \sigma\} \tag{30}$$

be the set of configurations visited at least ω-times. Then,

(a) *there exists $C > 0$ such that $\mathcal{Y}$-a.s., for all but finitely many N, for all $\omega \in \{2,\dots,N\}$*

$$|I_\omega| \le C^\omega(N^{1-\omega}r_N s \vee 1). \tag{31}$$

(b) *$\mathcal{Y}$-a.s., for all but finitely many N,*

$$|I_N| = 0. \tag{32}$$

Proof. The proof is very similar to the previous one: Lemma 4.2 can be used to show that the number of elements of I_ω with $\max\{i_k - i_{k-1} : 2 \le k \le \omega\} \ge \mathfrak{K}(N)$ is much smaller than the right-hand side of (31). Defining $\tilde{Z}_i = \mathbf{1}\{Y_N(i) \in I_\omega,\ i_1 = i,\ i_\omega - i_1 \le \omega\mathfrak{K}(N)\}$ and observing similarly as in Lemma 4.3 that $\mathbb{E}[\tilde{Z}_i] \sim cN^{1-\omega}$, the claim (a) can be proved by another application of Hoeffding's inequality.

Claim (b) follows from $\mathbb{P}[\exists i \le r_N s : Y_N(i) \in I_N] \le C^N sN^{-N+1}r_N$ and the Borel–Cantelli lemma. □

To prove Theorem 2.2, we need the following modification of Lemma 4.1.

Lemma 4.5. *Let $\gamma = \beta\beta_c$, that is $r_N = 2^N$. Then there exists a constant C such that for all $\xi < 1$*

$$\begin{aligned}\mathbb{P}\Big[\sum_{i\neq j=1}^{\xi r_N} \mathbf{1}\{Y_N(i) = Y_N(j)\} \geq \xi r_N\Big] &\leq C\xi, \\ \mathbb{P}[|I_\omega| \geq r_N \xi^{\omega/2}] &\leq C\xi^{\omega/2}.\end{aligned} \tag{33}$$

Proof. Following the same reasoning as in the proof of Lemma 4.1 one can show that $\mathbb{E}\big[\sum_{i\neq j=1}^{\xi r_N} \mathbf{1}\{Y_N(i) = Y_N(j)\}\big] \leq C\xi^2 r_N$. The first claim then follows from the Markov inequality.

The second claim can be obtained from $\mathbb{E}[|I_\omega|] \leq C(\xi r_N)^\omega 2^{-N(\omega-1)} \leq r_N \xi^\omega$. □

5. Comparison of two processes

We can now compare the clock process $\bar{S}_N^\gamma$ with the i.i.d. sum $\tilde{S}_N^\gamma$. We use Λ^0 and Λ^1 to denote the covariance matrices of X^0 and X^1:

$$\Lambda^0_{ij} = \mathbf{1}\{Y_N(i) = Y_N(j)\}, \qquad \Lambda^1_{ij} = \delta_{ij}. \tag{34}$$

For $h \in [0,1]$ we define the interpolating process $X_N^h(i) \equiv \sqrt{1-h}X_N^0(i) + \sqrt{h}X_N^1(i)$.

Let $\ell \in \mathbb{N}$, $0 = s_0 < \cdots < s_\ell = T$ and $u_1, \ldots, u_\ell > 0$ be fixed. For any Gaussian process X we define

$$\begin{aligned}F_N(X; \{s_i\}, \{u_i\}) &\equiv \mathbb{E}\Big[\exp\Big(-\sum_{k=1}^{\ell} \frac{u_k}{t_N} \sum_{i=s_{k-1}r_N}^{s_k r_N - 1} e_i \mathrm{e}^{\beta\sqrt{N}X(i)}\Big)\Big|X\Big](X) \\ &= \exp\Big(-\sum_{k=1}^{\ell} \sum_{i=s_{k-1}r_N}^{s_k r_N - 1} g\Big(\frac{u_k}{t_N}\mathrm{e}^{\beta\sqrt{N}X(i)}\Big)\Big).\end{aligned} \tag{35}$$

Note that $\mathbb{E}[F(X^0; \{s_i\}, \{u_i\})|\mathcal{Y}]$ is a joint Laplace transform of the distributions of the properly rescaled clock process at times s_i. The following proposition thus compares Laplace transforms of $\tilde{S}_N^\gamma$ and $\bar{S}_N^\gamma$.

Proposition 5.1.

(a) *If the assumptions of Theorem 2.1 are satisfied, then for all sequences $\{s_i\}$ and $\{u_i\}$,*

$$\lim_{N\to\infty} \mathbb{E}\big[F_N(X_N^0; \{s_i\}, \{u_i\})\big|\mathcal{Y}\big] - \mathbb{E}\big[F_N(X_N^1; \{s_i\}, \{u_i\})\big] = 0, \qquad \mathcal{Y}\text{-a.s.} \tag{36}$$

(b) *If the assumptions of Theorem 2.2 hold, then, in probability,*

$$\lim_{\xi\to 0}\lim_{N\to\infty} \mathbb{E}\big[F_N(X_N^0; \{\xi s_i\}, \{\xi^{-\frac{1}{\alpha}} u_i\})\big|\mathcal{Y}\big] - \mathbb{E}\big[F_N(X_N^1; \{\xi s_i\}, \{\xi^{-\frac{1}{\alpha}} u_i\})\big] = 0. \tag{37}$$

Proof. We use the well-known interpolation formula for functionals of two Gaussian processes due (probably) to Slepian and Kahane (see, e.g., [14])

$$\mathbb{E}[F_N(X_N^1) - F_N(X_N^0)|\mathcal{Y}] = \frac{1}{2}\int_0^1 \mathrm{d}h \sum_{\substack{i,j=1\\ i\neq j}}^{Tr_N} (\Lambda_{ij}^1 - \Lambda_{ij}^0)\mathbb{E}\Big[\frac{\partial^2 F_N(X_N^h)}{\partial X(i)\partial X(j)}\Big|\mathcal{Y}\Big]. \quad (38)$$

We will show that the integral in (38) converges to 0. To save on notation we assume that $\ell = 1$ and write $u = u_1$, $s = s_1$. Generalisation to larger ℓ is straightforward. The second derivative in (38) is 0 if at least one of i, j is larger than sr_N. For $i, j < sr_N$ the second derivative equals

$$\begin{aligned} &\frac{u^2\beta^2 N F_N(X_N^h)}{t_N^2} \prod_{\circ=i,j} \mathrm{e}^{\beta\sqrt{N}X_N^h(\circ)} g'\Big(\frac{u}{t_N}\mathrm{e}^{\beta\sqrt{N}X_N^h(\circ)}\Big) \\ &\quad\le \frac{u^2\beta^2 N}{t_N^2} \prod_{\circ=i,j} \mathrm{e}^{\beta\sqrt{N}X_N^h(\circ)} \exp\Big[-2g\Big(\frac{u}{t_N}\mathrm{e}^{\beta\sqrt{N}X_N^h(\circ)}\Big)\Big], \end{aligned} \quad (39)$$

where we used the fact that $g'(x) = (\ln(1+x))' = (1+x)^{-1} = \exp(-g(x))$, and omitted in the summation of $F_N(X_N^h)$ all terms different from i and j. To estimate the expected value of this expression we need the following technical lemma.

Lemma 5.1. *Let $c \in [0,1)$ and let U_1, U_2 be two standard normal variables with the covariance $\mathbb{E}[U_1U_2] = c$. For $u > 0$ define $\Xi_N(c) = \Xi_N(c,\beta,\gamma,u)$ and $\bar{\Xi}_N(c) = \bar{\Xi}_N(c,\beta,\gamma,u)$ by*

$$\Xi_N(c) = \frac{u^2\beta^2 N}{t_N^2} \prod_{\circ=1,2} \mathbb{E}\Big[\exp\Big\{\beta\sqrt{N}U_\circ - 2g\big(ut_N^{-1}\mathrm{e}^{\beta\sqrt{N}U_\circ}\big)\Big\}\Big] \quad (40)$$

and

$$\bar{\Xi}_N(c) = C(1-c)^{-1/2}(1\wedge u^2)N^{1/(1+c)}\exp\Big\{-\frac{\gamma^2 N}{\beta^2(1+c)}\Big\}, \quad (41)$$

where $C = C(\gamma,\beta)$ is a suitably chosen large constant. Then

$$\Xi_N(c) \le \bar{\Xi}_N(c). \quad (42)$$

Proof. Define $\kappa_\pm = \sqrt{2(1\pm c)}$. Since $c < 1$ both κ_+ and κ_- are positive. Let $\bar{U}_1$, $\bar{U}_2$ be two independent standard normal variables. Then U_1 and U_2 can be written as

$$U_1 = \frac{1}{2}(\kappa_+\bar{U}_1 + \kappa_-\bar{U}_2), \qquad U_2 = \frac{1}{2}(\kappa_+\bar{U}_1 - \kappa_-\bar{U}_2). \quad (43)$$

Hence, $U_1 + U_2 = \kappa_+\bar{U}_1$. Using $g(x) + g(y) = g(x+y+xy) \ge g(x+y)$ if $xy \ge 0$, and $\mathrm{e}^x + \mathrm{e}^{-x} \ge \mathrm{e}^{|x|}$ we get

$$\sum_{\circ=1,2} g(ut_N^{-1}\mathrm{e}^{\beta\sqrt{N}U_\circ}) \ge g\Big(ut_N^{-1}\exp\Big(\frac{\kappa_+\beta\sqrt{N}\bar{U}_1}{2} + \Big|\frac{\kappa_-\beta\sqrt{N}\bar{U}_2}{2}\Big|\Big)\Big). \quad (44)$$

Hence, $\Xi_N(c)$ is bounded from above by

$$\begin{aligned}&\frac{u^2\beta^2 N}{t_N^2}\int_{\mathbb{R}^2}\frac{\mathrm{d}y}{2\pi}\\&\times\exp\Big\{-\frac{y_1^2+y_2^2}{2}+\beta\sqrt{N}\kappa_+y_1-2g\big(ut_N^{-1}\mathrm{e}^{\kappa_+\beta\sqrt{N}y_1/2+\kappa_-\beta\sqrt{N}|y_2|/2}\big)\Big\}.\end{aligned}\tag{45}$$

Substituting $z_1=y_1-\beta\sqrt{N}\kappa_+$ and $z_2=y_2$ we get

$$\begin{aligned}&\frac{u^2\beta^2 N}{t_N^2}\mathrm{e}^{\beta^2\kappa_+^2N/2}\int_{\mathbb{R}^2}\frac{\mathrm{d}z}{2\pi}\exp\Big(-\frac{z_1^2+z_2^2}{2}\Big)\\&\times\exp\Big(-2g\Big(u\exp\Big\{\sqrt{N}\Big[\Big(\frac{\beta^2\kappa_+^2}{2}-\bar\gamma_N\Big)\sqrt{N}+\frac{\beta\kappa_+}{2}z_1+\frac{\beta\kappa_-}{2}|z_2|\Big]\Big\}\Big)\Big),\end{aligned}\tag{46}$$

where $\bar\gamma_N\equiv N^{-1}\log t_N=\gamma-\log N/(2\alpha N)$. Observe that $\exp(-2g(u\mathrm{e}^{\sqrt{N}x}))$ converges to the indicator function $\mathbf{1}_{x<0}$, as $N\to\infty$. The role of x will be played by the square bracket in the expression (46). Since $\kappa_+>\sqrt{2}$ and $\gamma/\beta^2<1$, this bracket is positive for 'typical' z_1, z_2 not far from 0. This means that the largest contribution to (46) comes again from non-typical values of z_1, z_2. We need another tilting:

$$z_1=\frac{1}{\sqrt{N}}\Big[v_1-\frac{\kappa_-}{\kappa_+}|v_2|-N\Big(\beta\kappa_+-\frac{2\bar\gamma_N}{\beta\kappa_+}\Big)\Big],\quad z_2=\frac{v_2}{\sqrt{N}}.\tag{47}$$

This substitution transforms the domain where the square bracket of (46) is negative into the half plane $v_1<0$: The expression inside the braces in (46) equals $\beta\kappa_+v_1/2$. Substituting (47) into $(z_1^2+z_2^2)/2$ produces an additional prefactor

$$N^{-1/\alpha}N^{1/(1+c)}\exp\Big(-\frac{(\beta^2\kappa_+^2-2\gamma)^2N}{2\beta^2\kappa_+^2}\Big).$$

Another prefactor N^{-1} comes from the Jacobian. The remaining terms can be bounded from above by

$$\begin{aligned}&C\int_{\mathbb{R}}\exp\Big\{-\frac{v_2^2}{2N}-\Big(\beta\kappa_--\frac{2\bar\gamma_N\kappa_-}{\beta\kappa_+^2}\Big)|v_2|\Big\}\mathrm{d}v_2\\&\times\int_{\mathbb{R}}\exp\Big\{\Big(\beta\kappa_+-\frac{2\bar\gamma_N}{\beta\kappa_+}\Big)v_1-2g(u\mathrm{e}^{\beta\kappa_+v_1/2})\Big\}\mathrm{d}v_1.\end{aligned}\tag{48}$$

Ignoring the quadratic term and using the facts that the parenthesis on the first line is always positive and $\gamma_N\to\gamma$ we can bound the first integral for N large by

$$C\Big(\beta\kappa_--\frac{2\gamma\kappa_-}{\beta\kappa_+^2}\Big)^{-1}\le C\kappa_-^{-1}\le C(1-c)^{-1/2}.\tag{49}$$

To bound the second integral observe that the integrand behaves as

$$\exp\{-2v_1\bar\gamma_N/\beta\kappa_+\}\quad\text{as}\quad v_1\to\infty,$$

and as

$$\exp\{(\beta\kappa_+-(2\bar\gamma_N/\beta\kappa_+))v_1\}\quad\text{as}\quad v_1\to-\infty.$$

Therefore, the second integral is bounded uniformly for all values of c. Moreover, as u increases, the second integral is $\mathrm{O}(u^{-2})$.

Putting everything together we get

$$\begin{aligned}\Xi_N(c) &\leq C(1-c)^{-\frac{1}{2}}\frac{u^2\beta^2 N}{t_N^2}(1\wedge u^{-2})\mathrm{e}^{\beta^2\kappa_+^2 N/2}N^{\frac{1}{1+c}-\frac{1}{\alpha}-1}\mathrm{e}^{-\frac{(\beta^2\kappa_+^2-2\gamma)^2 N}{2\beta^2\kappa_+^2}}\\ &= C(\gamma,\beta)(1-c)^{-1/2}N^{1/(1+c)}(1\wedge u^2)\exp\Big\{-\frac{\gamma^2 N}{\beta^2(1+c)}\Big\} = \bar{\Xi}_N(c).\end{aligned} \tag{50}$$

This finishes the proof of Lemma 5.1. □

We can now finish the proof of Proposition 5.1. First, note that for all $K > 0$

$$\lim_{N\to\infty} \mathrm{e}^{KN/2}\int_0^1 (1-h)^{-1/2}N^{1/(1+h)}\mathrm{e}^{-KN/(1+h)}\mathrm{d}h = C_K > 0. \tag{51}$$

Lemmas 4.1 and 5.1 imply that the absolute value of (38) can be bounded from above, $\mathcal{Y}$-a.s. for N large enough, by

$$CN^{-1}r_N\int_0^1 (1-h)^{-1/2}N^{1/(1+h)}\exp\Big\{-\frac{\gamma^2 N}{\beta^2(1+h)}\Big\}\mathrm{d}h \tag{52}$$

which converges to 0 as $N\to\infty$ using (51). This finishes the proof of Proposition 5.1(a).

In case (b) we can use Lemmas 4.5 and 5.1 to show that, out of a set with probability smaller than $C\xi$, (38) is bounded from above by

$$C\xi r_N\int_0^1 (1-h)^{-1/2}N^{1/(1+h)}\exp\Big\{-\frac{\gamma^2 N}{\beta^2(1+h)}\Big\}\mathrm{d}h, \tag{53}$$

which again converges to 0 after taking $N\to\infty$ and then $\xi\to 0$. □

6. Proofs of the main results

Proof of Theorem 2.1. Propositions 3.1 and 5.1(a) yield

$$\begin{aligned}\mathbb{E}\big[\exp\{-u\bar{S}_N^\gamma(s)\}\big|Y_N\big] &= \mathbb{E}\big[F_N(X_N^0;s,u)\big|Y_N\big]\\ &= \mathbb{E}\big[F_N(X_N^1;s,u)\big]+o(1) = \mathbb{E}[\mathrm{e}^{-uV_\alpha(\mathcal{K}s)}]+o(1).\end{aligned} \tag{54}$$

This implies the convergence of fixed time distributions of $\bar{S}_N^\gamma$ to those of $V_\alpha(\mathcal{K}\cdot)$. Analogous computation gives the convergence of more-dimensional marginals.

We still need to check tightness in the J_1-topology of the sequence $\bar{S}_N^\gamma$. For increasing processes it amounts to check (see, e.g., [12, Theorem 7.2, page 128]) that $\mathcal{Y}$-a.s.

$$\forall\eta>0, T>0 \quad \exists K \text{ such that } \mathbb{P}[\bar{S}_N(T)\geq K|\mathcal{Y}]\leq\eta \quad \forall N. \tag{55}$$

and

$$\forall\eta>0, T>0 \quad \exists\delta \text{ such that } \mathbb{P}[w_\delta^T(\bar{S}_N)\geq\eta|\mathcal{Y}]\leq\eta \quad \forall N, \tag{56}$$

where the modulus of continuity $w_\delta^T(f)$ is defined by

$$w_\delta^T(f) = \inf_{\{t_i\}} \max_i \sup \big\{|f(s) - f(t)| : s, t \in [t_{i-1}, t_i)\big\}, \tag{57}$$

where $\{t_i\}$ ranges over all partitions of the form $0 = t_0 < t_1 < \cdots < t_{n-1} < T \le t_n$ with $\min_{1\le i\le n}(t_i - t_{i-1}) > \delta$ and $n \ge 1$.

The condition (55) follows directly from the tightness of the fixed-time marginals, which is a consequence of the continuity of the limiting Laplace transform $e^{-T\mathcal{K}u^\alpha}$ at $u = 0$.

To prove (56) more work is necessary. First, we show that traps with energies 'much smaller' than $\gamma\sqrt{N}/\beta$ contribute hardly to the clock process. Recall the Definition (20) of B_N^ε and define

$$\bar{S}_N^\gamma(s,\varepsilon) = t_N^{-1} \sum_{i=0}^{\lfloor s r_N \rfloor} e_i \exp\big\{\beta\sqrt{N} H_N(Y_N(i))\big\} \mathbf{1}\{H_N(Y_N(i)) \le B_N^\varepsilon\}. \tag{58}$$

Lemma 6.1. *For every T and $\eta > 0$ there exists ε such that $\mathcal{Y}$-a.s., for all but finitely many N*

$$\mathbb{P}[\bar{S}_N^\gamma(T,\varepsilon) \ge \frac{\eta}{2}] \le \frac{\eta}{3}. \tag{59}$$

Proof. Using Definition (30) of I_ω it is possible to rewrite $\bar{S}_N^\gamma(T,\varepsilon)$ as

$$t_N^{-1} \sum_{\omega=1}^{\infty} \sum_{\sigma \in I_\omega \setminus I_{\omega+1}} \sum_{i=1}^{\omega} e'_{\sigma,i} e^{\beta\sqrt{N} H_N(\sigma)} \mathbf{1}\{H_N(\sigma) \le B_N^\varepsilon\} \equiv \sum_{\omega=1}^{\infty} \sum_{i=1}^{\omega} q_N^\varepsilon(\omega, i), \tag{60}$$

where $e'_{\sigma,i}$ are i.i.d. mean-one exponentials. By Lemma 4.4(b) we can restrict the sum to $\omega \le N$. For these ω we have, using the notation of Lemma 3.2,

$$\mathbb{P}\Big[q_N^\varepsilon(\omega,i) \ge \frac{\eta}{\omega 2^\omega}\Big] \le \frac{1 - \mathbb{E}[e^{-q_N^\varepsilon(\omega)}]}{1 - e^{-\eta/(2^\omega \omega)}} \le \frac{1 - (1 - r_N^{-1} f_N(\varepsilon))^{|I_\omega|}}{1 - e^{-\eta/(2^\omega \omega)}}. \tag{61}$$

For $\omega = 1$ this is bounded by $C f_N(\varepsilon)$ which can be made smaller than $\eta/2$ by choosing ε small enough. Moreover, using Lemma 4.4, $\mathcal{Y}$-a.s,

$$\sum_{\omega=2}^{N} \sum_{i=1}^{\omega} \mathbb{P}\Big[q_N^\varepsilon(\omega,i) \ge \frac{\eta}{\omega 2^\omega}\Big] \le \sum_{\omega=2}^{N} C f_N(\varepsilon) r_N^{-1} (N^{-\omega+1} r_N \vee 1) \omega^2 2^\omega. \tag{62}$$

which is smaller than $\eta/2$ for N large enough. □

Lemma 6.2. *For any fixed $\varepsilon > 0$, $\mathcal{Y}$-a.s,*

$$\lim_{N\to\infty} \mathbb{P}[\max\{H_N(\sigma) : \sigma \in I_2\} \ge B_N^\varepsilon | \mathcal{Y}] = 0. \tag{63}$$

Proof. By Lemma 4.4, $\mathcal{Y}$-a.s., $|I_2| \le T N^{-1} r_N$. Since the energies are i.i.d., the lemma follows by elementary arguments. □

We can now check (56). Fix $\eta > 0$ and $T > 0$. By Lemma 6.1 we can choose ε small such that $\mathbb{P}[\bar{S}_N^\gamma(T,\varepsilon) \geq \eta/2] \leq \eta/3$. Lemma 6.2 implies that $\mathbb{P}[\max\{H_N(\sigma) : \sigma \in I_2\} \geq B_N^\varepsilon | \mathcal{Y}] \leq \eta/3$ for all N large enough. Thus out of a set of probability smaller than $2\eta/3$, the contribution to the clock of the configurations with energies smaller than B_N^ε and of the configurations visited more than once is bounded by $\eta/2$. Out of this set the modulus of continuity $w_\delta^T(\bar{S}_N^\gamma)$ can be larger than η only if

$$\exists i,j : |i-j| \leq \delta r_N,\ Y_N(i), Y_N(j) \notin I_2 \text{ and } \min_{k\in\{i,j\}} \frac{e_k}{t_N} \mathrm{e}^{\beta\sqrt{N}H_N(Y_N(k))} \geq \frac{\eta}{2}. \quad (64)$$

However, the random variables $\{H_N(\sigma), \sigma \in I_1 \setminus I_2\}$ are i.i.d. An elementary calculation then shows that the probability of event in (64) can be made smaller than $\eta/3$ by choosing δ small. This finishes the proof of the tightness and thus of Theorem 2.1. □

Proof of Theorem 2.2. The theorem follows from Propositions 3.1 and 5.1(b) by a reasoning analogous to (54). □

Proof of Theorem 2.3. The proof of (a) is standard. It is sufficient to observe that $\mathbb{P}[\sigma_N(t_N) = \sigma_N(\theta t_N)|\mathcal{Y}]$ is very well approximated by $\mathbb{P}[\{\bar{S}_N^\gamma(t) : t \geq 0\} \cap [1,\theta] = \varnothing | \mathcal{Y}]$. The last probability converges, $\mathcal{Y}$-a.s, as $N \to \infty$, to $\mathbb{P}[\{V_\alpha(t) : t \geq 0\} \cap [1,\theta] = \varnothing]$ by the weak convergence of $\bar{S}_N^\gamma$ in the J_1-topology.

To imply claim (b) Theorem 2.2 is not sufficient. We need in addition the following estimate.

Lemma 6.3. *Let $\gamma = \beta\beta_c$, that is $r_N = 2^N$, and let $\hat{S}_N^\xi(t) = \xi^{-1/\alpha}\bar{S}_N^\gamma(\xi t)$. Then, for any $T > 0$ and $\eta > 0$ there exists $\delta > 0$ such that*

$$\limsup_{\xi\to 0} \limsup_{N\to\infty} \mathbb{P}\big[\mathbb{P}[w_\delta^T(\hat{S}_N^\xi) \geq \eta | \mathcal{Y}] \leq \eta\big] = 0. \quad (65)$$

Proof. We first set (see (58)), $\hat{S}_N^\xi(t,\varepsilon) = \xi^{-1/\alpha}\bar{S}_N^\gamma(\xi t, \xi^{1/\alpha}\varepsilon)$. We claim that, similarly as in Lemma 6.1, if ε is small enough, then

$$\mathbb{P}[\hat{S}_N^\xi(T,\varepsilon) \geq \eta | \mathcal{Y}] \leq \eta \quad (66)$$

holds with probability converging to 1 as $N \to \infty$ and $\xi \to 0$. The proof of this claim is analogous to the proof of Lemma 6.1; Lemma 4.5 is used instead of Lemmas 4.1 and 4.4.

Moreover, similarly to Lemma 6.2, $\mathbb{P}[\max\{H_N(\sigma) : \sigma \in I_2\} \geq B_N^{\xi^{1/\alpha}\varepsilon} | \mathcal{Y}]$ converges to 0 in probability as $N \to \infty$ and $\xi \to 0$, by Lemma 4.5 again. The proof then follows the same line as the proof of the tightness in Theorem 2.1. □

We now finish the proof of Theorem 2.3(b). As before,

$$\mathbb{P}[\sigma_N(\xi t_N) = \sigma_N(\xi\theta t_N) | \mathcal{Y}]$$

is well approximated by

$$\mathbb{P}[\{\hat{S}_N(t) : t \geq 0\} \cap [1,\theta] = \varnothing | \mathcal{Y}]. \quad (67)$$

The last probability can be bounded from above by

$$\mathbb{P}\Big[\bigcup_{i=0}^{K}\{\hat{S}_N(i\delta)\le 1\cap \hat{S}_N((i+1)\delta)\ge\theta\}\cup\{\hat{S}_N(K\delta)\le 1\}\Big|\mathcal{Y}\Big]. \tag{68}$$

The last quantity converges in probability, by Theorem 2.2, to

$$\mathbb{P}\Big[\bigcup_{i=0}^{K}\{V_\alpha(\mathcal{K}i\delta)\le 1\cap V_\alpha(\mathcal{K}(i+1)\delta)\ge\theta\}\cup\{\hat{S}_N(\mathcal{K}K\delta)\le 1\}\Big], \tag{69}$$

which can be made arbitrarily close to $\mathsf{Asl}_\alpha(\theta)$ by choosing δ small and K large. A lower bound on (67) can be obtained by considering the event

$$\bigcup_{i=0}^{\lfloor T\delta^{-1}\rfloor}\{\hat{S}_N(i\delta)\le 1-\eta\cap \hat{S}_N((i+1)\delta)\ge\theta+\eta\}\cap\{w_\delta^T(\hat{S}_N)\le\eta\}. \tag{70}$$

By Lemma 6.3, with probability converging to 1 as $N\to\infty$ and $\xi\to 0$, the conditional probability of the event on the right of the last expression is very close to 1, and the conditional probability of the union over i converges to a number that can be made arbitrarily close to $\mathsf{Asl}_\alpha(\theta)$ again. □

References

[1] Ben Arous, G., Bogachev, L.V., Molchanov, S.A.: Limit theorems for sums of random exponentials. Probab. Theory Related Fields **132**, 579–612 (2005).

[2] Ben Arous, G., Bovier, A., Černý, J.: Universality of the REM for dynamics of mean-field spin glasses. Comm. Math. Phys. **282**(3), 663–695 (2008).

[3] Ben Arous, G., Bovier, A., Gayrard, V.: Glauber dynamics of the random energy model. I. Metastable motion on the extreme states. Comm. Math. Phys. **235**(3), 379–425 (2003).

[4] Ben Arous, G., Bovier, A., Gayrard, V.: Glauber dynamics of the random energy model. II. Aging below the critical temperature. Comm. Math. Phys. **236**(1), 1–54 (2003).

[5] Ben Arous, G., Černý, J.: The arcsine law as a universal aging scheme for trap models. Comm. Pure Appl. Math. **61**(3), 289–329 (2008).

[6] Ben Arous, G., Černý, J., Mountford, T.: Aging in two-dimensional Bouchaud's model. Probab. Theory Related Fields **134**(1), 1–43 (2006).

[7] Ben Arous, G., Dembo, A., Guionnet, A.: Cugliandolo-Kurchan equations for dynamics of spin-glasses. Probab. Theory Related Fields **136**(4), 619–660 (2006).

[8] Ben Arous, G., Guionnet, A.: Symmetric Langevin spin glass dynamics. Ann. Probab. **25**(3), 1367–1422 (1997).

[9] Bouchaud, J.P.: Weak ergodicity breaking and aging in disordered systems. J. Phys. I (France) **2**(9), 1705–1713 (1992).

[10] Bouchaud, J.P., Dean, D.S.: Aging on Parisi's tree. J. Phys I(France) **5**, 265 (1995).

[11] Cugliandolo, L.F., Kurchan, J.: Analytical solution of the off-equilibrium dynamics of a long-range spin-glass. Phys. Rev. Lett. **71**, 173 (1993).

[12] Ethier, S.N., Kurtz, T.G.: Markov processes. John Wiley & Sons Inc., New York (1986).
[13] Hoeffding, W.: Probability inequalities for sums of bounded random variables. J. Amer. Statist. Assoc. **58**, 13–30 (1963).
[14] Ledoux, M., Talagrand, M.: Probability in Banach spaces, vol. 23. Springer-Verlag (1991).
[15] Matthews, P.: Mixing rates for a random walk on the cube. SIAM J. Algebraic Discrete Methods **8**(4), 746–752 (1987).

Jiří Černý
Departement Mathematik
ETH Zürich
Rämistrasse 101
CH-8092 Zürich, Switzerland
e-mail: `jiri.cerny@math.ethz.ch`

Progress in Probability, Vol. 62, 103–121

Spin Glass Identities and the Nishimori Line

Pierluigi Contucci, Cristian Giardinà and Hidetoshi Nishimori

Abstract. For a general spin glass model with asymmetric couplings we prove a family of identities involving expectations of generalized overlaps and magnetizations in the quenched state. Those identities hold pointwise in the Nishimori line and are reached at the rate of the inverse volume while, in the general case, they can be proved in integral average.

Mathematics Subject Classification (2000). 82B44.

Keywords. Spin glasses, sum rules, Nishimori line

1. Introduction and results

Overlap identities have played and continue to play a central role in spin glass statistical mechanics since the appearance of the Parisi solution [16] of the Sherrington–Kirkpatrick model [24]. The replica symmetry breaking theory contains indeed, as a built-in ansatz property, a family of identities for the overlap expectations that have been since then classified as replica equivalence and ultrametricity factorization properties. While the first is now largely understood for both mean field and short range finite-dimensional models [6, 7], the second is still an open conjecture even within the mean field cases. The ideas to obtain a rigorous proof of the identities trace back to the papers [2, 11] where the invariance property of stochastic stability and the role of the energy fluctuations for the spin glass quenched measure were introduced (see also [3] for a different and original derivation). The mentioned properties state a striking feature for the quenched measure on overlap expectations: the overlap moments do obey sum rules that reduce the distributional degrees of freedom and, at least in the mean field case, it is expected that the entire distribution be identified by the simple overlap distribution. Those relations are expected to hold everywhere but on isolated singularities and they can be in fact rigorously proved in β Riemann integral average [7, 25, 4].

We present in this paper a new family of identities in terms of generalized overlap and magnetization that hold when the interactions are not centered, which are proved in integral average. Moreover we find the remarkable result that in the Nishimori line they hold everywhere with respect to the parameters and are reached at the expected rate of the inverse volume.

We can illustrate the results proved in this paper considering the Edwards–Anderson spin glass model with non symmetric Gaussian interactions of variance δ^2 and non zero average μ. Considering the quenched measure over multiple copies of the system subject to the same disorder (for a precise definition see Section 2 the *link overlaps* and the *link magnetizations* fulfill an infinite family of identities when integrated in the inverse temperature β over arbitrary intervals. Those identities at the lowest moment read, for all μ and in the thermodynamic limit:

$$\int_{\beta_1}^{\beta_2} \mathrm{d}\beta \big| \beta^2 \left\langle -2q_{1,2}^2 + 8q_{1,2}q_{2,3} - 6q_{1,2}q_{3,4} \right\rangle \\ + \beta \left\langle 4q_{1,2}m_3 - 4q_{1,2}m_1 \right\rangle + \left\langle m_1^2 - m_1 m_2 \right\rangle \big| = 0, \tag{1}$$

$$\int_{\beta_1}^{\beta_2} \mathrm{d}\beta \big| \beta^2 (\langle q_{1,2}^2 \rangle - 6 \langle q_{1,2}q_{2,3} \rangle + 6 \langle q_{1,2}q_{3,4} \rangle - \langle q_{1,2} \rangle^2) \\ + \beta (2 \langle q_{1,2}m_1 \rangle - 4 \langle q_{1,2}m_3 \rangle + 2 \langle q_{1,2} \rangle \langle m_1 \rangle) \\ + \langle m_1 m_2 \rangle - \langle m_1 \rangle^2 \big| = 0. \tag{2}$$

Since we prove that the magnetization has vanishing fluctuation in μ-average in the infinite volume limit

$$\int_{\mu_1}^{\mu_2} \mathrm{d}\mu (\langle m^2 \rangle - \langle m \rangle^2) = 0, \tag{3}$$

then, when integrated over arbitrary boxes in the (β, μ) plane, the previous identities become

$$\int_{\mu_1}^{\mu_2} \mathrm{d}\mu \int_{\beta_1}^{\beta_2} \mathrm{d}\beta \big| \beta^2 \left\langle -2q_{1,2}^2 + 8q_{1,2}q_{2,3} - 6q_{1,2}q_{3,4} \right\rangle \big| = 0, \tag{4}$$

$$\int_{\mu_1}^{\mu_2} \mathrm{d}\mu \int_{\beta_1}^{\beta_2} \mathrm{d}\beta \big| \beta^2 (\langle q_{1,2}^2 \rangle - 6 \langle q_{1,2}q_{2,3} \rangle + 6 \langle q_{1,2}q_{3,4} \rangle - \langle q_{1,2} \rangle^2) \big| = 0\,, \tag{5}$$

thus reducing to the standard ones [20, 12, 2]. Notice the remarkable fact that the identities of type (1) have recently appeared in the theory of mean field diluted ferromagnets [1].

In the Nishimori line [18, 17],

$$\beta\delta^2 = \mu, \tag{6}$$

the identities (1) and (2) hold pointwise in β and μ and the identities (4) and (5) hold when integrated over μ on arbitrary intervals.

The plan of the paper is the following: in the next section we define the general class of models for which our result apply and we set the notations. In Section 3 we state the theorems, which are then proved in Section 4.

2. Definitions

We consider a disordered model of Ising configurations $\sigma_n = \pm 1$, $n \in \Lambda \subset \mathbb{Z}^d$ for some d-parallelepiped Λ of volume $|\Lambda|$. We denote Σ_Λ the set of all $\sigma = \{\sigma_n\}_{n\in\Lambda}$, and $|\Sigma_\Lambda| = 2^{|\Lambda|}$. In the sequel the following definitions will be used.

1. *Hamiltonian.*
   For every $\Lambda \subset \mathbb{Z}^d$ let $\{H_\Lambda(\sigma)\}_{\sigma\in\Sigma_N}$ be a family of $2^{|\Lambda|}$ *translation invariant (in distribution) Gaussian* random variables defined, in analogy with [22], according to the general representation

$$H_\Lambda(\sigma) \;=\; -\sum_{X\subset\Lambda} J_X\sigma_X \tag{7}$$

   where

$$\sigma_X = \prod_{i\in X}\sigma_i\,, \tag{8}$$

   ($\sigma_\varnothing = 0$) and the J's are independent Gaussian variables with mean

$$\mathsf{Av}(J_X) = \mu_X\,, \tag{9}$$

   and variance

$$\mathsf{Av}((J_X-\mu_X)^2) = \delta_X^2\,. \tag{10}$$

2. *Average and covariance matrix.*
   The Hamiltonian has average

$$\mathcal{B}_\Lambda(\sigma) \;:=\; \mathsf{Av}(H_\Lambda(\sigma)) \;=\; \sum_{X\subset\Lambda}\mu_X\sigma_X\,, \tag{11}$$

   and covariance matrix

$$\begin{aligned}\mathcal{C}_\Lambda(\sigma,\tau) &= \mathsf{Av}\big([H_\Lambda(\sigma)-\mathcal{B}_\Lambda(\sigma)][H_\Lambda(\tau)-\mathcal{B}_\Lambda(\tau)]\big)\\ &= \sum_{X\subset\Lambda}\delta_X^2\sigma_X\tau_X.\end{aligned} \tag{12}$$

   By the triangular inequality

$$|\mathcal{B}_\Lambda(\sigma)| \;\le\; \sum_{X\subset\Lambda}|\mu_X| \tag{13}$$

   for all σ and by the Schwarz inequality

$$|\mathcal{C}_\Lambda(\sigma,\tau)| \;\le\; \sqrt{\mathcal{C}_\Lambda(\sigma,\sigma)}\sqrt{\mathcal{C}_\Lambda(\tau,\tau)} \;=\; \sum_{X\subset\Lambda}\delta_X^2 \tag{14}$$

   for all σ and τ.

3. *Thermodynamic stability.*
The Hamiltonian (7) is thermodynamically stable if there exist constants $\bar{c}$ such that

$$\sup_{\Lambda\subset\mathbb{Z}^d}\frac{1}{|\Lambda|}\sum_{X\subset\Lambda}|\mu_X|\leq\bar{c}\,<\,\infty$$
$$\sup_{\Lambda\subset\mathbb{Z}^d}\frac{1}{|\Lambda|}\sum_{X\subset\Lambda}\delta_X^2\leq\bar{c}\,<\,\infty\,. \tag{15}$$

Thanks to the relations (13) and (14) a thermodynamically stable model fulfills the bounds

$$\mathcal{B}_\Lambda(\sigma)\leq\bar{c}\,|\Lambda|,$$
$$\mathcal{C}_\Lambda(\sigma,\tau)\leq\bar{c}\,|\Lambda|, \tag{16}$$

and has an order 1 normalized mean and covariance

$$b_\Lambda(\sigma):=\frac{1}{|\Lambda|}\mathcal{B}_\Lambda(\sigma),$$
$$c_\Lambda(\sigma,\tau):=\frac{1}{|\Lambda|}\mathcal{C}_\Lambda(\sigma,\tau)\,. \tag{17}$$

4. *Random partition function.*

$$\mathcal{Z}(\beta)=\sum_{\sigma\in\Sigma_\Lambda}\mathrm{e}^{-\beta H_\Lambda(\sigma)}\,. \tag{18}$$

5. *Random free energy.*

$$-\beta\mathcal{F}(\beta)=\mathcal{A}(\beta)\;:=\;\ln\mathcal{Z}(\beta)\,. \tag{19}$$

6. *Random internal energy.*

$$\mathcal{U}(\beta)\;:=\;\frac{\sum_{\sigma\in\Sigma_\Lambda}H_\Lambda(\sigma)\mathrm{e}^{-\beta H_\Lambda(\sigma)}}{\sum_{\sigma\in\Sigma_\Lambda}\mathrm{e}^{-\beta H_\Lambda(\sigma)}}\,. \tag{20}$$

7. *Quenched free energy.*

$$-\beta F(\beta)\;:=\;A(\beta)=\mathsf{Av}\big(\mathcal{A}(\beta)\big)\,. \tag{21}$$

8. *R-product random Gibbs-Boltzmann state.*

$$\Omega(-)=\sum_{\sigma^{(1)},\ldots,\sigma^{(R)}}(-)\,\frac{\mathrm{e}^{-\beta[H_\Lambda(\sigma^{(1)})+\cdots+H_\Lambda(\sigma^{(R)})]}}{[\mathcal{Z}(\beta)]^R}\,. \tag{22}$$

9. *Quenched equilibrium state.*

$$\langle-\rangle:=\mathsf{Av}\big(\Omega(-)\big)\,. \tag{23}$$

10. *Observables.*
For any smooth bounded function $G(b_\Lambda,c_\Lambda)$ (without loss of generality we consider $|G|\leq 1$ and no assumption of permutation invariance on G is made) of the mean and covariance matrix entries we introduce the random (with respect to $<->$) R-dimensional vector of elements $\{m_k\}$ (called *generalized*

magnetization) and the $R \times R$ matrix of elements $\{q_{k,l}\}$ (called *generalized overlap*) by the formula

$$\langle G(m,q)\rangle := \mathsf{Av}\big(\Omega(G(b_\Lambda, c_\Lambda))\big), \tag{24}$$

e.g.,

$$\begin{aligned} G(b_\Lambda, c_\Lambda) &= b_\Lambda(\sigma^1)c_\Lambda(\sigma^{(1)},\sigma^{(2)})c_\Lambda(\sigma^{(2)},\sigma^{(3)}) \\ \langle m_1 q_{1,2} q_{2,3}\rangle & \qquad\qquad\qquad\qquad\qquad\qquad\qquad\qquad (25) \\ &= \mathsf{Av}\Big(\sum_{\sigma^{(1)},\sigma^{(2)},\sigma^{(3)}} b_\Lambda(\sigma^1)c_\Lambda(\sigma^{(1)},\sigma^{(2)})c_\Lambda(\sigma^{(2)},\sigma^{(3)})\,\frac{\mathrm{e}^{-\beta[\sum_{i=1}^3 H_\Lambda(\sigma^{(i)})]}}{[\mathcal{Z}(\beta)]^3}\Big). \end{aligned}$$

3. Theorems

To state our results we introduce the random variable J' defined by

$$J_X = J'_X + \mu_X\,,$$

and deform uniformly the averages μ_X with a parameter μ defined by

$$\mu\mu'_X = \mu_X\,,$$

in such a way that

$$J_X = J'_X + \mu\mu'_X\,.$$

Our results can be summarized in the following theorems.

3.1. Identities in β-average

For every observable G of the kind considered in the previous section, we define

$$\begin{aligned} f_1(\beta,\mu) &= \sum_{l=1}^{R}\langle (m_l - m_{R+1})\,G\rangle \\ &\quad - \beta\Big(\Big\langle \sum_{\substack{k,l=1\\k\neq l}}^{R} G\,q_{l,\,k} - 2RG\sum_{l=1}^{R} q_{l,\,R+1} + R(R+1)G\,q_{R+1,\,R+2}\Big\rangle\Big) \end{aligned} \tag{26}$$

and

$$\begin{aligned} f_2(\beta,\mu) &= \langle m_{R+1}G\rangle - \langle m_1\rangle\,\langle G\rangle \\ &\quad - \beta\Big(\sum_{k=1}^{R+1}\langle G\,q_{k,\,R+1}\rangle - (R+1)\,\langle G\,q_{R+1,\,R+2}\rangle - \langle G\rangle\,(\langle q_{1,1}\rangle - \langle q_{1,2}\rangle)\Big). \end{aligned} \tag{27}$$

We then have the following

Theorem 3.1. *The quenched equilibrium state of a thermodynamically stable Hamiltonian fulfills, for every observable G and every temperature interval* $[\beta_1, \beta_2]$ *the following identities in the thermodynamic limit*

$$\lim_{\Lambda \nearrow \mathbb{Z}^d} \int_{\beta_1}^{\beta_2} \mathrm{d}\beta \, |f_1(\beta, \mu)| = 0, \tag{28}$$

$$\lim_{\Lambda \nearrow \mathbb{Z}^d} \int_{\beta_1}^{\beta_2} \mathrm{d}\beta \, |f_2(\beta, \mu)| = 0. \tag{29}$$

Remark 3.1. The two previous relations when applied to $G(m,q) = \beta q_{1,2} + m_1$ and $R = 2$ yield the identities mentioned in the introduction, formulae (1) and (2) – here it is assumed that $q_{1,1} = 1$, as it happens for the Edwards–Anderson model.

3.2. Identities in (β, μ)-average

We also define

$$g_1(\beta,\mu) = -\beta\left(\left\langle \sum_{\substack{k,l=1 \\ k\neq l}}^{R} G\,q_{l,\,k} - 2RG \sum_{l=1}^{R} q_{l,\,R+1} + R(R+1)G\,q_{\,R+1,\,R+2} \right\rangle\right) \tag{30}$$

and

$$\begin{aligned} g_2(\beta,\mu) = -\beta\Bigg(& \sum_{k=1}^{R+1} \langle G\,q_{\,k,\,R+1}\rangle \\ & - (R+1)\,\langle G\,q_{\,R+1,\,R+2}\rangle - \langle G\rangle\,(\langle q_{1,1}\rangle - \langle q_{1,2}\rangle)\Bigg). \end{aligned} \tag{31}$$

Theorem 3.2. *The quenched equilibrium state of a thermodynamically stable Hamiltonian fulfills, for every observable G and every set* $[\beta_1, \beta_2] \times [\mu_1, \mu_2]$, *the following identities in the thermodynamic limit:*

$$\lim_{\Lambda \nearrow \mathbb{Z}^d} \int_{\mu_1}^{\mu_2} \mathrm{d}\mu \int_{\beta_1}^{\beta_2} \mathrm{d}\beta \, |g_1(\beta, \mu)| = 0, \tag{32}$$

$$\lim_{\Lambda \nearrow \mathbb{Z}^d} \int_{\mu_1}^{\mu_2} \mathrm{d}\mu \int_{\beta_1}^{\beta_2} \mathrm{d}\beta \, |g_2(\beta, \mu)| = 0. \tag{33}$$

Remark 3.2. Two similar families of identities have been proved in [7] for the centered case $\mathsf{Av}(J_X) = 0$ for all X. This theorem generalizes the old result and reduces to it when the observable G doesn't depend on m.

3.3. Identities pointwise

Theorem 3.3. *In the space of parameters* $(\mu_X, \delta_X^2)_{X\in\Lambda}$ *there exists a region called the 'Nishimori manifold'*

$$\mu_X = \beta\delta_X^2, \tag{34}$$

where the identities of Theorem 3.1 *hold pointwise, namely*

$$\lim_{\Lambda \nearrow \mathbb{Z}^d} f_1(\beta, \mu) = 0, \tag{35}$$

$$\lim_{\Lambda \nearrow \mathbb{Z}^d} f_2(\beta, \mu) = 0. \tag{36}$$

4. Proofs

Theorem 3.1 is proved in the lemmas of Subsections 4.1 and 4.2. The proof uses only elementary methods like concentration and classical inequalities. Along the same line Theorem 3.2 is proved in Subsection 4.3. Theorem 3.3 is proved in Subsection 4.4 making use of an exact computation on the Nishimori manifold.

Let $h(\sigma) = |\Lambda|^{-1} H_\Lambda(\sigma)$ denote the Hamiltonian per particle. We consider the quantity

$$\sum_{l=1}^{R} \{ \left\langle h(\sigma^{(l)})\, G \right\rangle - \left\langle h(\sigma^{(l)}) \right\rangle \langle G \rangle \} = \delta_1 G + \delta_2 G, \tag{37}$$

where

$$\delta_1 G = \sum_{l=1}^{R} \Big\{ \mathsf{Av}\big(\Omega[h(\sigma^{(l)})\, G] - \Omega[h(\sigma^{(l)})]\Omega[G]\big) \Big\}, \tag{38}$$

$$\delta_2 G = \sum_{l=1}^{R} \Big\{ \mathsf{Av}\big(\Omega[h(\sigma^{(l)})]\Omega[G]\big) - \mathsf{Av}\big(\Omega[h(\sigma^{(l)})]\big)\, \mathsf{Av}\big(\Omega[G]\big) \Big\}. \tag{39}$$

We are going to show that both $\delta_1 G$ and $\delta_2 G$ vanish (in β average) in the thermodynamic limit. This implies, by a simple application of integration by parts, the relations (28) and (29).

4.1. Stochastic stability bound

We follow the method of stochastic stability as developed in [6].

Lemma 4.1. *For any inverse temperature interval* $[\beta_1, \beta_2]$ *one has*

$$\int_{\beta_1}^{\beta_2} \mathsf{Av}\big(\Omega(h^2) - \Omega^2(h)\big) \mathrm{d}\beta \le \frac{\bar{c}(2 + \beta_1 + \beta_2)}{|\Lambda|}. \tag{40}$$

Proof. The variance of the Hamiltonian per particle $h(\sigma)$ with respect to the Boltzmann state is nothing but (minus) the derivative of the average of $h(\sigma)$ up to a factor $1/|\Lambda|$, i.e.,

$$\mathsf{Av}\big(\Omega(h^2) - \Omega^2(h)\big) = -\frac{1}{|\Lambda|} \frac{\mathrm{d}}{\mathrm{d}\beta} \mathsf{Av}\big(\Omega(h)\big). \tag{41}$$

Application of integration by parts (for its general form, see formula (50) below) yields

$$\begin{aligned}\mathsf{Av}(\Omega(h)) &= \frac{1}{|\Lambda|}\mathsf{Av}\Big(\sum_{X\subset\Lambda} J_X\Omega(\sigma_X)\Big)\\ &= \frac{1}{|\Lambda|}\mathsf{Av}\Big(\sum_{X\subset\Lambda}\mu_X\Omega(\sigma_X)\Big) + \frac{1}{|\Lambda|}\sum_{X\subset\Lambda}\beta\delta_X^2\left[1-\mathsf{Av}(\Omega^2(\sigma_X))\right]\\ &\le (1+\beta)\bar{c}\,, \end{aligned} \tag{42}$$

where thermodynamic stability condition, Eq. (16), has been used in the last inequality. The lemma statement follows from integration of Eq. (41) over an arbitrary inverse temperature interval $[\beta_1,\beta_2]$ and the use of fundamental theorem of calculus together with the bound (42). □

Lemma 4.2. *For every bounded observable G, see Definition* 10 *of Section* 2, *we have that for every interval $[\beta_1,\beta_2]$ in the thermodynamic limit one has*

$$\int_{\beta_1}^{\beta_2}|\delta_1 G|\mathrm{d}\beta = 0. \tag{43}$$

Proof. From the definition of $\delta_1 G$, Eq. (38), we have

$$\int_{\beta_1}^{\beta_2}|\delta_1 G|\mathrm{d}\beta \le \int_{\beta_1}^{\beta_2}\sum_{l=1}^{R}\left|\mathsf{Av}\left(\Omega[h(\sigma^{(l)})\,G]-\Omega[h(\sigma^{(l)})]\Omega[G]\right)\right|\mathrm{d}\beta \tag{44}$$

$$\le \int_{\beta_1}^{\beta_2}\sum_{l=1}^{R}\sqrt{\mathsf{Av}\left(\left\{\Omega[h(\sigma^{(l)})\,G]-\Omega[h(\sigma^{(l)})]\Omega[G]\right\}^2\right)}\mathrm{d}\beta \tag{45}$$

$$\le R\int_{\beta_1}^{\beta_2}\sqrt{\mathsf{Av}\left(\Omega[h^2(\sigma)]-\Omega^2[h(\sigma)]\right)}\mathrm{d}\beta \tag{46}$$

$$\le R\sqrt{\beta_2-\beta_1}\sqrt{\int_{\beta_1}^{\beta_2}\mathsf{Av}\left(\Omega[h^2(\sigma)]-\Omega^2[h(\sigma)]\right)\mathrm{d}\beta} \tag{47}$$

$$\le R\sqrt{\beta_2-\beta_1}\sqrt{\frac{\bar{c}(2+\beta_1+\beta_2)}{|\Lambda|}}\,, \tag{48}$$

where (44) follows from triangular inequality, (45) is obtained by applying Jensen inequality on the measure $\mathsf{Av}(-)$, (46) comes from application of the Schwarz inequality to the measure $\Omega(-)$ and boundedness of G, (47) is again Jensen inequality on the measure $\frac{1}{\beta_2-\beta_1}\int_{\beta_1}^{\beta_2}(-)\mathrm{d}\beta$ and finally (48) comes from Lemma 4.1. □

Remark 4.1. The previous lemma is related to a general property of disordered systems which is known as 'stochastic stability' (see [2, 6]). It says that the equilibrium state in a spin glass model is invariant under a suitable class of perturbations in all temperature intervals of continuity. The result presented in [6] for the case of zero average couplings holds with the absolute value outside the integral in beta with a vanishing rate of the inverse volume. Here instead we got the stronger result for the absolute value inside the integral but with a weaker vanishing rate of the square root inverse volume.

Lemma 4.3. *The following expression holds:*

$$\delta_1 G = \sum_{l=1}^{R} \langle (m_l - m_{R+1})\, G\rangle - \beta \Big\langle G\Big[\sum_{\substack{k,l=1\\ k\neq l}}^{R} q_{l,\,k} - 2R\sum_{l=1}^{R} q_{l,\,R+1} + R(R+1)\, q_{R+1,\,R+2}\Big]\Big\rangle . \tag{49}$$

Proof. For each replica l $(1 \le l \le R)$, we evaluate separately the two terms in the sum of the right side of Eq. (38) by using integration by parts (generalized Wick formula) for correlated Gaussian random variables, $x_1, x_2, \dots, x_n$ with means $\mathsf{Av}(x_i)$ and covariances $\mathsf{Av}\big((x_i - \mathsf{Av}(x_i))(x_j - \mathsf{Av}(x_j))\big)$, namely

$$\begin{aligned}\mathsf{Av}\big(x_i\, \psi(x_1, \dots, x_n)\big) &= \mathsf{Av}(x_i)\, \mathsf{Av}(\psi(x_1, \dots, x_n)) \\ &\quad + \sum_{j=1}^{n} \mathsf{Av}\big((x_i - \mathsf{Av}(x_i)(x_j - \mathsf{Av}(x_j))\big)\ \mathsf{Av}\Big(\frac{\partial \psi(x_1, \dots, x_n)}{\partial x_j}\Big) .\end{aligned} \tag{50}$$

It is convenient to denote by $p\,(R)$ the Gibbs-Boltzmann weight of R copies of the system

$$p_R\,(\sigma^1, \dots, \sigma^R) = \frac{\mathrm{e}^{-\beta\,[\sum_{k=1}^{R} H_\Lambda(\sigma^{(k)})\,]}}{[\mathcal{Z}(\beta)]^R}\,, \tag{51}$$

so that we have

$$\begin{aligned}-\frac{1}{\beta}\frac{\mathrm{d}p_R\,(\sigma^1, \dots, \sigma^R)}{\mathrm{d}H(\tau)} &= p_R\,(\sigma^1, \dots, \sigma^R) \sum_{k=1}^{R} \delta_{\sigma^{(k)},\,\tau} \\ &\quad - R\, p_{R+1}\,(\sigma^1, \dots, \sigma^R, \tau)\,.\end{aligned} \tag{52}$$

We obtain

$$\begin{aligned}\mathsf{Av}\big(\Omega(h(\sigma^{(l)})\, G)\big) &= \frac{1}{|\Lambda|}\, \mathsf{Av}\Big(\sum_{\sigma^{(1)},\dots,\sigma^{(r)}} G\, H_\Lambda(\sigma^{(l)})\, p_R\,(\sigma^1, \dots, \sigma^R)\Big) \\ &= \mathsf{Av}\Big(\sum_{\sigma^{(1)},\dots,\sigma^{(r)}} G\, b_\Lambda(\sigma^{(l)})\, p_R\,(\sigma^1, \dots, \sigma^R)\Big) \\ &\quad + \mathsf{Av}\Big(\sum_{\sigma^{(1)},\dots,\sigma^{(r)}} \sum_{\tau} G\, c_\Lambda(\sigma^{(l)}, \tau)\, \frac{\mathrm{d}p_R\,(\sigma^1, \dots, \sigma^R)}{\mathrm{d}H(\tau)}\Big) \end{aligned} \tag{53}$$

$$= \langle m_l\, G\rangle - \beta \Big[\sum_{k=1}^{R} \langle G\, q_{l,\,k}\rangle - R\,\langle G\, q_{l,\,R+1}\rangle \Big], \tag{54}$$

where in (53) we made use of the integration by parts formula and (54) is obtained by (52). Analogously, the other term reads

$$\begin{aligned}
&\mathsf{Av}\big(\Omega(h(\sigma^{(l)}))\,\Omega(G)\big) = \\
&= \frac{1}{|\Lambda|}\,\mathsf{Av}\Big(\sum_{\sigma^{(l)}}\sum_{\tau^{(1)},\dots,\tau^{(R)}} G\,H_\Lambda(\sigma^{(l)})\,p_{R+1}\,(\sigma^l,\tau_1,\dots,\tau_R)\Big) \\
&= \mathsf{Av}\Big(\sum_{\sigma^{(l)}}\sum_{\tau^{(1)},\dots,\tau^{(R)}} G\,b_\Lambda(\sigma^{(l)})\,p_{R+1}\,(\sigma^l,\tau_1,\dots,\tau_R)\Big) \\
&\quad + \mathsf{Av}\Big(\sum_{\sigma^{(l)}}\sum_{\tau^{(1)},\dots,\tau^{(R)}}\sum_{\gamma} G\,c_\Lambda(\sigma^{(l)},\gamma)\,\frac{\mathrm{d}p_{R+1}\,(\sigma^l,\tau_1,\dots,\tau_R)}{\mathrm{d}H(\gamma)}\Big) \qquad (55) \\
&= \langle m_{R+1}\,G\rangle - \beta\left[\sum_{k=1}^{R+1}\langle G\,q_{k,\,R+1}\rangle - (R+1)\,\langle G\,q_{R+1,\,R+2}\rangle\right]. \qquad (56)
\end{aligned}$$

Inserting (54) and (56) in Eq. (38) and summing over l we obtain the expression (49). □

4.2. Selfaveraging bound

The selfaveraging of the free energy is a well-established property of spin glass models. The vanishing of the fluctuations with respect to the disorder of the free energy can be obtained either by martingales arguments [21, 7] or by concentration of measure [25, 13]. Here we follow the second approach. Our formulation applies to both mean field and finite-dimensional models and, for instance, includes the non summable interactions in finite dimensions [14] and the p-spin mean field model as well as the REM and GREM models.

Lemma 4.4. *The disorder fluctuation of the free energy satisfies the following inequality: for all $x > 0$*

$$\mathbb{P}\left(|\mathcal{A} - \mathsf{Av}(\mathcal{A})| \geq x\right) \;\leq\; 2\exp\left(-\frac{x^2}{2\bar{c}|\Lambda|}\right). \qquad (57)$$

The free energy is then a self averaging quantity, i.e.,

$$V(\mathcal{A}) \;=\; \mathsf{Av}(\mathcal{A}^2) - \big(\mathsf{Av}(\mathcal{A})\big)^2 \;\leq\; 4\,\bar{c}\,\beta^2\,|\Lambda|. \qquad (58)$$

Proof. Consider an $s > 0$. By the Markov inequality, one has

$$\begin{aligned}
\mathbb{P}\{\mathcal{A} - \mathsf{Av}(\mathcal{A}) \geq x\} &= \mathbb{P}\{\exp[s(\mathcal{A} - \mathsf{Av}(\mathcal{A}))] \geq \exp(sx)\} \\
&\leq \mathsf{Av}\big(\exp[s(\mathcal{A} - \mathsf{Av}(\mathcal{A}))]\big)\ \exp(-sx). \qquad (59)
\end{aligned}$$

To bound the generating function

$$\mathsf{Av}\big(\exp[s(\mathcal{A} - \mathsf{Av}(\mathcal{A}))]\big) \qquad (60)$$

one introduces, for a parameter $t \in [0,1]$, the following interpolating function:

$$\phi(t) = \ln \mathsf{Av}_1\{\exp(s\ \mathsf{Av}_2\{\ln Z(t)\})\}\ , \qquad (61)$$

where $\mathsf{Av}_1\{-\}$ and $\mathsf{Av}_2\{-\}$ denote expectation with respect to two independent copies $X_1(\sigma)$ and $X_2(\sigma)$ of the random variable $X(\sigma)$, which is a *centered* Gaussian process with the same covariance as the Hamiltonian $H_\Lambda(\sigma)$, and the partition function $Z(t)$ is

$$Z(t) = \sum_{\sigma \in \Sigma_\Lambda} \mathrm{e}^{-\beta\sqrt{t}X_1(\sigma)-\beta\sqrt{1-t}X_2(\sigma)-\beta\mathcal{B}_\Lambda(\sigma)} . \tag{62}$$

Indeed, since $H_\Lambda(\sigma) = X(\sigma) + \mathcal{B}_\Lambda(\sigma)$, it is immediate to verify (see Definition (19)) that

$$\phi(0) = s \ \mathsf{Av}(\mathcal{A}) , \tag{63}$$

and

$$\phi(1) = \ln \mathsf{Av}(\mathrm{e}^{s\ \mathcal{A}}). \tag{64}$$

This implies that

$$\mathsf{Av}\big(\exp[s(\mathcal{A} - \mathsf{Av}(\mathcal{A}))]\big) = \mathrm{e}^{\phi(1)-\phi(0)} = \mathrm{e}^{\int_0^1 \phi'(t)dt}. \tag{65}$$

On the other hand, the derivative with respect to t can be easily bounded. Defining

$$K(t) = \exp(s \ \mathsf{Av}_2\{\ln Z(t)\}) \tag{66}$$

and

$$p^t(\sigma) = \frac{\mathrm{e}^{-\beta\sqrt{t}X_1(\sigma)-\beta\sqrt{1-t}X_2(\sigma)-\beta\mathcal{B}_\Lambda(\sigma)}}{Z(t)} \tag{67}$$

one has

$$\phi'(t) = \frac{\mathsf{Av}_1\Big\{K(t)\ s\ \mathsf{Av}_2\Big\{\sum_\sigma p^t(\sigma)\left[\frac{1}{2\sqrt{t}}X_1(\sigma) - \frac{1}{2\sqrt{1-t}}X_2(\sigma)\right]\Big\}\Big\}}{\mathsf{Av}_1\{K(t)\}} . \tag{68}$$

Applying the integration by parts formula (50), a simple computation gives

$$\begin{aligned}
&\sum_\sigma \mathsf{Av}_1 \left\{K(t)\ \mathsf{Av}_2\Big\{p^t(\sigma)\frac{1}{\sqrt{t}}\ X_1(\sigma)\Big\}\right\} \\
&\quad = s\sum_{\sigma,\tau} \mathsf{Av}_1 \left\{K(t)\, \mathcal{C}_\Lambda(\sigma,\tau)\ \mathsf{Av}_2\Big\{p^t(\tau)\Big\}\ \mathsf{Av}_2\Big\{p^t(\sigma)\Big\}\right\} \\
&\qquad + \mathsf{Av}_1 \left\{K(t)\ \mathsf{Av}_2\Big\{\sum_\sigma \mathcal{C}_\Lambda(\sigma,\sigma)p^t(\sigma)\Big\}\right\} \\
&\qquad - \mathsf{Av}_1 \left\{K(t)\ \mathsf{Av}_2\Big\{\sum_{\sigma,\tau} \mathcal{C}_\Lambda(\sigma,\tau)p^t(\sigma)p^t(\tau))\Big\}\right\}
\end{aligned}$$

and

$$\mathsf{Av}_1\left\{K(t)\ \mathsf{Av}_2\Big\{\sum_{\sigma} p^t(\sigma)\frac{1}{\sqrt{1-t}}\ X_2(\sigma)\Big\}\right\}$$
$$= \mathsf{Av}_1\left\{K(t)\ \mathsf{Av}_2\Big\{\sum_{\sigma} \mathcal{C}(\sigma,\sigma)_\Lambda p^t(\sigma)\Big\}\right\}$$
$$- \mathsf{Av}_1\left\{K(t)\ \mathsf{Av}_2\Big\{\sum_{\sigma,\tau} \mathcal{C}(\sigma,\tau)_\Lambda p^t(\sigma)p^t(\tau))\Big\}\right\}.$$

Taking the difference of the previous two expressions one finds

$$\phi'(t) = \frac{s^2}{2}\frac{\sum_{\sigma,\tau}\mathsf{Av}_1\left\{K(t)\ \mathcal{C}_\Lambda(\sigma,\tau)\ \mathsf{Av}_2\{p^t(\tau)\}\ \mathsf{Av}_2\{p^t(\sigma)\}\right\}}{\mathsf{Av}_1\{K(t)\}}. \tag{69}$$

Using the thermodynamic stability condition (16), this yields

$$|\phi'(t)| \leq \frac{s^2}{2}|\Lambda|\bar{c} \tag{70}$$

from which it follows

$$\mathsf{Av}\big(\exp[s(\mathcal{A}-\mathsf{Av}(\mathcal{A}))]\big) \leq \exp\left(\frac{s^2}{2}|\Lambda|\bar{c}\right). \tag{71}$$

Inserting this bound into the inequality (59) and optimizing over s one finally obtains

$$\mathbb{P}\left(\mathcal{A}-\mathsf{Av}(\mathcal{A}) \geq x\right)\ \leq\ \exp\left(-\frac{x^2}{2\bar{c}|\Lambda|}\right). \tag{72}$$

The proof of inequality (57) is completed by observing that one can repeat a similar computation for $\mathbb{P}\left(\mathcal{A}-\mathsf{Av}(\mathcal{A}) \leq -x\right)$. The result for the variance (58) is then immediately proved using the identity

$$\mathsf{Av}\big((\mathcal{A}-\mathsf{Av}(\mathcal{A}))^2\big) = 2\int_0^\infty x\ \mathbb{P}(|\mathcal{A}-\mathsf{Av}(\mathcal{A})| \geq x)\mathrm{d}x. \tag{73}$$

□

Lemma 4.5. *The internal energy is self averaging almost everywhere in β, i.e., defining $u = \mathcal{U}/|\Lambda|$ and $V(u) = \mathsf{Av}(u^2) - (\mathsf{Av}(u))^2$ it holds in the thermodynamic limit*

$$\int_{\beta_1}^{\beta_2} V(u)\mathrm{d}\beta\ \rightarrow\ 0. \tag{74}$$

Proof. The result is obtained in two steps which use general theorems of measure theory. First from Lemma 4.4 we obtain the convergence to zero almost everywhere (in β) of the variance of the internal energy, then thanks to a bound on the variance of the internal energy we apply the Lebesgue dominated convergence theorem which gives the lemma statement. The sequence of convex functions $\mathcal{A}(\beta)/|\Lambda|$ converges a.e. (in J) to the limiting value $a(\beta)$ of its average and the convergence is self averaging in the sense of Lemma 4.4. By general convexity arguments [22] it follows that the sequence of the derivatives $\mathcal{A}'(\beta)/|\Lambda|$ converges to $u(\beta) = a'(\beta)$ almost everywhere in β and also that the convergence is self averaging. In fact the vanishing of the variance of a sequence of convex functions is inherited, in all points in which the derivative exists (which is almost everywhere for a convex function), to the sequence of its derivatives (see [23, 19]). From Lemma 4.4 we have then

$$V(u) \;\to\; 0 \quad \beta\text{-a.e.} \tag{75}$$

In order to obtain the convergence in β-average we use the Lebesgue dominated convergence theorem. In fact we prove that the sequence of variances of u is uniformly bounded (in every interval $[\beta_1, \beta_2]$) by an integrable function of β. A lengthy but simple computation which uses again integration by parts gives

$$\begin{aligned}
\mathsf{Av}(\mathcal{U}^2) &= \mathsf{Av}\Big(\sum_{X,Y\subset\Lambda} J_X J_Y \Omega(\sigma_X)\Omega(\sigma_Y)\Big)\\
&= \sum_{X,Y\subset\Lambda} \mu_X\mu_Y\Omega(\sigma_X)\Omega(\sigma_Y) + \sum_{X\subset\Lambda} \delta_X^2\Omega^2(\sigma_X)\\
&\quad + 2\beta \sum_{X,Y\subset\Lambda} \mu_X\delta_Y^2 \left[\Omega(\sigma_X\sigma_Y)\Omega(\sigma_Y) + \Omega(\sigma_X) - 2\Omega(\sigma_X)\Omega^2(\sigma_Y)\right]\\
&\quad + \beta^2 \sum_{X,Y\subset\Lambda} \delta_X^2\delta_Y^2 \,\mathrm{Av}\big[1 - \Omega^2(\sigma_X) - \Omega^2(\sigma_Y) + 6\Omega^2(\sigma_X)\Omega^2(\sigma_Y)\\
&\quad - 6\Omega(\sigma_X)\Omega(\sigma_Y)\Omega(\sigma_X\sigma_Y) + \Omega^2(\sigma_X\sigma_Y)\big]
\end{aligned} \tag{76}$$

from which

$$V(u) \;\le\; |\Lambda|^{-2}\,\mathsf{Av}(\mathcal{U}^2) \;\le\; \bar{c}^2(2 + 4\beta + 14\beta^2). \tag{77}$$

From this follows (74). □

Lemma 4.6. *For every bounded observable G, see Definition* 10 *of Section* 2*, we have that for every interval $[\beta_1, \beta_2]$ in the thermodynamic limit*

$$\int_{\beta_1}^{\beta_2} |\delta_2 G|\,\mathrm{d}\beta = 0. \tag{78}$$

Proof. From the definition of $\delta_2 G$, Eq. (39), we have

$$\int_{\beta_1}^{\beta_2} |\delta_2 G|\mathrm{d}\beta \le \int_{\beta_1}^{\beta_2} \sum_{l=1}^{R} \Big|\mathsf{Av}(\Omega[h(\sigma^{(l)})]\Omega[G]) - \mathsf{Av}(\Omega[h(\sigma^{(l)})])\,\mathsf{Av}(\Omega[G])\Big|\mathrm{d}\beta \tag{79}$$

$$\le \int_{\beta_1}^{\beta_2} \sum_{l=1}^{R} \sqrt{\mathsf{Av}(\Omega[h^2(\sigma^{(l)})]) - \big(\mathsf{Av}(\Omega[h(\sigma^{(l)})])\big)^2}\mathrm{d}\beta \tag{80}$$

$$\le R\int_{\beta_1}^{\beta_2} \sqrt{V(u)}\mathrm{d}\beta \tag{81}$$

$$\le R\sqrt{\beta_2 - \beta_1}\sqrt{\int_{\beta_1}^{\beta_2} V(u)\mathrm{d}\beta}, \tag{82}$$

where (79) follows from triangular inequality, (80) is obtained by applying the Schwarz inequality to the measure $\mathsf{Av}(-)$ and boundedness of G, (82) is Jensen inequality on the measure $\frac{1}{\beta_2-\beta_1}\int_{\beta_1}^{\beta_2}(-)\mathrm{d}\beta$. The statement (78) follows then using the result of the previous lemma. □

Lemma 4.7. *The following expression holds:*

$$\delta_2 G = \sum_{l=1}^{R}\left(\langle m_{R+1}G\rangle - \langle m_l\rangle\,\langle G\rangle\right) \tag{83}$$
$$-\beta R\Bigg[\sum_{k=1}^{R+1}\langle G\,q_{k,\,R+1}\rangle - (R+1)\,\langle G\,q_{R+1,\,R+2}\rangle - \langle G\rangle\,(\langle q_{1,1}\rangle - \langle q_{1,2}\rangle)\Bigg].$$

Proof. In order to obtain $\delta_2 G$ we are left with the explicit evaluation of the other term in (39) which simply gives

$$\mathsf{Av}(\Omega(h(\sigma^{(l)})))\ \mathsf{Av}(\Omega(G)) = \frac{1}{|\Lambda|}\,\mathsf{Av}\bigg(\sum_{\sigma^{(l)}} H_\Lambda(\sigma^{(l)})\,p_1(\sigma)\bigg)\,\langle G\rangle$$
$$= \mathsf{Av}\bigg(\sum_{\sigma^{(l)}} b_\Lambda(\sigma^{(l)})\,p_1(\sigma)\bigg)\,\langle G\rangle + \mathsf{Av}\bigg(\sum_{\sigma^{(l)}}\sum_{\gamma} c_\Lambda(\sigma^{(l)},\gamma)\,\frac{\mathrm{d}p_1(\sigma)}{\mathrm{d}H_\Lambda(\gamma)}\bigg)\,\langle G\rangle$$
$$= \langle m_l\rangle\,\langle G\rangle - \beta\,\langle G\rangle\,[\langle q_{1,1}\rangle - \langle q_{1,2}\rangle]. \tag{84}$$

Inserting (56) and (84) in Eq. (39) and summing over l we obtain (83). □

4.3. Vanishing fluctuations of the generalized magnetization

Lemma 4.8. *For every interval* $[\mu_1,\mu_2]$, *in the thermodynamic limit*

$$\int_{\mu_1}^{\mu_2}\mathrm{d}\mu(\langle m^2\rangle - \langle m\rangle^2) = 0. \tag{85}$$

Proof. The proof that the generalized magnetization has vanishing fluctuation follows the strategy that has been pursued so far to control fluctuations of the internal energy. We have

$$\langle m^2\rangle - \langle m\rangle^2 = \mathsf{Av}(\Omega(b_\Lambda(\sigma)^2) - (\Omega(b_\Lambda(\sigma))^2) \tag{86}$$
$$+\,\mathsf{Av}((\Omega(b_\Lambda(\sigma))^2) - \big(\mathsf{Av}(\Omega(b_\Lambda(\sigma)))\big)^2 \tag{87}$$

and we observe that generalized magnetization is related to the pressure by

$$\frac{\partial}{\partial\mu}\left(\frac{\mathcal{A}}{|\Lambda|}\right) = \frac{\beta}{\mu}\frac{1}{|\Lambda|}\Omega\bigg(\sum_{X\in\Lambda}\mu_X\sigma_X\bigg) = \frac{\beta}{\mu}\,\Omega(b_\Lambda(\sigma)). \tag{88}$$

The fluctuations w.r.t. the Gibbs state (r.h.s. of Eq. (86)) are easily controlled by a stochastic stability argument:

$$\int_{\mu_1}^{\mu_2} d\mu \, \mathsf{Av}\big(\Omega(b_\Lambda(\sigma)^2) - (\Omega(b_\Lambda(\sigma))^2\big)$$
$$= \frac{1}{|\Lambda|} \int_{\mu_1}^{\mu_2} d\mu \, \frac{\mu^2}{\beta^2} \, \mathsf{Av}\Big(\frac{\partial}{\partial \mu} \frac{\beta}{\mu} \Omega(b_\Lambda(\sigma))\Big) \,, \tag{89}$$

ma where the right-hand side can be bounded, integrating by parts in μ, by $\beta^{-1}3(\mu_2 - \mu_1)$.

The fluctuations w.r.t. the disorder (Eq. (87)) are bounded by the same argument of Lemma 4.5. Indeed from self-averaging of the pressure per particle

$$V\left(\frac{\mathcal{A}}{|\Lambda|}\right) \le \frac{c}{|\Lambda|} \to 0 \,,$$

and convexity of finite volume pressure

$$\frac{\partial^2}{\partial \mu^2} \frac{\mathcal{A}}{|\Lambda|} = |\Lambda| [\Omega(b_\Lambda(\sigma)^2) - (\Omega(b_\Lambda(\sigma))^2] \ge 0$$

one deduces that also the sequence of derivatives (88) is self-averaging in μ-average. Hence

$$\int_{\mu_1}^{\mu_2} d\mu \left[\mathsf{Av}((\Omega(b_\Lambda(\sigma))^2) - \big(\mathsf{Av}(\Omega(b_\Lambda(\sigma))\big)^2\right]$$
$$= \int_{\mu_1}^{\mu_2} d\mu V \left(\frac{\partial}{\partial \mu}\left(\frac{\mathcal{A}}{|\Lambda|}\right)\right) \longrightarrow 0. \tag{90}$$

Combining together (89) and (90) completes the proof of the lemma. □

4.4. Nishimori manifold

Our strategy here is to prove the vanishing of fluctuations of the Hamiltonian per particle $h(\sigma)$. This implies, following the same reasoning of the previous sections, the vanishing of correlations with a generic bounded function G using the Schwarz inequality:

$$|\langle hG\rangle - \langle h\rangle \langle G\rangle| \le \sqrt{\langle h^2\rangle - \langle h\rangle^2} \sqrt{\langle G^2\rangle - \langle G\rangle^2} \,. \tag{91}$$

Lemma 4.9. *On the Nishimori manifold*

$$\mu_X = \beta \delta_X^2 \tag{92}$$

the random internal energy per particle is selfaveraging. More precisely the following result holds:

$$\langle h^2\rangle - \langle h\rangle^2 \le \frac{\bar{c}}{|\Lambda|}. \tag{93}$$

Proof. This is proved by an explicit computation of both terms in formula (93). Let us first consider $\mathsf{Av}\,(\Omega(J_X\sigma_X))$. We first write the definition explicitly:

$$\begin{aligned}&\mathsf{Av}\,(\Omega(J_X\sigma_X))\\&=\int_{-\infty}^{\infty}\prod_Y\Big(\mathrm{d}J_Y\frac{1}{\sqrt{2\pi}\delta_Y}\exp\Big(-\frac{(J_Y-\mu_Y)^2}{2\delta_Y^2}\Big)\Big)\cdot\frac{\sum_\sigma J_X\sigma_X\mathrm{e}^{\beta\sum_Z J_Z\sigma_Z}}{\sum_\sigma \mathrm{e}^{\beta\sum_Z J_Z\sigma_Z}}.\end{aligned}\tag{94}$$

We apply the gauge transformation

$$J_X\to J_X\tau_X,\quad \sigma_i\mapsto\sigma_i\tau_i,$$

for all $i\in\Lambda$ and $X\subset\Lambda$, where τ_i is a 'gauge' variable fixed to 1 or -1 at each $i\in X$ and $\tau_X=\prod_{i\in X}\tau_i$. This change of variables leaves the integral and sums in the above equation invariant. Then, only the J_Y in the exponent for the Gaussian weight changes:

$$\begin{aligned}&\mathsf{Av}\Big(\Omega(J_X\sigma_X)\Big)\\&=\int_{-\infty}^{\infty}\prod_Y\Big(\mathrm{d}J_Y\frac{1}{\sqrt{2\pi}\delta_Y}\exp\Big(-\frac{(J_Y\tau_Y-\mu_Y)^2}{2\delta_Y^2}\Big)\Big)\cdot\frac{\sum_\sigma J_X\sigma_X\mathrm{e}^{\beta\sum_Z J_Z\sigma_Z}}{\sum_\sigma \mathrm{e}^{\beta\sum_Z J_Z\sigma_Z}}\\&=\int_{-\infty}^{\infty}\prod_Y\Big(\mathrm{d}J_Y\frac{1}{\sqrt{2\pi}\delta_Y}\exp\Big(-\frac{J_Y^2+\mu_Y^2}{2\delta_Y^2}\Big)\Big)\\&\qquad\cdot\exp\Big(\sum_Y\frac{J_Y\mu_Y\tau_Y}{\delta_Y^2}\Big)\cdot\frac{\sum_\sigma J_X\sigma_X\mathrm{e}^{\beta\sum_Z J_Z\sigma_Z}}{\sum_\sigma \mathrm{e}^{\beta\sum_Z J_Z\sigma_Z}}.\end{aligned}\tag{95}$$

Since this expression holds for any assignment of ±1 to τ_i, we may sum it up over all possible $\{\tau_i\}_i$ and divide the result by $2^{|\Lambda|}$,

$$\begin{aligned}&\mathsf{Av}\Big(\Omega(J_X\sigma_X)\Big)\\&=\frac{1}{2^{|\Lambda|}}\int_{-\infty}^{\infty}\prod_Y\Big(\mathrm{d}J_Y\frac{1}{\sqrt{2\pi}\delta_Y}\exp\Big(-\frac{J_Y^2+\mu_Y^2}{2\delta_Y^2}\Big)\Big)\sum_\tau\mathrm{e}^{\sum_Y J_Y\mu_Y\tau_Y/\delta_Y^2}\\&\quad\cdot\frac{\sum_\sigma J_X\sigma_X\mathrm{e}^{\beta\sum_Z J_Z\sigma_Z}}{\sum_\sigma \mathrm{e}^{\beta\sum_Z J_Z\sigma_Z}}.\end{aligned}\tag{96}$$

The sum over τ and the sum over σ in the denominator cancel each other for NL ($\beta=\mu_Y/\delta_Y^2$), and we have a simplified expression

$$\begin{aligned}&\mathsf{Av}\Big(\Omega(J_X\sigma_X)\Big)\\&=\frac{1}{2^{|\Lambda|}}\int_{-\infty}^{\infty}\prod_Y\Big(\mathrm{d}J_Y\frac{1}{\sqrt{2\pi}\delta_Y}\exp\Big(-\frac{J_Y^2+\mu_Y^2}{2\delta_Y^2}\Big)\Big)\sum_\sigma J_X\sigma_X\mathrm{e}^{\beta\sum_Z J_Z\sigma_Z}\\&=\frac{1}{2^{|\Lambda|}}\sum_\sigma\int_{-\infty}^{\infty}\prod_Y\Big(\mathrm{d}J_Y\frac{1}{\sqrt{2\pi}\delta_Y}\exp\Big(-\frac{J_Y^2+\mu_Y^2}{2\delta_Y^2}\Big)\Big)J_X\sigma_X\mathrm{e}^{\,\sum_Z J_Z\sigma_Z\mu_Z/\delta_Z^2}\end{aligned}\tag{97}$$

For given $\{\sigma_i\}_i$, let us change the integral variable as $J_Y \to J_Y\sigma_Y$. Then σ disappears completely and the integral is just for the average of J_X

$$\begin{aligned}
&\mathsf{Av}\Big(\Omega(J_X\sigma_X)\Big)\\
&= \frac{1}{2^{|\Lambda|}}\sum_{\sigma}\int_{-\infty}^{\infty}\prod_Y\Big(\mathrm{d}J_Y\frac{1}{\sqrt{2\pi}\delta_Y}\exp\Big(-\frac{J_Y^2+\mu_Y^2}{2\delta_Y^2}\Big)\Big)J_X\,\mathrm{e}^{\,\sum_Z J_Z\mu_Z/\delta_Z^2}\\
&= \frac{1}{2^{|\Lambda|}}\cdot 2^{|\Lambda|}\cdot 1\cdot\int_{-\infty}^{\infty}\mathrm{d}J_X\frac{1}{\sqrt{2\pi}\delta_X}J_X\exp\Big(-\frac{(J_X-\mu_X)^2}{2\delta_X^2}\Big)\\
&= \mathsf{Av}(J_X).
\end{aligned} \tag{98}$$

From the previous computation we obtain the final result for the quenched internal energy on the NL:

$$\mathsf{Av}\Big(\Omega(H_\Lambda(\sigma))\Big) = \mathsf{Av}\Big(\Omega\Big(\sum_X J_X\sigma_X\Big)\Big) = \mathsf{Av}\Big(\sum_X J_X\Big) = \sum_X \mu_X. \tag{99}$$

The other term in the variance (93) is evaluated similarly:

$$\begin{aligned}
\mathsf{Av}(\Omega(H_\Lambda(\sigma)^2)) &= \mathsf{Av}\Big(\Omega\Big(\sum_{X,Y} J_XJ_Y\sigma_X\sigma_Y\Big)\Big)\\
&= \mathsf{Av}\Big(\sum_{X,Y} J_XJ_Y\Big)\\
&= \sum_{X\neq Y}\mu_X\mu_Y + \sum_X(\mu_X^2+\delta_X^2)\\
&= \sum_{X,Y}\mu_X\mu_Y + \sum_X\delta_X^2.
\end{aligned} \tag{100}$$

Therefore, using (99) and (100) in the expression for the variance of h one finds

$$\begin{aligned}
\langle h(\sigma)^2\rangle - \langle h(\sigma)\rangle^2 &= \frac{1}{|\Lambda|^2}\mathsf{Av}\Big(\Omega(H_\Lambda(\sigma)^2)\Big) - \frac{1}{|\Lambda|^2}\{\mathsf{Av}\Big(\Omega(H_\Lambda(\sigma))\Big)\}^2\\
&= \frac{1}{|\Lambda|^2}\sum_X\delta_X^2 \le \frac{\bar{c}}{|\Lambda|}.
\end{aligned} \tag{101}$$

□

Acknowledgement

P.C. and C.G. thank the Tokyo Institute of Technology for the hospitality during the period July–August 2007 in which this work was developed and Michel Talagrand for interesting discussions.

References

[1] Agliari, E., Barra, A., Camboni, F.: Criticality in diluted ferromagnet. J. Stat. Mech. Theory Exp. **2008**(10), P10003, 21 pp. (2008).

[2] Aizenman, M., Contucci, P.: On the stability of the quenched state in mean-field spin-glass models. J. Stat. Phys. **92**(5-6), 765–783 (1998).

[3] Barra, A.: Irreducible free energy expansion and overlaps locking in mean field spin glasses. J. Stat. Phys. **123**(3), 601–614 (2006).

[4] Bovier, A.: Statistical Mechanics of Disordered Systems, *MaPhySto Lecture Notes*, vol. 10. MaPhySto, Aarhus (2001).

[5] Contucci, P.: Replica equivalence in the Edwards-Anderson model. J. Phys. A **36**(43), 10 961–10 966 (2003). Special issue: Statistical physics of disordered systems: from real materials to optimization and codes.

[6] Contucci, P., Giardinà, C.: Spin-glass stochastic stability: a rigorous proof. Ann. Henri Poincaré **6**(5), 915–923 (2005).

[7] Contucci, P., Giardinà, C.: The Ghirlanda-Guerra identities. J. Stat. Phys. **126**(4-5), 917–931 (2007).

[8] Edwards, S., Anderson, P.W.: Theory of spin glasses. J. Phys. F **5**, 965–974 (1975).

[9] Franz, S., Mézard, M., Parisi, G., Peliti, L.: Measuring equilibrium properties in aging systems. Phys. Rev. Lett. **81**(9), 1758–1761 (1998).

[10] Franz, S., Mézard, M., Parisi, G., Peliti, L.: The response of glassy systems to random perturbations: a bridge between equilibrium and off-equilibrium. J. Statist. Phys. **97**(3-4), 459–488 (1999).

[11] Ghirlanda, S., Guerra, F.: General properties of overlap probability distributions in disordered spin systems. Towards Parisi ultrametricity. J. Phys. A **31**(46), 9149–9155 (1998).

[12] Guerra, F.: About the overlap distribution in mean field spin glass models. Int. Jou. Mod. Phys. B **10**(13-14), 1675–1684 (1996). Memorial issue for H. Umezawa.

[13] Guerra, F., Toninelli, F.L.: The infinite volume limit in generalized mean field disordered models. Markov Process. Related Fields **9**(2), 195–207 (2003). Inhomogeneous random systems (Cergy-Pontoise, 2002).

[14] Khanin, K.M., Sinaĭ, Y.G.: Existence of free energy for models with long-range random Hamiltonians. J. Statist. Phys. **20**(6), 573–584 (1979).

[15] Marinari, E., Parisi, G., Ricci-Tersenghi, F., Ruiz-Lorenzo, J.J., Zuliani, F.: Replica symmetry breaking in short-range spin glasses: theoretical foundations and numerical evidences. J. Statist. Phys. **98**(5-6), 973–1047 (2000).

[16] Mézard, M., Parisi, G., Virasoro, M.A.: Spin glass theory and beyond, *World Scientific Lecture Notes in Physics*, vol. 9. World Scientific Publishing Co. Inc., Teaneck, NJ (1987).

[17] Nishimori, H.: Internal Energy, Specific Heat and Correlation Function of the Bond-Random Ising Model. Prog. Theor. Phys. **66**(4), 1169–1181 (1981).

[18] Nishimori, H.: Statistical physics of spin glasses and information processing, *International Series of Monographs on Physics*, vol. 111. Oxford University Press, New York (2001). An introduction, Translated from the 1999 Japanese original.

[19] Orlandini, E., Tesi, M.C., Whittington, S.G.: Self-averaging in the statistical mechanics of some lattice models. J. Phys. A **35**(19), 4219–4227 (2002).
[20] Parisi, G.: On the probability distribution of the overlap in spin glasses. Internat. J. Modern Phys. B **18**(4-5), 733–743 (2004).
[21] Pastur, L.A., Shcherbina, M.V.: Absence of self-averaging of the order parameter in the Sherrington-Kirkpatrick model. J. Statist. Phys. **62**(1-2), 1–19 (1991).
[22] Ruelle, D.: Statistical mechanics: Rigorous results. W.A. Benjamin, Inc., New York-Amsterdam (1969).
[23] Shcherbina, M.: On the replica symmetric solution for the Sherrington-Kirkpatrick model. Helv. Phys. Acta **70**(6), 838–853 (1997).
[24] Sherrington, D., Kirkpatrick, S.: Solvable model of a spin-glass. Phys. Rev. Lett. **35**(26), 1792–1796 (1975).
[25] Talagrand, M.: Spin glasses: a challenge for mathematicians, *Ergebnisse der Mathematik und ihrer Grenzgebiete. 3. Folge.*, vol. 46. Springer-Verlag, Berlin (2003). Cavity and mean field models.

Pierluigi Contucci
Dipartimento di Matematica
Università di Bologna
I-40127 Bologna, Italy
e-mail: contucci@dm.unibo.it

Cristian Giardinà
Eindhoven University of Technology & EURANDOM
P.O. Box 513
NL-5600 MB Eindhoven, The Netherlands
e-mail: c.giardina@tue.nl

Hidetoshi Nishimori
Department of Physics
Tokyo Institute of Technology
Oh-okayama, Meguru-ku
Tokyo 152-8551, Japan
e-mail: nishimori@phys.titech.ac.jp

Progress in Probability, Vol. 62, 123–142

Self-averaging Identities for Random Spin Systems

Luca De Sanctis and Silvio Franz

Abstract. We provide a systematic treatment of self-averaging identities, whose validity is proven in integral average, for dilute spin glasses. The method is quite general, and as a special case recovers the Ghirlanda–Guerra identities, which are therefore proven, together with their extension, to be valid in dilute spin glasses. We focus on dilute spin glasses, but the results hold in all models enjoying stability with respect to the perturbations we introduce; although such a stability is believed to hold for several models, we do not classify them here.

Mathematics Subject Classification (2000). 82B44; 82B20.

Keywords. Diluted spin glasses, Ghirlanda–Guerra, self-averaging

1. Introduction

Despite many years of intense work, and the much awaited proof of the validity of the Parisi ansatz for the free-energy of the Sherrington–Kirkpatrick (SK) and related models, the mathematical comprehension of the thermodynamics of mean field spin glasses remains largely incomplete. We know from theoretical physics that in fully connected models all the properties of the low temperature spin glass phase can be encoded in the probability distribution of the overlap between two different copies of the system. The analysis of Parisi et al. predicts an ultrametric organization of the phases (see [12] and references therein). So far a rigorous proof (or disproof) of ultrametricity and, more in general, the analysis of the structure of Gibbs measures at low temperature, turned out to be a very difficult task. A step in this direction was performed by Ghirlanda and Guerra in [8]. They found a simple and elegant way, based on the self-averaging of the internal energy, to prove a remarkable property of the overlaps. Given s replicas, the Gibbs measure must be such that when one adds a further replica this is either identical to one of these, or statistically independent of them; each case occurring with the same probability. More generally, various constraints on the distribution of the different overlaps have been found in the same spirit [2, 14]. Such features have found several applications [16, 4] in the rigorous analysis of spin glass models. For example, the property of non-negativity of the overlap, which in some models plays a role in turning the cavity free-energy into a rigorous lower bound, turns out to be a consequence of the Ghirlanda–Guerra self-averaging identities [16]. In the same way these identities have a role in the rigorous analysis of spin glasses close to the critical temperature [1].

In more general spin-glass systems, like finite-dimensional systems or spin systems on random graphs, the statistics of the overlap is not enough to fully characterize the low temperature spin glass phase. For instance, in diluted models the statistics of the local cavity fields, or equivalently of all the multi-overlaps, is necessary to describe the low temperature thermodynamic properties. In this paper, we analyse two families of identities for the local fields and multi-overlap distributions that are a consequence of self-averaging relations. We will see that the first of the two families is a consequence of the self-averaging with respect to the Gibbs measure or, equivalently, of stochastic stability, as the two phenomena turn out to be equivalent. The other family of identities is instead a consequence of self-averaging with respect to the global measure (quenched after Gibbs). The second family contains the first. The form of the identities we obtain is due only to the form of the perturbations we introduce and does not depend on the specific form of the Hamiltonian. However, the self-averaging at the basis of the results does not necessarily hold for all Hamiltonians. So we will stick with the case of dilute spin glass (Viana–Bray model) for which the self-averaging is assured, but our method shows that the same identities we find hold whenever the self-averaging is true. We do not provide a classification of all the models exhibiting self-averaging when perturbed. A second reason to use the example of spin models on sparse random graphs (dilute spin glass models) is that we expect that our results could provide hints for progresses in the mathematical analysis of the low temperature phases. Diluted mean field spin glasses have, in recent time, attracted a lot of attention in statistical physics due to the intrinsic interest of spin glasses where each spin interacts with a finite number of variables, but more importantly because fundamental problems in computer science, such as the random K-SAT and graph coloring, the random X-OR-SAT, tree reconstruction [11] and others, admit a formulation in terms of spin glass systems on random graphs. The cavity approach to these problems has led in many cases to results believed to be exact, albeit for the moment several rigorous proofs are still lacking.

One of the two families of the identities that we will find appeared already in [7] to discuss free energy bounds in diluted models with non-Poissonian connectivity. In [7] such a family of identities was shown to be a consequence of self-averaging of certain random polynomial function of some spins variables, and the self-averaging was deduced from the convexity of the perturbed free energy. Here we re-derive this family of identities with a different strategy, employing stochastic stability of the free energy with respect to suitable perturbations of the Hamiltonian, and we show that the stability implies self-averaging. The use of stochastic stability makes our results valid even when a measure different from the Gibbs one is considered, provided certain conditions hold (we will briefly hint at this when introducing Random Multi-Overlap Structures). Moreover, our method shows that the constraints we find are valid whenever stochastic stability or self-averaging hold, whichever turns out to be easier to study, and in the physics literature many models have been investigated from at least one of the two points of view. We also exhibit a second family of new identities, which contains the first family and follows from the self-averaging with respect to the quenched expectation.

For both families we will start with the two-spin model, perturbed with a two-spin random field. This provides identities involving squared overlaps. The same identical method, reproduced for generic p-spin interactions, yields the same identities involving the pth power of the overlaps. Therefore, considering Hamiltonians and perturbations with all p-spin interactions we will conclude that the identities hold for all regular functions of the overlaps.

Let us stress here that the basic tool we employ is the introduction of suitable perturbations, such that the pressure (minus the free energy divided by the temperature) is convex in all the perturbing parameters. This guarantees the existence of the derivatives with respect to the perturbing parameters only almost everywhere. Therefore the relations we find are valid only when one integrates (with Lebesgue measure) back against the perturbing parameters over any given intervals. We summarize this by recalling that our relations hold in "integral average".

2. The notations

We will deal with the stereotypical dilute spin glass model, the Viana–Bray (VB), for which we are about to describe the notations we need to derive our results in the next two sections. Let α, β be non-negative real numbers (degree of connectivity and inverse temperature respectively); P_ζ be a Poisson random variable of mean ζ; $\{i_\nu\}, \{j_\nu\}$, etc. be independent identically distributed random variables, uniformly distributed over the points $\{1, \dots, N\}$; $\{J_\nu\}, J$, etc. be independent identically distributed random variables, with symmetric distribution; $\mathcal{J}$ be the set of all the quenched random variables above. The map $\sigma\colon i \to \sigma_i,\ i \in \{1, \dots, N\}$ is a spin configuration from the configuration space $\Sigma = \{-1, 1\}^N$; $\pi_\zeta(\cdot)$ is the Poisson measure of mean ζ. The Hamiltonian of the Viana–Bray model is defined as

$$H_N(\sigma, \alpha; \mathcal{J}) = -\sum_{\nu=1}^{P_{\alpha N}} J_\nu \sigma_{i_\nu} \sigma_{j_\nu} \,. \tag{1}$$

We will limit to the case $J = \pm 1$, without loss of generality [10]. We follow the usual basic definitions and notations of thermodynamics for the partition function Z_N, the pressure p_N, the free energy per site f_N and its thermodynamic limit f, so to have in general

$$Z_N(\beta,\alpha) \equiv Z(H_N;\beta,\alpha;\mathcal{J}) = \sum_{\{\sigma\}} \exp(-\beta H_N(\sigma,\alpha;\mathcal{J}))\,, \tag{2}$$

$$\begin{aligned} p_N(\beta,\alpha) &= -\beta f_N(\beta,\alpha) = \frac{1}{N}\mathbb{E}\ln Z_N(\beta,\alpha)\,,\\ f(\beta,\alpha) &= \lim_{N\to\infty} f_N(\beta,\alpha)\,. \end{aligned} \tag{3}$$

The Boltzmann–Gibbs average of an observable $\mathcal{O}\colon \Sigma \to \mathbb{R}$ is

$$\Omega(\mathcal{O}) = Z_N(\beta,\alpha;\mathcal{J})^{-1} \sum_{\{\sigma\}} \mathcal{O}(\sigma)\exp(-\beta H_N(\sigma,\alpha;\mathcal{J}))\,, \tag{4}$$

and $\langle \mathcal{O} \rangle = \mathbb{E}\Omega(\mathcal{O})$ is the global average, where $\mathbb{E}$ denotes the average with respect to the quenched variables. When dealing with more than one configuration, we need the product measure of the needed copies of Ω, which will be denoted again by Ω.

The multi-overlaps $q_{1,\dots,m}\colon \Sigma^m \to [-1,1]$, where we use the notation $\Sigma^n = \Sigma^{(1)} \times \cdots \times \Sigma^{(n)}$, among the "replicas" $\Sigma^{(r_1)} \ni \sigma^{(r_1)}, \dots, \Sigma^{(r_n)} \ni \sigma^{(r_n)}$ is defined by

$$q_{r_1,\dots,r_n} = \frac{1}{N}\sum_{i=1}^{N} \sigma_i^{(r_1)} \cdots \sigma_i^{(r_n)}\,, \tag{5}$$

but sometimes we will just write q_n; q_1 can be identified with the magnetization m

$$m = \frac{1}{N}\sum_{i=1}^{N} \sigma_i\,.$$

Note that

$$\begin{aligned} \mathbb{E}[\Omega(\sigma_{i_1})]^{2n} &= \langle q_{1,\dots,2n} \rangle\,,\\ \mathbb{E}\Omega(\sigma_{i_1}) &= \mathbb{E}\Omega(m) = \langle m \rangle\,, \end{aligned} \tag{6}$$

and that

$$\begin{aligned} \mathbb{E}[\Omega(\sigma_{i_1}\cdots\sigma_{i_p})]^{2n} &= \langle q^p_{1,\dots,2n} \rangle\,,\\ \mathbb{E}\Omega(\sigma_{i_1}\cdots\sigma_{i_p}) &= \mathbb{E}\Omega(m^p) = \langle m^p \rangle\,, \end{aligned} \tag{7}$$

for all integers n and p.

3. Stochastic stability and self-averaging of the Gibbs measure

In the study of finite connectivity models it emerged that in a suitable probability space it is possible to formulate an exact variational principle for the computation of the free energy. This was obtained with the introduction of Random Multi-Overlap Structures (RaMOSt). We refer to [6] for details. The RaMOSt approach is based on the use of generic random weights to average the "cavity" part and the relative "internal correction" in the free energy (these are the numerator and the denominator of the trial free energy G_N introduced in (10). See [6] for details).

Here we are not interested in a detailed discussion of the RaMOSt approach, but we study the effect of a perturbation to the measure of our model, which does not need to be the Gibbs measure. That is why we introduce this more general weighting scheme, although the reader may keep in mind the Gibbs measure as a guiding example.

3.1. Random multi-overlap structures

The proper framework for the calculation of the free energy per spin is that of the Random Multi-Overlap Structures (RaMOSt, see [6] for more details).

Definition 3.1. Given a probability space $\{\Omega, \mu(d\omega)\}$, a *Random Multi-Overlap Structure* $\mathcal{R}$ is a triple $(\tilde{\Sigma}, \{\tilde{q}_{2n}\}, \xi)$ where

- $\tilde{\Sigma}$ is a discrete space;
- $\xi\colon \tilde{\Sigma} \to \mathbb{R}_+$ is a system of random weights, such that $\sum_{\gamma\in\tilde{\Sigma}} \xi_\gamma \leq \infty$ μ-almost surely;
- $\tilde{q}_{2n}\colon \tilde{\Sigma}^{2n} \to \mathbb{R}$, $n \in \mathbb{N}$ is a positive semi-definite *Multi-Overlap Kernel* (equal to 1 on the diagonal of $\tilde{\Sigma}^{2n}$, so that by the Schwarz inequality $|\tilde{q}| \leq 1$).

A RaMOSt needs to be equipped with N independent copies of a random field $\{\tilde{h}^i_\gamma(\alpha; \tilde{\mathcal{J}})\}_{i=1}^N$ and with another random field $\hat{H}_\gamma(\alpha N; \mathcal{J}')$ such that

$$\frac{\mathrm{d}}{\mathrm{d}\alpha}\mathbb{E}\ln \sum_{\gamma\in\tilde{\Sigma}} \xi_\gamma \exp(-\beta\tilde{h}^i_\gamma(\alpha;\tilde{\mathcal{J}})) = 2\sum_{n>0} \frac{1}{2n}\tanh^{2n}(\beta)(1-\langle\tilde{q}_{2n}\rangle), \tag{8}$$

$$\frac{\mathrm{d}}{\mathrm{d}\alpha}\mathbb{E}\ln \sum_{\gamma\in\tilde{\Sigma}} \xi_\gamma \exp(-\beta\hat{H}_\gamma(\alpha N;\mathcal{J}')) = \sum_{n>0} \frac{1}{2n}\tanh^{2n}(\beta)(1-\langle\tilde{q}^2_{2n}\rangle). \tag{9}$$

The quenched variables in $\tilde{h}$ and $\hat{H}$ are independent one another and independent of those in the weights ξ. These two fields just introduced are employed in the definition of the trial pressure

$$G_N(\mathcal{R};\beta,\alpha) = \frac{1}{N}\mathbb{E}\ln \frac{\sum_{\gamma,\sigma} \xi_\gamma \exp(-\beta\sum_{i=1}^N \tilde{h}^i_\gamma(\alpha;\tilde{\mathcal{J}})\sigma_i)}{\sum_\gamma \xi_\gamma \exp(-\beta\hat{H}_\gamma(\alpha N;\mathcal{J}'))}. \tag{10}$$

Notice that the expectation $\langle\cdot\rangle$ here is not necessarily the quenched Gibbs one: it is the generic one of the RaMOSt.

The reason why this is the proper framework for the calculation of the free energy is explained by the next [6]

Theorem 3.1 (Extended Variational Principle). *Taking the infimum for each N separately of the trial function $G_N(\mathcal{R};\beta,\alpha)$ over the space of all RaMOSt's, the resulting sequence tends to the limiting pressure $-\beta f(\beta,\alpha)$ of the VB model as N tends to infinity:*

$$-\beta f(\beta,\alpha) = \lim_{N\to\infty} \inf_{\mathcal{R}} G_N(\mathcal{R};\beta,\alpha)\,. \tag{11}$$

A RaMOSt $\mathcal{R}$ is said to be optimal if $G(\mathcal{R};\beta,\alpha) = -\beta f(\beta,\alpha)$ for all β,α. Recall that we denote by Ω the measure associated to the RaMOSt weights ξ.

The Boltzmann RaMOSt [6] is optimal, and constructed by thinking of a reservoir of M spins τ

$$\Sigma = \{-1,1\}^M \ni \tau\,, \quad \xi_\tau = \exp(-\beta H_M(\tau,\alpha,\mathcal{J}))\,,$$

$$\tilde{q}_{1,\ldots,2n} = \frac{1}{M}\sum_{k=1}^{M} \tau_k^{(1)}\cdots\tau_k^{(2n)}$$

with

$$\tilde{h}^i_\tau(\alpha,\tilde{\mathcal{J}}) = \sum_{\nu=1}^{\tilde{P}_{2\alpha}} \tilde{J}^i_\nu \tau_{k^i_\nu}\,, \quad \hat{H}_\tau(\alpha N,\mathcal{J}') = -\sum_{\nu=1}^{\hat{P}_{\alpha N}} \hat{J}_\nu \tau_{k_\nu}\tau_{l_\nu}\,,$$

with $\tilde{J}$, $\hat{J}$ all independent copies of J, and with the random site variables k_ν, l_ν all mutually independent and uniformly distributed over $\{1,\ldots M\}$.

Let $c_i = 2\cosh(\beta\tilde{h}^i_\tau(\alpha,\tilde{\mathcal{J}}))$. It is possible to show [6] that optimal RaMOSt's enjoy the same factorization property enjoyed by the Boltzmann RaMOSt and described in the next theorem.

Theorem 3.2 (Factorization of optimal RaMOSt's, [6]**).** *The following Cesàro limit is linear in N and $\bar{\alpha}$*

$$\text{Clim}_M\, \mathbb{E}\ln\Omega_M\{c_1\cdots c_N \exp[-\beta\hat{H}_\tau(\bar{\alpha},\mathcal{J}')]\} = N(-\beta f + \alpha A) + \bar{\alpha}A\,, \tag{12}$$

where

$$A = \sum_{n=1}^{\infty}\frac{1}{2n}\mathbb{E}\tanh^{2n}(\beta J)(1-\langle q^2_{2n}\rangle)\,, \tag{13}$$

and the equality holds in integral average, i.e., once both sides are integrated against α over any given interval.

This factorization property is called *invariance with respect to the cavity step*, or *Quasi-Stationarity*, and it is found in the hierarchical Parisi ansatz as well. When $\bar{\alpha}$ is zero, the theorem above states the factorization of the cavity fields, and it is possible to show that from this property one can deduce the family of identities we will discuss in the next Subsection [3]. When one removes instead the cavity terms $c_1,\ldots,c_N$ from the previous theorem, the statement becomes what is usually referred to as Stochastic Stability. We will show that the latter too implies the same family of identities. We will have in mind the case of a small perturbation of our spin system, but what we find holds for more general RaMOSt's, provided the previous theorem holds, that is for Quasi-Stationary RaMOSt's.

3.2. The first family of identities

We will now prove a lemma that expresses the stability of the Gibbs measure of our model against a macroscopic but small stochastic perturbation. In different terms, the lemma expresses the linear response of the free energy to the connectivity shift the perturbation consists of. The lemma we are about to prove will be used to show that from stochastic stability one can deduce a certain self-averaging which in turn imposes a family of constraints on the distribution of the overlaps.

Lemma 3.1. *Let* Ω, $\langle\cdot\rangle$ *be the usual infinite volume Gibbs and quenched Gibbs expectations at inverse temperature* β, *associated with the Hamiltonian* $H_N(\sigma,\alpha;\mathcal{J})$, $N\to\infty$. *Then the following equality (understood to be in the thermodynamic limit) holds*

$$\mathbb{E}\ln\Omega\exp\left(\beta'\sum_{\nu=1}^{P_{\alpha'}}J'_\nu\sigma_{i'_\nu}\sigma_{j'_\nu}\right)=\alpha'\sum_{n=1}^{\infty}\frac{1}{2n}\tanh^{2n}(\beta')(1-\langle q_{2n}^2\rangle)\,, \tag{14}$$

in integral average with respect to the degree of connectivity. The random variables $P_{\alpha'}$, $\{J'_\nu\}$, $\{i'_\nu\}$, $\{j'_\nu\}$ *are independent copies of the analogous random variables in the Hamiltonian contained in* Ω.

Notice that, in distribution

$$\beta\sum_{\nu=1}^{P_{\alpha N}}J_\nu\sigma_{i_\nu}\sigma_{j_\nu}+\beta'\sum_{\nu=1}^{P_{\alpha'}}J'_\nu\sigma_{i'_\nu}\sigma_{j'_\nu}\sim\beta\sum_{\nu=1}^{P_{\alpha N+\alpha'}}J''_\nu\sigma_{i_\nu}\sigma_{j_\nu} \tag{15}$$

where $\{J''_\nu\}$ are independent copies of J with probability $\alpha N/(\alpha N+\alpha')$ and independent copies of $J\beta'/\beta$ with probability $\alpha'/(\alpha N+\alpha')$. In the right-hand side above, the quenched random variables will be collectively denoted by $\mathcal{J}''$. Notice also that the sum of Poisson random variables is a Poisson random variable with mean equal to the sum of the means, and hence we can write

$$A_t\equiv\mathbb{E}\ln\Omega\exp\left(\beta'\sum_{\nu=1}^{P_{\alpha' t}}J'_\nu\sigma_{i'_\nu}\sigma_{j'_\nu}\right)=\mathbb{E}\ln\frac{Z_N(\alpha_t;\mathcal{J}'')}{Z_N(\alpha;\mathcal{J})}\,, \tag{16}$$

where we defined, for $t\in[0,1]$,

$$\alpha_t=\alpha+\alpha'\frac{t}{N} \tag{17}$$

so that $\alpha_t\to\alpha\ \forall\ t$ as $N\to\infty$.

Proof. Let us compute the t-derivative of A_t, as defined in (16)

$$\frac{\mathrm{d}}{\mathrm{d}t}A_t=\mathbb{E}\sum_{m=1}^{\infty}\frac{\mathrm{d}}{\mathrm{d}t}\pi_{\alpha' t}(m)\ln\sum_\sigma\exp\left(\beta'\sum_{\nu=1}^{m}J'_\nu\sigma_{i'_\nu}\sigma_{j'_\nu}\right). \tag{18}$$

Using the following elementary property of the Poisson measure

$$\frac{\mathrm{d}}{\mathrm{d}t}\pi_{t\zeta}(m)=\zeta(\pi_{t\zeta}(m-1)-\pi_{t\zeta}(m)) \tag{19}$$

we get

$$\begin{aligned}\frac{\mathrm{d}}{\mathrm{d}t}A_t &= \alpha'\mathbb{E}\sum_{m=0}^{\infty}[\pi_{\alpha' t}(m-1)-\pi_{\alpha' t}(m)]\ln\sum_{\sigma}\exp\Big(\beta'\sum_{\nu=1}^{m}J'_\nu\sigma_{i'_\nu}\sigma_{j'_\nu}\Big)\\
&= \alpha'\mathbb{E}\ln\sum_{\sigma}\exp(\beta' J'\sigma_{i'_m}\sigma_{j'_m})\exp\Big(\beta'\sum_{\nu=1}^{P_{\alpha' t}}J'_\nu\sigma_{i'_\nu}\sigma_{j'_\nu}\Big)\\
&\quad - \alpha'\mathbb{E}\ln\sum_{\sigma}\exp\Big(\beta'\sum_{\nu=1}^{P_{\alpha' t}}J'_\nu\sigma_{i'_\nu}\sigma_{j'_\nu}\Big)\\
&= \alpha'\mathbb{E}\ln\Omega_t\exp(\beta' J'\sigma_{i'_m}\sigma_{j'_m}),\end{aligned}$$

where the average Ω_t is associated with the Hamiltonian plus the t-dependent weights in the exponential in (16). Now use the following identity

$$\exp(\beta' J'\sigma_i\sigma_j) = \cosh(\beta' J') + \sigma_i\sigma_j\sinh(\beta' J')$$

to get

$$\frac{\mathrm{d}}{\mathrm{d}t}A_t = \alpha'\mathbb{E}\ln\Omega_t[\cosh(\beta' J')(1+\tanh(\beta' J')\sigma_{i'_m}\sigma_{j'_m})]\,. \tag{20}$$

We now expand the logarithm in power series and see that, when $N\to\infty$, as $\alpha_t\to\alpha$ the result does not depend on t, wherever the expectation Ω_t is continuous as a function of the parameter t. From the comments that preceded the current proof, formalized in (15)–(16)–(17), this is the same as assuming that Ω is regular as a function of α, because $J''\to J$ in the sense that in the large N limit J'' can only take the usual values ±1 since the probability of being $\pm\beta'/\beta$ becomes zero. Therefore integrating against t from 0 to 1 is the same as multiplying by 1. Due to the symmetric distribution of J, the expansion of the logarithm yields the right-hand side of (14), where the odd powers are missing. □

Notice that the left-hand side of (14) can be written, according to our notations, $\mathbb{E}\ln\Omega\exp(-\beta'\hat{H}_\sigma(\alpha',\mathcal{J}'))$. We want now to consider the statement of Lemma 3.1 in the case of two independent perturbations, assuming we are always in the thermodynamic limit. The consequent generalization obtained by using twice the fundamental theorem of calculus simply gives (in integral average)

$$\mathbb{E}\ln\Omega[\exp(-\beta'_1\hat{H}_\sigma(\alpha'_1;\mathcal{J}'_1)-\beta'_2\hat{H}_\sigma(\alpha'_2;\mathcal{J}'_2))] = (\alpha'_1+\alpha'_2)A\,, \tag{21}$$

where A again does not depend, in the thermodynamic limit, on α'_1,α'_2, and has the same form as the right-hand side of (14) although the explicit form of A is not important here. In the equation above, assumed to be taken in the thermodynamic limit, Ω is the Gibbs measure associated with the unperturbed Hamiltonian of the original model, and the same holds for the averages appearing in A, just like in the previous lemma. Clearly at this point we have

$$\frac{\partial^2}{\partial\alpha'_1\partial\alpha'_2}\mathbb{E}\ln\Omega[\exp(-\beta'_1\hat{H}_\sigma(\alpha'_1;\mathcal{J}'_1)-\beta'_2\hat{H}_\sigma(\alpha'_2;\mathcal{J}'_2))] = 0\,,$$

whenever the derivatives exist, i.e., with the possible exception of a zero measure set, by convexity. A simple computation yields

$$\frac{\partial^2}{\partial\alpha_1'\partial\alpha_2'}\mathbb{E}\ln\Omega[\exp(-\beta_1'\hat{H}_\sigma(\alpha_1';\mathcal{J}_1')-\beta_2'\hat{H}_\sigma(\alpha_2';\mathcal{J}_2'))]=0$$
$$=\mathbb{E}\ln\Omega'[\exp(\beta_1'J_1'\sigma_{i_1}\sigma_{j_1}+\beta_2'J_2'\sigma_{i_2}\sigma_{j_2}]$$
$$-\mathbb{E}\ln\Omega'[\exp(\beta_1'J_1'\sigma_{i_1}\sigma_{j_1})]\Omega'[\exp(\beta_2'J_2'\sigma_{i_2}\sigma_{j_2})].$$

Every time a derivative with respect to a perturbing parameter is taken, the relative perturbation is added to the weights of the measure Ω, which hence is replaced by a perturbed measure denoted by Ω'. If the perturbation is small (like in our case, as explained in the previous lemma) it disappears from the measure in the thermodynamic limit. Recall that this holds with the usual limitation, i.e., only in integral average, because each derivative exists only almost everywhere, and meaningful equalities are hence proven only under integration over any given interval. Hence both in the equation above and in the next calculation β_1',β_2' are not in the measure Ω', and we get

$$\frac{\partial^2}{\partial(\beta_1'J_1)\partial(\beta_2'J_2)}\mathbb{E}\ln\Omega'[\exp(\beta_1'J_1'\sigma_{i_1}\sigma_{j_1}+\beta_2'J_2'\sigma_{i_2}\sigma_{j_2}]$$
$$=\mathbb{E}\Omega''(\sigma_{i_1}\sigma_{j_1})-\mathbb{E}\Omega''(\sigma_{i_1})\Omega''(\sigma_{j_1})=0\,, \quad (22)$$

again in integral average with respect to all the parameters $\alpha_1',\alpha_2',\beta_1',\beta_2'$. These new derivatives with respect to β_1',β_2' again introduce further perturbations in the weights, which is why we used the notation Ω'', but as usual in the thermodynamic limit they vanish (in integral average). The first line of (22) gives us the generator of a family of relations that we will obtain by means of an expansion in powers of β_1',β_2'. The second line of the equation formulates the self-averaging (with respect to the Gibbs measure) implied by stochastic stability.

So we proceed starting from the next lemma and the next theorem, summarizing what we just discussed. We want here to remind ourselves of the presence of the perturbations, vanishing only in the thermodynamic limit and only with probability one in the space of all parameters, by denoting any perturbed Gibbs expectation with Ω', independently of the perturbations.

Lemma 3.2. *Let Ω' be the Gibbs measure including two independent perturbations of the form*

$$\hat{H}_\sigma(\alpha';\mathcal{J}')=\sum_{\nu=1}^{P_{\alpha'}}J_\nu'\sigma_{i_\nu}\sigma_{j_\nu}$$

with parameters $\alpha_1',\alpha_2',\beta_1',\beta_2'$ like in (21). *Then, recalling that m is the magnetization, the following self-averaging* (*with respect to the Gibbs measure*) *identity*

$$\mathbb{E}\{\Omega'(m^2)-[\Omega'(m)]^2\}=0 \quad (23)$$

holds in the thermodynamic limit in integral average with respect to the perturbing parameters $\alpha_1',\alpha_2',\beta_1',\beta_2'$.

We will see again that in the first line of equation (22) the expression remains zero even without the derivative. In fact, the generator of the identities we want to prove is expressed in the following

Theorem 3.3. *In the thermodynamic limit the following identity holds*

$$\begin{aligned}&\mathbb{E}\ln\Omega'(\exp(\beta_1'J_1'\sigma_{i_1}\sigma_{j_1}+\beta_2'J_2'\sigma_{i_2}\sigma_{j_2}))\\&\quad=\mathbb{E}\ln\Omega'(\exp(\beta_1'J_1'\sigma_{i_1}\sigma_{j_1}))+\mathbb{E}\ln\Omega'(\exp(\beta_2'J_2'\sigma_{i_2}\sigma_{j_2}))\,,\end{aligned}\tag{24}$$

in integral average with respect to $\alpha_1', \alpha_2', \beta_1', \beta_2'$.

The relations we will derive are a simple consequence of this theorem, and fomalized in the next

Corollary 3.1. *In the thermodynamic limit, in integral average with respect to the perturbing parameters* $\alpha_1', \alpha_2', \beta_1', \beta_2'$, *we have*

$$\sum_{a=0}^{\min\{r,s\}}(-)^{a+1}\frac{(2r+2s-a-1)!}{a!(2r-a)!(2s-a)!}\langle q_{2r}^2 q_{2s}^2\rangle_a' = 0 \quad \forall\, r,s\in\mathbb{N}\,,$$

where the subscript a *in the global average* $\langle\cdot\rangle_a' = \mathbb{E}\Omega_a'$ *means that precisely* a *replicas are in common among those in* q_{2r} *and those in* q_{2s}.

Proof. The following shorthand will be employed

$$t_1=\tanh(\beta_1'J_1')\,,\quad t_2=\tanh(\beta_2'J_2')\,,$$

$$\Omega_1=\Omega'(\sigma_{i_1}\sigma_{j_1})\,,\quad \Omega_2=\Omega'(\sigma_{i_2}\sigma_{j_2})\,,\quad \Omega_{12}=\Omega'(\sigma_{i_1}\sigma_{j_1}\sigma_{i_2}\sigma_{j_2})$$

and

$$W=\Omega'(\exp(\beta_1'J_1'\sigma_{i_1}\sigma_{j_1}+\beta_2'J_2'\sigma_{i_2}\sigma_{j_2}))\,.$$

Observe that, if we let $\delta=1,2$,

$$\frac{\partial}{\partial\beta J_\delta'}=(1-t_\delta^2)\frac{\partial}{\partial t_\delta}\,.\tag{25}$$

Now,

$$\ln W=\ln(1+t_1\Omega_1+t_2\Omega_2+t_1t_2\Omega_{12})+\ln\cosh\beta J_1'+\ln\cosh\beta J_2'$$

and

$$\begin{aligned}&\ln(1+t_1\Omega_1+t_2\Omega_2+t_1t_2\Omega_{12})\\&\quad=\sum_{n=1}^{\infty}\sum_{l=0}^{n}\sum_{m=0}^{l}\frac{(-)^{n+1}}{n}\binom{n}{l}\binom{l}{m}t_1^{n-l+m}t_2^{n-m}\Omega_1^m\Omega_2^{l-m}\Omega_{12}^{n-l}\\&\quad=\sum_{n,l,m}(-)^{n+1}\frac{(n-1)!}{(n-l)!(l-m)!m!}t_1^{n-l+m}t_2^{n-m}\Omega_1^m\Omega_2^{l-m}\Omega_{12}^{n-l}\,.\end{aligned}$$

The derivatives in (22) kill the two terms with the hyperbolic cosines, and from (25) we know that we can replace the derivatives with respect to $\beta J_\delta'$ with the derivatives with respect to t_δ, $\delta=1,2$. Notice that the logarithm just expanded is zero for $t_1 = 0$ and for $t_2 = 0$, therefore as its derivative like in (22) is zero, the logarithm itself is zero. This is why Theorem 3.3 holds, being (24) just the integral of the second line in (22).

Thanks to (6), if we put

$$n - l + m = r\,,\ n - m = s\,,\ n - l = a$$

we get

$$\sum_{r,s} \mathbb{E}[t_1^r t_2^s] \sum_{a=0}^{\min\{r,s\}} (-)^{a+1} \frac{(r+s-a-1)!}{a!(r-a)!(s-a)!} \langle q_r^2 q_s^2 \rangle'_a = 0$$

where, recall, $\langle \cdot \rangle_a$ means that a replicas are in common among those in q_r and those in q_s. Hence the statement of the theorem to be proven

$$\sum_{a=0}^{\min\{2r,2s\}} (-)^{a+1} \frac{(2r+2s-a-1)!}{a!(2r-a)!(2s-a)!} \langle q_{2r}^2 q_{2s}^2 \rangle'_a = 0\,,$$

where only the terms with an even number of replicas in each overlap survive because of the symmetry of the variables J'_1, J'_2 in t_1, t_2. □

3.3. Generalization to smooth functions of multi-overlaps

The fact that in our formulas we always got the square power of the overlaps is due to the fact that the Hamiltonian has 2-spin interactions. Everything we did so far could then be reproduced in the case of p-spin interactions, and we would obtain the same relations just derived, except the overlaps would appear with the power p instead of 2. Clearly the perturbation needed in this case is a p-spin perturbation too. More in general, we could consider a Hamiltonian consisting of the sum (over p) of p-spin Hamiltonians for any integer p. Then we could perturb each of the p-spin Hamiltonians with its proper small p-spin perturbation, and add all these perturbations to the system. Clearly we have to make sure that all the terms in this whole Hamiltonian are weighted with sufficiently small weights so to have the necessary convergence. More explicitly, the perturbed Hamiltonian is

$$H_N(\sigma, \alpha; \mathcal{J}) = -\sum_p \left[a_p \sum_{\nu=1}^{P_{\alpha N}^{(p)}} J_\nu \sigma_{i_\nu^1} \cdots \sigma_{i_\nu^p} + b_p \lambda_p \sum_{\nu=1}^{P'^{(p)}_{\alpha'}} J'_\nu \sigma_{j_\nu^1} \cdots \sigma_{j_\nu^p} \right],$$

where $\sum_p |a_p|^2 = \sum_p |b_p|^2 = 1$, the notation for all the quenched variables is the usual one, and $\{\lambda_p\}$ are the independent perturbing real parameters.

It is not surprising then that we can state

Corollary 3.2. *The following constraints hold in integral average with respect to the set of all the perturbing parameters*

$$\sum_{a=0}^{\min\{2r,2s\}} (-)^{a+1} \frac{(2r+2s-a-1)!}{a!(2r-a)!(2s-a)!} \langle q_{2r}^m q_{2s}^n \rangle'_a = 0 \quad \forall\, r, s, m, n \in \mathbb{N}$$

in the thermodynamic limit.

Again, this corollary can be seen as a consequence of a self-averaging property, namely

$$\mathbb{E}\Omega'(\sigma_{i_1^1}\cdots\sigma_{i_1^m}\sigma_{j_1^1}\cdots\sigma_{j_1^n}) - \mathbb{E}[\Omega'(\sigma_{i_1^1}\cdots\sigma_{i_1^m})\Omega'(\sigma_{j_1^1}\cdots\sigma_{j_1^n})] = 0\,. \tag{26}$$

Therefore we can replace each overlap by any smooth function of the relative replicas in the statement of the corollaries.

As a last remark, notice that in [7] the strategy consisted of using the fact that the second derivative of the free energy with respect to the "perturbing inverse temperatures" (β_1', β_2' in our case) is bounded to deduce self-averaging, and from the latter the identities. Here we used the connectivity as opposed to the inverse temperature to analyze stochastic stability, we then showed that the latter implies the self-averaging of [7]. So we obtained a comparison between self-averaging and stochastic stability (of a quite general validity), both providing a precious factorization of the Gibbs measure, and we also obtained that if any of the two is given, we know how to derive the same constraints.

4. Self-averaging of the quenched-Gibbs measure

Roughly speaking, if a convex random function does not fluctuate much, then its derivative does not fluctuate much either, with the exception of bad cases. This is well explained in Proposition 4.3 of [15] and Lemma 8.10 of [5]. We are not interested in general theorems here, in our case the convex function is the free energy density, and we only need to know that it is self-averaging (in the sense that the random free energy density does not fluctuate around its quenched expectation, in the thermodynamic limit). In the case of finite connectivity random spin systems like the VB model, a detailed proof of this can be found in [10]. The derivative of the free energy density (times $-\beta$) with respect to $-\beta$ is, in full generality, the expectation of the internal energy density $u_N = H_N/N$. Like in [9] and in Section 2 of [8], we have therefore this further self-averaging (in integral average with respect to β)

$$\lim_{N\to\infty} [\langle u_N^2\rangle - \langle u_N\rangle^2] = 0$$

which implies (due to Schwarz inequality)

$$\lim_{N\to\infty} \langle u_N^{(1)}\phi_s\rangle = \lim_{N\to\infty} \langle u_N\rangle\langle\phi_s\rangle \tag{27}$$

for any bounded function ϕ_s of s replicas, and $u_N^{(1)}$ is the internal energy density in the configuration space of the replica 1. More precisely, the spin-configuration space is $\{-1,1\}^N = \Sigma$, and we consider a bounded function ϕ_s of s replicas, i.e., $\phi_s\colon \Sigma^s \to \mathbb{R}$. The product space Σ^s ("the space of the replicas") is equipped with the product Gibbs measure ("replica measure") Ω, but the quenched variables are the same in each factor of the product space, and this means that the measure $\langle\cdot\rangle = \mathbb{E}\Omega(\cdot)$ on the product space Σ^s is not a product measure. So $f_N^{(1)}$ is the free energy in the space which is the first factor in the product space Σ^s. Notice that Σ has the cardinality of the continuum in the thermodynamic limit $N\to\infty$.

Now we want to perturb the Hamiltonian

$$-\beta H_N(\sigma) \longmapsto -\beta H_N(\sigma) + \beta' \sum_{\nu=1}^{P'\alpha} J'_\nu \sigma_{i'_\nu} \sigma_{j'_\nu} \, ,$$

and consider the derivative with respect to the perturbing parameter, as we did in the previous section in order to obtain an expansion in powers of β' with coefficients not depending on β' in the thermodynamic limit. As before, recall that we always need to take derivatives (with respect to the inverse temperature this time), which by convexity exist only almost everywhere, and therefore we will obtain equality only integrating back over a given interval [16].

We are going to prove, first of all, the following theorem:

Theorem 4.1. *For a given bounded function ϕ_s of s replicas, the following relation, holding in integral average with respect to the inverse temperature β, constrains the distribution of the 4-overlap*

$$\frac{s(s+1)(s+2)}{3!}\langle q^2_{1,s+1,s+2,s+3}\phi_s\rangle - \frac{s(s+1)}{2!}\sum_{a}^{2,s}\langle q^2_{1,a,s+1,s+2}\phi_s\rangle$$
$$+ s\sum_{a<b}^{2,s}\langle q^2_{1,a,b,s+1}\phi_s\rangle - \sum_{a<b<c}^{2,s}\langle q^2_{1,a,b,c}\phi_s\rangle = \langle q^2_{1234}\rangle\langle\phi_s\rangle \, .$$

The proof is straightforward but long, and it will be splitted into several steps.

Let us consider the right-hand side of (27). Put $t = \tanh(\beta')$, $q_0 = 1$, and let us just indicate the number of replicas in the overlaps, rather than denumerating them all. The presence of the perturbation implies that the pressure and the free energy are functions, $\tilde{p}_N(\beta, \beta', \alpha)$ and $\tilde{f}_N(\beta, \beta', \alpha)$ respectively, of both β and β', and recall also that according to our notations $\tilde{p}_N(\beta, \beta', \alpha) = -\beta\tilde{f}_N(\beta, \beta', \alpha)$. Let us prove the next

Lemma 4.1. *The derivative of the (perturbed) pressure $\tilde{p}_N(\beta, \beta', \alpha)$ with respect to the perturbing parameter β' has the following form as a series in powers of $t = \tanh(\beta')$*

$$\partial_{\beta'}\tilde{p}_N(\beta, \beta', \alpha) = -\alpha\sum_{n=0}^{\infty} t^{2n+1}(\langle q^2_{2n}\rangle - \langle q^2_{2n+2}\rangle) \, . \tag{28}$$

Proof. We have

$$\partial_{\beta'}\tilde{p}_N(\beta, \beta', \alpha) = -\sum_{m=1}^{\infty}\pi_\alpha(m)\sum_{\nu=1}^{m}\langle J'_\nu\sigma_{i'_\nu}\sigma_{j'_\nu}\rangle_m = -\sum_{m=1}^{\infty}m\pi_\alpha(m)\langle J'_m\sigma_{i'_m}\sigma_{j'_m}\rangle_m$$
$$= -\alpha\sum_{m=1}^{\infty}\pi_\alpha(m-1)\langle J'_m\sigma_{i'_m}\sigma_{j'_m}\rangle_m$$

where the subscript m indicates that the variable P'_α has been fixed to m. It is easy to see that

$$\langle J'_m \sigma_{i'_m} \sigma_{j'_m} \rangle_m = \mathbb{E} \frac{\Omega(J'_m \sigma_{i'_m} \sigma_{j'_m} \exp(\beta J'_m \sigma_{i'_m} \sigma_{j'_m}))_{m-1}}{\Omega(\exp(\beta J'_m \sigma_{i'_m} \sigma_{j'_m}))_{m-1}} \,. \tag{29}$$

Hence

$$\partial_{\beta'} \tilde{p}_N(\beta, \beta', \alpha) = -\alpha \mathbb{E} J' \frac{t+w}{1+tw} \,, \quad w \equiv \Omega(\sigma_{i'_m} \sigma_{j'_m}) \,, \tag{30}$$

according to the usual notations. Now a simple expansion (that we will explicitly write in the next lemma) of $(1+tw)^{-1}$ in powers of t yields

$$\partial_{\beta'} \tilde{p}_N(\beta, \beta', \alpha) = -\alpha \sum_{n=0}^{\infty} t^{2n+1} (\langle q_{2n}^2 \rangle - \langle q_{2n+2}^2 \rangle) \,. \tag{31}$$

So the lemma is proven and we have an expression for the right-hand side of (27), if we just multiply the average of the multi-overlaps by the average of ϕ_s. □

Let us now consider the left-hand side of (27), recalling that ϕ_s is a function of s replicas, that indices in the spins indicate which factor of the product space Σ^s (which replica) the spin belongs to, and that the energy density is assumed to be taken in the first replica.

Lemma 4.2. *Recalling that* $w \equiv \Omega(\sigma_{i'_m} \sigma_{j'_m})$, *we have*

$$\begin{aligned}
\langle u_N^{(1)} \phi_s \rangle = -\alpha t \mathbb{E} \Big\{ \Omega \Big[& \phi_s (1 + J' t^{-1} \sigma_{i_1}^{(1)} \sigma_{j_1}^{(1)}) \\
& \times \Big(1 + J' \sum_a^{2,s} \sigma_{i_1}^{(a)} \sigma_{j_1}^{(a)} t + \sum_{a<b}^{2,s} \sigma_{i_1}^{(a)} \sigma_{i_1}^{(b)} \sigma_{j_1}^{(a)} \sigma_{j_1}^{(b)} t^2 \\
& + J' \sum_{a<b<c}^{2,s} \sigma_{i_1}^{(a)} \sigma_{i_1}^{(b)} \sigma_{i_1}^{(c)} \sigma_{j_1}^{(a)} \sigma_{j_1}^{(b)} \sigma_{j_1}^{(c)} t^3 + \cdots \Big) \Big] \\
& \times \Big(1 - J' stw + \frac{s(s+1)}{2!} t^2 w^2 - J' \frac{s(s+1)(s+2)}{3!} t^3 w^3 \\
& + \frac{s(s+1)(s+2)(s+3)}{4!} t^4 w^4 - \cdots \Big) \Big\} .
\end{aligned}$$

Proof. From the proof of the previous lemma, in particular equations (29)–(30), and by definition of replica measure, we immediately get

$$\langle u^{(1)} \phi_s \rangle = -\alpha \mathbb{E} \frac{\Omega[J' \sigma_{i_1}^{(1)} \sigma_{j_1}^{(1)} \exp(\beta J' (\sigma_{i_1}^{(1)} \sigma_{j_1}^{(1)} + \cdots + \sigma_{i_1}^{(s)} \sigma_{j_1}^{(s)})) \phi_s]}{\Omega^s (\exp(\beta J' \sigma_{i_1} \sigma_{j_1}))} \,, \tag{32}$$

which we rewrite as

$$\langle u^{(1)} \phi_s \rangle = -\alpha \mathbb{E}\, t \frac{\Omega[(1 + J' t^{-1} \sigma_{i_1}^{(1)} \sigma_{j_1}^{(1)}) \prod_{a=2}^{s} (1 + J' t \sigma_{i_1}^{(a)} \sigma_{j_1}^{(a)}) \phi_s]}{(1+J'tw)^s} \,. \tag{33}$$

Let us write explicitly the power expansion of the denominator, which we omitted in the previous lemma

$$\frac{1}{(1+J'tw)^s} = 1 - J'stw + \frac{s(s+1)}{2!}t^2w^2 - J'\frac{s(s+1)(s+2)}{3!}t^3w^3 + \frac{s(s+1)(s+2)(s+3)}{4!}t^4w^4 \cdots .$$

It is also clear that

$$\prod_{a=2}^{s}(1+J't\sigma_{i_1}^{(a)}\sigma_{j_1}^{(a)}) = 1 + J'\sum_{a}^{2,s}\sigma_{i_1}^{(a)}\sigma_{j_1}^{(a)}t + \sum_{a<b}^{2,s}\sigma_{i_1}^{(a)}\sigma_{i_1}^{(b)}\sigma_{j_1}^{(a)}\sigma_{j_1}^{(b)}t^2 + J'\sum_{a<b<c}^{2,s}\sigma_{i_1}^{(a)}\sigma_{i_1}^{(b)}\sigma_{i_1}^{(c)}\sigma_{j_1}^{(a)}\sigma_{j_1}^{(b)}\sigma_{j_1}^{(c)}t^3 + \cdots .$$

Gathering all the ingredients completes the proof of the lemma. □

We are now able to compare the two sides of (27), and see what the self-averaging of the internal energy density in the thermodynamic limit brings.

Equating the expressions computed in the last two lemmas gives

$$\begin{aligned}\sum_{n=0}^{\infty} t^{2n}(\langle q_{2n}^2\rangle - \langle q_{2n+2}^2\rangle)\langle\phi_s\rangle = \mathbb{E}\Big\{\Omega\Big[\phi_s(1+Jt^{-1}\sigma_{i_1}^{(1)}\sigma_{j_1}^{(1)}) \\ \Big(1 + J'\sum_{a}^{2,s}\sigma_{i_1}^{(a)}\sigma_{j_1}^{(a)}t + \sum_{a<b}^{2,s}\sigma_{i_1}^{(a)}\sigma_{i_1}^{(b)}\sigma_{j_1}^{(a)}\sigma_{j_1}^{(b)}t^2 \\ + J'\sum_{a<b<c}^{2,s}\sigma_{i_1}^{(a)}\sigma_{i_1}^{(b)}\sigma_{i_1}^{(c)}\sigma_{i_1}^{(a)}\sigma_{i_1}^{(b)}\sigma_{j_1}^{(c)}t^3 \\ + \cdots + J'^{s-1}t^{s-1}\sigma_{i_1}^{(2)}\cdots\sigma_{i_1}^{(s)}\sigma_{j_1}^{(2)}\cdots\sigma_{j_1}^{(s)}\Big)\Big] \\ \Big(1 - J'stw + \frac{s(s+1)}{2!}t^2w^2 - J'\frac{s(s+1)(s+2)}{3!}t^3w^3 \\ + \frac{s(s+1)(s+2)(s+3)}{4!}t^4w^4 - \cdots\Big)\Big\}. \end{aligned} \tag{34}$$

The equality holds for any smooth function ϕ_s (typical interesting information is obtained for $\phi_s \equiv 1$ or $\phi_{s=2n} = q_{2n}^2$), so that we get equalities between expressions involving averages of (squared) overlaps.

Let us see in detail what information we can get from the lowest orders.

Denote by $\mathbb{E}(\cdot|\mathcal{A}_s)$ the conditional expectation with respect to the sigma-algebra $\mathcal{A}_s$ generated by the overlaps of s replicas. Let us show that the usual [8] Ghirlanda–Guerra identities for the overlap hold in our quite general case too:

Proposition 4.1. *The Ghirlanda–Guerra relation holds*

$$\mathbb{E}(q_{a,s+1}^2|\mathcal{A}_s) = \frac{1}{s}\langle q_{12}^2\rangle + \frac{1}{s}\sum_{b\neq a} q_{ab}^2 \tag{35}$$

in integral average with respect to the inverse temperature β.

Proof. In the expansion (34), where only the terms of even order survive due to the symmetry of the variables J, at the lowest order in t one gets

$$\begin{aligned}\langle\phi_s\rangle - \langle q_{12}^2\rangle\langle\phi_s\rangle &= \langle\phi_s\rangle - s\mathbb{E}[\omega(\sigma_{i_1}^{(1)}\sigma_{j_1}^{(1)})w\phi_s] + \sum_a^{2,s}\mathbb{E}[\Omega(\sigma_{i_1}^{(1)}\sigma_{i_1}^{(a)}\sigma_{j_1}^{(1)}\sigma_{j_1}^{(a)})\phi_s] \\ &= \langle\phi_s\rangle - s\langle q_{1,s+1}^2\phi_s\rangle + \sum_a^{2,s}\langle q_{1a}^2\phi_s\rangle\,,\end{aligned}$$

which is precisely what is stated in (35), (see [16]), immediately completing the proof of the proposition. □

So the usual Ghirlanda–Guerra identities for 2-overlaps are recovered and proven to hold in dilute spin glasses too.

At the next order we get instead

$$\begin{aligned}\langle q_{12}^2\rangle\langle\phi_s\rangle - \langle q_{1234}^2\rangle\langle\phi_s\rangle &= \sum_{a<b}^{2,s}\langle q_{ab}^2\phi_s\rangle + \frac{s(s+1)}{2!}\langle q_{s+1,s+2}^2\phi_s\rangle \\ &- s\sum_a^{2,s}\langle q_{a,s+1}^2\phi_s\rangle - \frac{s(s+1)(s+2)}{3!}\langle q_{1,s+1,s+2,s+3}^2\phi_s\rangle \\ &+ \frac{s(s+1)}{2!}\sum_a^{2,s}\langle q_{1,a,s+1,s+2}^2\phi_s\rangle - s\sum_{a<b}^{2,s}\langle q_{1,a,b,s+1}^2\phi_s\rangle + \sum_{a<b<c}^{2,s}\langle q_{1,a,b,c}^2\phi_s\rangle\,. \end{aligned}\tag{36}$$

Now consider the four 2-overlaps terms. A simple generalization of the usual Ghirlanda–Guerra relations [8] to the case when two replicas are added to a previously assigned set of other replicas, tells us that these terms cancel out. Let us check that explicitly.

Corollary 4.1. *Relation* (35) *implies*

$$\mathbb{E}(q_{s+1,s+2}^2|\mathcal{A}_s) = \frac{2}{s+1}\langle q_{12}^2\rangle + \frac{2}{s(s+1)}\sum_{a<b}^{1,s} q_{ab}^2\,, \tag{37}$$

under the same conditions, i.e., in integral average with respect to β.

Proof. Let us re-write (35) in the case of $s+1$ given replicas

$$\mathbb{E}(q_{s+1,s+2}^2|\mathcal{A}_{s+1}) = \frac{1}{s+1}\langle q_{12}^2\rangle + \frac{1}{s+1}\sum_b^{1,s} q_{b,s+1}^2\,.$$

Now use

$$\mathbb{E}(\mathbb{E}(\,\cdot\,|\mathcal{A}_{s+1})|\mathcal{A}_s) = \mathbb{E}(\,\cdot\,|\mathcal{A}_s) \tag{38}$$

to get

$$\begin{aligned}\mathbb{E}(q^2_{s+1,s+2}|\mathcal{A}_s) &= \frac{1}{s+1}\langle q^2_{12}\rangle + \frac{1}{s+1}\sum_b^{1,s}\mathbb{E}(q^2_{b,s+1}|\mathcal{A}_s)\\ &= \frac{1}{s+1}\langle q^2_{12}\rangle + \frac{1}{s+1}\left(\langle q^2_{12}\rangle + \frac{1}{s}\sum_b^{1,s}\sum_{c\neq b}^{1,s} q^2_{bc}\right).\end{aligned}$$

That is

$$\mathbb{E}(q^2_{s+1,s+2}|\mathcal{A}_s) = \frac{2}{s+1}\langle q^2_{12}\rangle + \frac{2}{s(s+1)}\sum_{a<b}^{1,s} q^2_{ab}\,, \tag{39}$$

which is what we wanted to prove. □

Proof of Theorem 4.1 (*end*). Now with (35) and (37) in our hands, let us take the three 2-overlap terms in the right-hand side of (36)

$$\frac{s(s+1)}{2}\langle q^2_{s+1,s+2}\phi_s\rangle = s\langle q^2_{12}\rangle\langle\phi_s\rangle + \sum_{a<b}^{1,s}\langle q^2_{ab}\phi_s\rangle\,,$$

$$\begin{aligned}-s\sum_a^{2,s}\langle q^2_{a,s+1}\phi_s\rangle &= -s\sum_a^{1,s}\langle q^2_{a,s+1}\phi_s\rangle + s\langle q^2_{1,s+1}\phi_s\rangle\\ &= -s\langle q^2_{12}\rangle\langle\phi_s\rangle - \sum_a^{1,s}\sum_{b\neq a}^{1,s}\langle q^2_{ab}\phi_s\rangle + \langle q^2_{12}\rangle\langle\phi_s\rangle + \sum_a^{2,s}\langle q^2_{1a}\phi_s\rangle\,,\end{aligned}$$

$$\sum_{a<b}^{2,s}\langle q^2_{ab}\phi_s\rangle = \sum_{a<b}^{1,s}\langle q^2_{ab}\phi_s\rangle - \sum_a^{2,s}\langle q^2_{1a}\phi_s\rangle\,.$$

The sum of these three terms clearly reduces to $\langle q^2_{12}\rangle\langle\phi_s\rangle$, which is precisely what we find in the left-hand side of (36). The 2-overlap terms thus cancel out from (36). We are hence left with a new relation for 4-overlaps:

$$\begin{aligned}\frac{s(s+1)(s+2)}{3!}\langle q^2_{1,s+1,s+2,s+3}\phi_s\rangle - \frac{s(s+1)}{2!}\sum_a^{2,s}\langle q^2_{1,a,s+1,s+2}\phi_s\rangle\\ + s\sum_{a<b}^{2,s}\langle q^2_{1,a,b,s+1}\phi_s\rangle = \langle q^2_{1234}\rangle\langle\phi_s\rangle + \sum_{a<b<c}^{2,s}\langle q^2_{1,a,b,c}\phi_s\rangle\,,\end{aligned}$$

and the proof of Theorem 4.1 is now complete. □

We report for the sake of completeness the general expression of the generic order in the power series expansion (34). From the explicit calculation in Lemma 4.2

we get

$$\begin{aligned}
&\langle q_{2n}^2\rangle\langle\phi_s\rangle - \langle q_{2n+2}^2\rangle\langle\phi_s\rangle \\
&= \sum_{m=2n-s+1}^{2n}\sum_{l=0}^{s-1}\sum_{a_1<\cdots<a_l}^{2,s}(-)^m\binom{s+m+1}{m} \\
&\qquad \times \mathbb{E}[w^m\Omega(\phi_s\sigma_{i_1}^{a_1}\cdots\sigma_{i_1}^{a_l}\sigma_{j_1}^{a_1}\cdots\sigma_{j_1}^{a_l})]\delta_{2n,m+l} \\
&\quad + \sum_{m=2n-s+2}^{2n+1}\sum_{l=0}^{s-1}\sum_{a_1<\cdots<a_l}^{2,s}(-)^m\binom{s+m+1}{m} \\
&\qquad \times \mathbb{E}[w^m\Omega(\phi_s\sigma_{i_1}^{1}\sigma_{j_1}^{1}\sigma_{i_1}^{a_1}\cdots\sigma_{i_1}^{a_l}\sigma_{j_1}^{a_1}\cdots\sigma_{j_1}^{a_l})]\delta_{2n,m+l-1}
\end{aligned}$$

which becomes, denoting by $x\wedge y$ the minimum between x and y,

$$\begin{aligned}
\langle q_{2n}^2\rangle\langle\phi_s\rangle - \langle q_{2n+2}^2\rangle\langle\phi_s\rangle &= \sum_{l=0}^{2n\wedge s-1}\sum_{a_1<\cdots<a_l}^{2,s}(-)^{2n-l}\binom{2n+s-l+1}{2n-l} \\
\times\left[\langle\phi_s q_{a_1\cdots a_l}^2 q_{s+1\cdots s+2n-l}^2\rangle - \frac{2n-l+s+2}{2n-l+1}\langle\phi_s q_{1a_1\cdots a_l}^2 q_{s+1\cdots s+2n-l+1}^2\rangle\right]. \quad (40)
\end{aligned}$$

In both the expressions above the term for $l=0$ is understood to be one.

The right-hand side of (40), due to the presence of $1+Jt^{-1}\sigma$ in the right-hand side of (34) – along with the symmetry of J, makes the expansion somewhat recursive. This means that at each order we find some terms already found in the previous order. More precisely, we claim without proving that at each $2n$th order of the expansion, all the terms involving $2m$-overlaps with $2m\le 2n$ cancel out thanks to a repeated use of (38) with the relations coming from the lower orders. Hence from the $2n$th order we get new relations involving $2n+2$-overlaps only. This is what we explicitly verified only for 4-overlaps in the previous pages. More explicitly, if we re-write the difference in the right-hand side of (40) as

$$\langle q_{2n}^2\rangle\langle\phi_s\rangle - \langle q_{2n+2}^2\rangle\langle\phi_s\rangle = c_{2n} - d_{2n+2}\,,$$

we have

$$\langle q_{2n}^2\rangle\langle\phi_s\rangle = c_{2n}\,,\quad \langle q_{2n+2}^2\rangle\langle\phi_s\rangle = d_{2n+2}\,,\quad c_{2n}=d_{2n}\,.$$

So that the final formula becomes

$$\langle q_{2n}^2\rangle\langle\phi_s\rangle = \sum_{l=0}^{2n\wedge s-1}\sum_{a_1<\cdots<a_l}^{2,s}(-)^{2n-l}\binom{2n+s-l+1}{2n-l}\langle q_{a_1\cdots a_l}^2 q_{s+1\cdots s+2n-l}^2\phi_s\rangle\,,$$

and holds in integral average, as always.

4.1. Generalization to smooth functions of multi-overlaps

Just like for the family of identities discussed in the previous section, we started our analysis with the most natural quantity: the energy of our model with 2-spin interactions. And so we got again some relations for the squared multi-overlaps. But we already know how to generalize these formulas to smooth functions of the overlaps. We can consider p-spin interactions, and the procedure would provide us with the same relations for the pth power of the overlaps. Then, as already explained, we can take a convergent sum over all integers p of p-spin Hamiltonians, and consider the self-averaging of the desired one among them. The perturbed Hamiltonian is again

$$H_N(\sigma,\alpha;\mathcal{J}) = -\sum_p \left[a_p \sum_{\nu=1}^{P^{(p)}_{\alpha N}} J_\nu \sigma_{i^1_\nu} \cdots \sigma_{i^p_\nu} + b_p \lambda_p \sum_{\nu=1}^{P'^{(p)}_{\alpha'}} J'_\nu \sigma_{j^1_\nu} \cdots \sigma_{j^p_\nu} \right],$$

where $\sum_p |a_p|^2 = \sum_p |b_p|^2 = 1$, the notation for all the quenched variables is the usual one, and $\{\lambda_p\}$ are the independent perturbing real parameters. As a side remark, we just point out that (like in [8]) in the case of this second family of identities it is not necessary to consider a Hamiltonian consisting of the sum of all possible p-spin Hamiltonians: only the perturbation must be so.

Let us remind once again that all the identities we provided hold in integral average with respect to all the variables used to compute derivatives. If more than one variable is used in some derivatives, the relations hold under integral performed simultaneously against all the variables, over the Cartesian product of the chosen intervals of variation of each variable.

Concluding remarks

Notice that while we derived our identities having as reference diluted spin glasses, all that matters in the derivation are the properties of the perturbing Hamiltonian, and they are therefore generically valid (in integral average with respect to the perturbing parameters).

The Ghirlanda–Guerra identities for the overlap have been useful to prove non-trivial properties of mean-field spin glasses. For instance Talagrand could prove [16] that for all models where the identities are valid, the overlap is positive with probability one. This positivity property is important as it enters in the Guerra free-energy bounds in spin systems without spin reversal symmetry, like in the case of odd-spin interactions. Unfortunately the derivation of Talagrand for the overlap does not extend immediately to the multi-overlap case. We believe however that the self-averaging identity will be useful in the mathematical analysis of diluted spin models.

Acknowledgement

LDS thanks Fabio Lucio Toninelli and Anton Bovier for useful discussions.

References

[1] Agostini, A., Barra, A., De Sanctis, L.: Positive-Overlap Transition and Critical Exponents in Mean Field Spin Glasses, J. Stat. Mech. Theory Exp. **2006**(11), P11015, 13 pp. (2006).

[2] Aizenman, M., Contucci, P.: On the stability of the quenched state in mean field spin glass models, J. Stat. Phys. **92**(5-6), 765–783 (1998).

[3] Barra, A., De Sanctis, L.: Stability properties and probability distribution of multi-overlaps in dilute spin glasses, J. Stat. Mech. Theory Exp. **2007**(8), P08025, 20 pp. (2007).

[4] Bovier, A.: Statistical Mechanics of Disordered Systems, A mathematical perspective, Cambridge University Press, Cambridge, 2006.

[5] Bovier, A., Gayrard, V.: Hopfield Models as Generalized Random Mean Field Models, ArXiv:cond-mat/9607103 (1997), in A. Bovier and P. Picco Eds., Mathematical Aspects of Spin Glasses and Neural Networks, Progress in Probability **41** Birkhäuser, Boston-Basel-Berlin, 1998.

[6] De Sanctis, L.: Random Multi-Overlap Structures and Cavity Fields in Diluted Spin Glasses, J. Stat. Phys. **117** 785–799 (2004).

[7] Franz, S., Leone, M., Toninelli, F.L.: Replica bounds for diluted non-Poissonian spin systems, J. Phys. A **36** 10967 (2003).

[8] Ghirlanda, S., Guerra, F.: General properties of overlap distributions in disordered spin systems. Towards Parisi ultrametricity, J. Phys. A, **31** 9149–9155 (1998).

[9] Guerra, F.: About the overlap distribution in mean field spin glass models, Int. J. Mod. Phys. B **10**, 1675–1684 (1996).

[10] Guerra, F., Toninelli, F.L.: The high temperature region of the Viana–Bray diluted spin glass model, J. Stat. Phys.**115**(1-2), 531–555 (2004).

[11] Mézard, M., Montanari, A.: Reconstruction on trees and spin glass transition, J. Stat. Phys. **124**(6), 1317–1350 (2006).

[12] Mézard, M., Parisi, G., Virasoro, M.A.: Spin glass theory and beyond, World Scientific, Singapore, 1987.

[13] Mézard, M., Parisi, G., Zecchina, R.: Analytic and Algorithmic Solution of Random Satisfiability Problems, Science **247**, 812–815 (2002).

[14] Parisi, G.: On the probabilistic formulation of the replica approach to spin glasses, ArXiv:cond-mat/9801081.

[15] Talagrand, M.: The Sherrington–Kirkpatrick model: a challenge for mathematicians, Probab. Theory Related Fields **110**, 109–176 (1998).

[16] Talagrand, M.: Spin glasses: a challenge for mathematicians. Cavity and Mean field models, Springer Verlag, Heidelberg, 2003.

Luca De Sanctis
Scuola Normale Superiore
Centro di Ricerca Matematica
P.zza dei Cavalieri 3
I-56100 Pisa, Italy
e-mail: `luca.desanctis@sns.it`

Silvio Franz
Université Paris-Sud Paris 11
F-91405 Orsay Cedex, France
e-mail: `silvio.franz@lptms.u-psud.fr`

Progress in Probability, Vol. 62, 143–157

Chaos in Mean-field Spin-glass Models

Tommaso Rizzo

Abstract. Physicists understand mean-field spin-glass models as possessing a complex free-energy landscape with many equilibrium states. The problem of chaos concerns the evolution of this landscape upon changing the external parameters of the system and is considered relevant for the interpretation of important features of real spin-glass and for understanding the performance of numerical algorithms. The subject is strongly related to that of constrained systems which is considered by mathematicians the natural framework for proving rigorously some of the most peculiar properties of Parisi's replica-symmetry-breaking solution of mean-field spin-glass models, notably ultrametricity. Many aspects of the problems turned out to possess an unexpected level of difficulty and are still open. We present the results of the physics literature on the subject and discuss the main unsolved problems from a wider perspective.

Mathematics Subject Classification (2000). 82-06, 82D30.

Keywords. Spin-glass, chaos, constrained systems, replica method.

Introduction

The chaos problem is a very old one in spin-glass theory and is intimately related to the disordered nature of these systems. Physicists understand these models as possessing a complex free energy landscape with many states, and the problem is essentially the evolution of this landscape upon changing the external fields, say temperature or magnetic field, or upon switching on random perturbations. It has received a lot of attention over the years in connection with many different problems [1]. It is believed that this framework is suitable to understand the surprising rejuvenation and memory effects observed in the dynamics of real spin-glasses [1]. Furthermore, a detailed understanding of this problem would probably shed light on the success of the Parallel Tempering procedure, which is nowadays considered an essential ingredient to achieve thermalization in numerical spin-glass simulations [3]. The problem, especially the so-called temperature chaos problem, presents amazing technical and conceptual difficulties which have not yet been solved completely. On the other hand the chaos problem pertains to the general problem of constrained systems. This seems the natural framework to rigorously prove the predictions of Parisi's Replica-Symmetry-Breaking, in particular those concerning the distribution of the overlap and ultrametricity. In this context a number of problems are still open; even statements that were originally expected to be at hand have eluded a rigorous prove up to now and were dubbed obnoxious by M. Talagrand [14].

The Sherrington–Kirkpatrick (SK) model is a system of N Ising spins $\{s_i\}$ with the following Hamiltonian

$$H_J(\{\sigma\}) = -\sum_{i<j} J_{ij}\sigma_i\sigma_j \tag{1}$$

where the coupling J_{ij} are random variables with zero mean and variance $\overline{J_{ij}^2} = 1/N$. A measure of the correlation between the relevant configurations at *different* temperatures is given by the function

$$P_J^{T_1T_2}(q) = \frac{\sum_{\{\tau\}\{\sigma\}} \delta(q - \sum_i \sigma_i\tau_i)\exp[-\beta_1 H_J(\{\sigma\}) - \beta_2 H_J(\{\tau\})]}{\sum_{\{\tau\}\{\sigma\}} \exp[-\beta_1 H_J(\{\sigma\}) - \beta_2 H_J(\{\tau\})]} \tag{2}$$

where $\beta_i \equiv T_i^{-1}$. This function gives the probability of observing an overlap q if we extract two configurations according to their Boltzmann weights from systems with the same Hamiltonian but different temperatures.

A peculiar feature of the SK model according to Parisi theory [2] is that if the two systems have the same temperature the function $P_J^{T_1T_1}(q)$ in the low temperature phase has a support between $-q_{\rm EA}$ and $q_{\rm EA}$, where $q_{\rm EA}$ is the so-called Edwards–Anderson parameter. In particular the disorder average $P(q) = \overline{P_J(q)}$ is given by $P(q) = \mathrm{d}x/\mathrm{d}q$ where $q(x)$ is a continuous function between zero and $q_{\rm EA}$.

The problem of chaos in temperature concerns the function $P_J^{T_1T_2}(q)$. In particular, we say that there is chaos if

$$P_J^{T_1T_2}(q) = \delta(q), \tag{3}$$

i.e., if $P_J^{T_1T_2}(q)$ has a support concentrated on $q = 0$ and that there is no chaos otherwise. In other words, the problem deals with the correlations between the relevant thermodynamic configurations of systems with different external parameters. As such it is related to the disordered nature of these systems being trivial in non-disordered models like a ferromagnet.

The problem of chaos can be also addressed considering different spin-glass models. In particular we can consider the case of "soft" spins, *i.e.*, with a distribution of values $\rho(\sigma)$ not just concentrated on ± 1 as in the Ising case and continuous spins with a spherical global constraint $\sum_i \sigma_i^2 = N$.

Models with different Hamiltonians can be also considered, in particular p-spin models whose Hamiltonian contains random terms involving the product of more than two spins. In general spin-glass models are classified according to the nature of the function $P(q)$ in the low temperature phase; we talk of one-step

Replica-Symmetry-Breaking (1RSB) models when this function has a support concentrated on just three values, $\{0, q_{\mathrm{EA}}, -q_{\mathrm{EA}}\}$ while we talk of full-RSB models (like the SK model) when $P(q)$ has a non-zero support between $-q_{\mathrm{EA}}$ and q_{EA}.

Furthermore, different types of external field variations can be considered. Notably we talk of chaos in a magnetic field in the case of two systems with different magnetic fields and of *bond chaos* in the case of models with slightly different random Hamiltonians. There is, however, an important difference between these last two cases and the problem of chaos in temperature. Indeed the Boltzmann weight of a given configuration is modified by a truly random amount when a magnetic field is switched on or when the coupling between the spins is sligthly modified, while when the temperature is changed the modification is determined by the original weight.

The problem of chaos in temperature in the various mean-field models has been investigated intensively by the author over the years. Today we know that there are full-RSB models that do not display chaos in temperature [4] and full-RSB models that do have chaos including notably the SK model [5]. Similarly there are 1RSB models that do have chaos in temperature and models that do not [6].

On technical grounds the main result has been to obtain a consistent analytical picture for the absence of chaos [7]. As we will see absence of chaos in full-RSB models has been connected to the possibility of solving the variational equations that appear in the replica treatment of the problem with some special non-chaotic solutions. Much as the Parisi solution, these solutions encode a lot of information on the correlations between the configurations at different temperatures, namely ultrametricity and the distribution of the weights of the physical states. Therefore the problem is essentially to check whether the variational equations of the given model admit this type of solutions. If these solutions do not exist it seems naturally to conclude that $P_J^{T_1T_2} = \delta(q)$, in this case a complete solution of the chaos problem amounts to the computation of the large deviations of $P_J(q)$ and this represents the main open problem.

The computation of the large deviations of $P^{T_1T_2}$ is essentially related to the study of two coupled systems, indeed one expects that if chaos is present,

$$P_J^{T_1T_2}(q) \propto \exp\left[-N\left(F_{12}(T_1, T_2, q) - F(T_1) - F(T_2)\right)\right], \tag{4}$$

i.e., the large deviations are given by the difference between the free energy per spin of the coupled systems, whose allowed configurations are those that satisfy the constraint $q - \sum_i \sigma_i \tau_i$,

$$F_{12}(T_1, T_2, q) = -\frac{1}{N}\Big(\ln \sum_{\{\tau\}\{\sigma\}} \delta(q - \sum_i \sigma_i \tau_i) \exp[-\beta_1 H_J(\{\sigma\}) - \beta_2 H_J(\{\tau\})]\Big)$$

and the free energy *per spin* of the two systems in the absence of the coupling

$$F(T) = -\frac{1}{N}\Big(\ln \sum_{\{\tau\}} \exp[-\beta H_J(\{\tau\})]\Big). \tag{5}$$

The main object of study has been up to now the constrained free energy functional $F(T_1, T_2, q)$. This functional must be larger than or equal to the free energies of the two unconstrained systems. If it is larger one can safely conclude that the function $P^{T_1 T_2}(q)$ is a delta function in $q = 0$ while if it is zero for some values of $q \neq 0$ one expects that $P^{T_1 T_2}(q)$ has support on these values although the actual determination of $P^{T_1 T_2}(q)$ requires the computation of subleading contributions. The latter case however is rather subtle, because the function $P^{T_1 T_2}(q)$ could still converge to a delta function in the large N limit with subexponential large deviations. This unpleasant phenomenon happens indeed in the spherical SK model [15]. In the following we will consider the constrained free energy functional $F(T_1, T_2, q)$ averaged over the disorder. It is usually assumed that this quantity (and correspondingly the large deviations) does not fluctuate in the large N limit.

1. Chaos in temperature

We consider a system composed of two copies (replicas) of a SK spin glass model constrained to have fixed values of their mutual overlaps. This model was first studied in the case of replicas of the same temperature [10]; here we shall use the generalization to replicas with two different temperatures below the critical one [8]. Denoting by S_i^r the ith spin of the rth replica, we fix a constraint

$$q_c = \frac{1}{N} \sum_{i=1}^{N} S_i^1 S_i^2 \tag{6}$$

where N is the total number of spins. Including the temperature difference, the Hamiltonian of the system reads

$$H = -\sum_{i<j} J_{ij} (\beta_1 S_i^1 S_j^1 + \beta_2 S_i^2 S_j^2). \tag{7}$$

So we take the same realization of the quenched $\{J_{ij}\}$ for the two systems. They are chosen with Gaussian probability, zero mean and variance $\frac{1}{N}$. The partition function is restricted to those spin configurations that satisfy (6). The constraint (6) is implemented introducing a Lagrange multiplier ϵ

$$Z = \sum_{\{S_i^1, S_i^2\}} \int_{-i\infty}^{i\infty} \frac{d\epsilon}{2\pi} \exp\left[-H - \epsilon\left(\sum_{i=1}^{N} S_i^1 S_i^2 - N q_c\right)\right]. \tag{8}$$

Instead of fixing the constraint q_c we can consider the partition function corresponding to the following Hamiltonian

$$H(\epsilon) = -\sum_{i<j} J_{ij} (\beta_1 S_i^1 S_j^1 + \beta_2 S_i^2 S_j^2) - \epsilon \sum_{i=1}^{N} S_i^1 S_i^2 \tag{9}$$

this corresponds to systems coupled by a forcing term which selects configurations with higher overlap. In the thermodynamic limit the two descriptions are obtained one from the other by a Legendre transformation, in particular defining $F = -\ln Z$ the following relation holds:

$$\epsilon = \frac{\partial F(q_c)}{\partial q_c} . \tag{10}$$

Introducing replicas to average over the disorder we obtain via standard manipulation the average partition function to the power n

$$\begin{aligned} \overline{Z^n} = \mathsf{S.P.} \exp & -\left[\frac{N}{4}\beta_1^2 \operatorname{Tr} Q_1^2 + \frac{N}{4}\beta_2^2 \operatorname{Tr} Q_2^2 + \frac{N}{2}\beta_1\beta_2 \operatorname{Tr} P^2 \right. \\ & \left. - N \ln Z[\hat{Q}] - N q_c \sum_\alpha \epsilon_\alpha - \frac{N}{2}\sum_\alpha \left(\frac{\epsilon_\alpha^2}{\beta_1\beta_2} - 2P_{\alpha\alpha}\epsilon_\alpha\right)\right] \end{aligned} \tag{11}$$

$$\begin{aligned} Z[\hat{Q}] = & \sum_{\{S_\alpha^1, S_\alpha^2\}} \exp\left[\frac{1}{2}\beta_1^2 \sum_{\alpha\beta} Q_{1\alpha\beta} S_\alpha^1 S_\beta^1 \right. \\ & \left. + \frac{1}{2}\beta_2^2 \sum_{\alpha\beta} Q_{2\alpha\beta} S_\alpha^2 S_\beta^2 + \beta_1\beta_2 \sum_{\alpha\beta} P_{\alpha\beta} S_\alpha^1 S_\beta^2\right], \end{aligned} \tag{12}$$

where by S.P. we mean the value computed at the saddle point with respect to the set $\{\epsilon_\alpha\}$ and to the order parameter which is a $2n \times 2n$ matrix

$$\hat{Q} = \begin{pmatrix} Q_1 & P \\ P^t & Q_2 \end{pmatrix} .$$

The saddle-point (SP) equations then read

$$\begin{aligned} Q_{1\alpha\beta} &= \langle S_\alpha^1 S_\beta^1 \rangle & Q_{2\alpha\beta} &= \langle S_\alpha^2 S_\beta^2 \rangle \\ P_{\alpha\beta} &= \langle S_\alpha^1 S_\beta^2 \rangle + \frac{1}{\beta_1\beta_2}\epsilon_\alpha \delta_{\alpha\beta} & \epsilon_\alpha &= \beta_1\beta_2 (P_{\alpha\alpha} - q_c), \end{aligned} \tag{13}$$

where the square brackets mean average taken with respect to the Hamiltonian

$$H = \frac{1}{2}\beta_1^2 \sum_{\alpha\beta} Q_{1\alpha\beta} S_\alpha^1 S_\beta^1 + \frac{1}{2}\beta_2^2 \sum_{\alpha\beta} Q_{2\alpha\beta} S_\alpha^2 S_\beta^2 + \beta_1\beta_2 \sum_{\alpha\beta} P_{\alpha\beta} S_\alpha^1 S_\beta^2 . \tag{14}$$

The SP equation in the unconstrained case can be obtained by setting ϵ to zero and neglecting the last equation in (13). We eliminate the Lagrange multipliers $\{\epsilon_\alpha\}$ in (11) replacing their saddle point values. This gives for the ϵ-dependent term in $\overline{Z^n}$

$$-N q_c \sum_\alpha \epsilon_\alpha - \frac{N}{2}\sum_\alpha \left(\frac{\epsilon_\alpha^2}{\beta_1\beta_2} - 2P_{\alpha\alpha}\epsilon_\alpha\right) = N\frac{\beta_1\beta_2}{2} \sum_\alpha (P_{\alpha\alpha} - q_c)^2 . \tag{15}$$

To solve the model we need a variational ansatz for the matrices Q_1, Q_2, and P. We choose each of them to be a Parisi hierarchical matrix; in particular this fixes $P_{\alpha\alpha} = p_d$ and $\epsilon_\alpha = \epsilon$ for any replica index α. In the $n \to 0$ limit the matrices are parameterized with three functions defined on the interval $[0, 1]$

$$Q_1 \equiv (0, q_1(x)), \quad Q_2 \equiv (0, q_2(x)), \quad P \equiv (p_d, p(x))$$

and the constrained free energy reads:

$$
\begin{aligned}
F(q_c,T_1,T_2) = &-\frac{\beta_1^2}{4}-\frac{\beta_2^2}{4}+\frac{\beta_1^2}{4}\left(-\int_0^1 q_1^2(x)\mathrm{d}x+2q_1(1)\right)\\
&+\frac{\beta_1\beta_2}{2}\left(p_d^2-\int_0^1 p^2(x)\mathrm{d}x\right)+\frac{\beta_2^2}{4}\left(-\int_0^1 q_2^2(x)\mathrm{d}x+2q_2(1)\right)\\
&-\frac{\beta_1\beta_2}{2}(p_d-q_c)^2-f(0,0,0).
\end{aligned}
\tag{16}
$$

The function $f(x,y_1,y_2)$ obeys through the following differential equation:

$$
\begin{aligned}
-\frac{\partial f}{\partial x} = &\frac{\dot{q}_1}{2}\left(\frac{\partial^2 f}{\partial y_1^2}+x\left(\frac{\partial f}{\partial y_1}\right)^2\right)+\frac{\dot{q}_2}{2}\left(\frac{\partial^2 f}{\partial y_2^2}+x\left(\frac{\partial f}{\partial y_2}\right)^2\right)\\
&+\dot{p}\left(\frac{\partial^2 f}{\partial y_1\partial y_2}+x\frac{\partial f}{\partial y_1}\frac{\partial f}{\partial y_2}\right)
\end{aligned}
\tag{17}
$$

with initial condition:

$$
f(1,y_1,y_2)=\ln[2\mathrm{e}^{\delta}\cosh[\beta_1 y_1+\beta_2 y_2]+2\mathrm{e}^{-\delta}\cosh[\beta_1 y_1-\beta_2 y_2]] \tag{18}
$$

$$
\delta=\beta_1\beta_2(p_d-p(1)). \tag{19}
$$

Notice that the initial condition depends on the difference $p_d - p(1)$.

1.1. The $\beta_1 = \beta_2$ case

When $T_1 = T_2$ we distinguish two cases depending on the value of the constraint q_c. For all values of q_c inside the support of the overlap distribution function $P(q)$ we have solutions with $\delta = 0$ and therefore $\epsilon = 0$. As a consequence for $0 \le q_c \le q_{\mathrm{EA}}$ we have $\Delta F = 0$ as it is to be expected since the function $P(q)$ is non-trivial. The solutions of the SP equations satisfy $p_d = q_c$ and are:

$$
\begin{aligned}
&q_1(x)=q_2(x)=p(x)=q_{\mathrm{Parisi}}(2x), && 0\le x\le \frac{1}{2}x_{\mathrm{Parisi}}(p_d),\\
&q_1(x)=q_2(x)=p(x)=p_d && \frac{1}{2}x_{\mathrm{Parisi}}(p_d)\le x\le x_{\mathrm{Parisi}}(p_d),\\
&q_1(x)=q_2(x)=q_{\mathrm{Parisi}}(x),\quad p(x)=p_d, && x_{\mathrm{Parisi}}(p_d)\le x\le 1.
\end{aligned}
\tag{20}
$$

These solutions were first proposed in [10] for the truncated model. In [4] it has been proven that they exist in any models with Replica-symmetry breaking (RSB) by noticing that they are a permutation of the standard Parisi solution. Quite interestingly they are also a crucial ingredient in Talagrand's proof of the correctness of the Parisi free energy [13].

When q_c lies outside the support of the function $P(q)$ there are off-equilibrium solutions of the SP equations with $\delta \neq 0$ and $q_1(x) = q_2(x) = p(x)$ [10]. These solutions can be obtained by solving the SP equations perturbatively in $\delta q = q_c - q_{\mathrm{EA}}(T)$ and $\tau = T - T_c$.

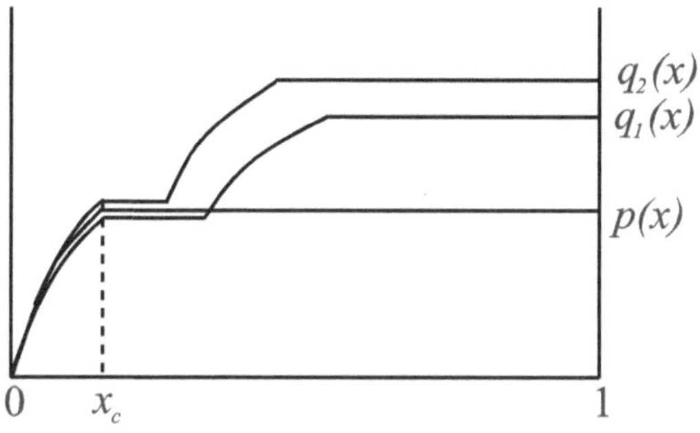

FIGURE 1. The non-chaotic ansatz. The real solutions are expected to pick up small corrections induced by chaos. The actual difference between the three functions in the small-x region is much smaller than in the figure.

1.2. The $\beta_1 \neq \beta_2$ case

In [7] an analytical picture for the absence of chaos in spin-glasses was proposed based on the possible existence of specific solutions of the SP equations. The structure of these solutions is depicted in Figure 1: in the small-x region the three functions $q_1(x)$, $q_2(x)$ and $p(x)$ are all different till they reach the point x_c where $p(x) = p_d = q_c$; then for x greater than x_c, $p(x)$ remains constant while $q_1(x)$ and $q_2(x)$ after an intermediate plateau are connected continuously to the corresponding free Parisi solutions. These solutions may exist for values of the constraint from zero to a maximum value where the two plateaus of the function at the higher temperature merge ($T_1 > T_2$ in Figure 1). These solutions reduce to eq. (20) in the limit $T_1 \to T_2$ and are intrinsically non-chaotic due to the relation $\delta = 0$ [7]. However, their existence must be checked explicitly by attempting to solve the SP equations with the ansatz $\delta = 0$.

It has been proven that these solutions exist in the spherical spin-glass models with multi-p spin interaction [5]. The Hamiltonian of the model is defined [11] as

$$H = \sum_{p=2}^{\infty} \sum_{i_1<i_2<\ldots<i_p} J_{i_1 i_2 \ldots i_p} S_{i_1} S_{i_2} \ldots S_{i_p} + h \sum_i S_i .$$

The J's are independent Gaussian random variables with zero mean and variance $\left\langle J^2_{i_1 i_2 \ldots i_p} \right\rangle = (p-1)! J_p^2 N^{1-p}$. The spins are subjected to the spherical constraint $\sum_i S_i^2 = N\sigma$. The variational free energy can be expressed introducing the function

$$f(q) = \sum_{p=2}^{\infty} \frac{1}{p} J_p^2 q^p$$

and it reads

$$2\beta F_n = -\beta^2 \sum_{ab} [f(Q_{ab}) + H^2 Q_{ab}] - \text{Tr} \ln Q - n$$

where the variational parameter is an $n \times n$ matrix Q_{ab} with diagonal $q_d = \sigma$. In this context it has been argued that the non-chaotic solutions imply ultrametricity between the equilibrium states at different temperatures so that the Parisi tree is essentially the same at all temperatures. Furthermore, arguments have been advanced in order to show that the reference free energies of the clusters of states do not change with temperature, much as in the generalized random-energy model. Perfect ultrametricity is a special property of those models due to the fact that the functions $q_1(x)$, $q_2(x)$ and $p(x)$ are all equal in the small-x region. In general, however, we believe that the physical interpretation of the solutions proposed in [7] is similar. Indeed, we think that the non-chaotic solutions proposed in [7] are the mathematical implementation of the old idea of a unique Parisi tree which bifurcates when lowering the temperature plus the additional hypothesis that the ordering of the reference free energies of the clusters of states does not change when lowering the temperature.

The problem was studied in the SK using a perturbative expansion of the constrained free energy near the critical temperature where the order parameter is small. In [7] it was found that if one plugs the non-chaotic ansatz into the expression of the free energy the following free energy cost is obtained:

$$\Delta F(q_c) = -\frac{q_c^3 u}{6\omega}\left(1 - \frac{v}{\omega^2}\right)\Delta T^2, \tag{21}$$

where ω, u and v are respectively the coefficients of the terms of the permutational invariants $\operatorname{Tr} Q^2$, $\sum_{ab} Q_{ab}^4$ and $\operatorname{Tr} Q^4$ in the expansion of the free energy in powers of Q. Remarkably in the SK model we have $\omega = u = v = 1$ and the above free energy cost is zero. The relationship $1 - v/\omega^2 = 0$ holds also in the case of continuous distribution of spin but could possibly fail for other models.

In order to assess if the free energy cost vanishes also at higher orders in the SK model a high-order computer-assisted perturbative expansion was used [5]. The procedure is conceptually straightforward: one tries to extremize the constrained free energy with the non-chaotic ansatz of Figure 1 assuming that near the critical temperature the functions $q_1(x)$, $q_2(x)$, $p(x)$ are small and the x regions where they vary are also small. It was found that up to that relatively high order the SP equations admit a non-chaotic solution with $\delta = 0$ for values of q_c between 0 and some maximum value near the self-overlap of the states at the higher temperature. However at ninth order in the free energy it is impossible to find a solution of the variational equations with $\delta = 0$: *the non-chaotic solutions don't exist in the SK model.* Accordingly there must be a free-energy cost in imposing the constraint, which one expects to be of ninth order.

Since off-equilibrium solutions must have $\delta \neq 0$, things become very complicated; indeed, as explained in [7], when $\delta \neq 0$ the three functions $q_1(x)$, $q_2(x)$ and $p(x)$ are coupled in the whole small-x region, *i.e.*, if one of them is varying in some interval of x the others must vary too. It seems reasonable to expect that off-equilibrium solutions have the same structure of the non chaotic solutions of Figure 1 plus small corrections of order δ, which is expected to be of the seventh order. Therefore it was tried to solve the variational equations with a complicated ansatz made up of five x regions on each of whom the functions $q_1(x)$, $q_2(x)$ and $p(x)$ have a different expansion. Unfortunately, in spite of the very high number of parameters of this ansatz, the authors of [5] were not able to solve the variational equations explicitly, and this is still a problem; a possible reason for this failure is that the variation of the variational functions have a non-analytic contribution of the form ΔT^{α} with a non-integer exponent α. This is precisely what happens in the case of bond chaos that has been recently studied by Aspelmeier [12]. Nevertheless an estimate of the free-energy cost was obtained plugging the non-chaotic ansatz directly into the expression of the free energy and extremizing with respect to the variational parameters. The following estimate for the free-energy cost of constraining two copies of a SK spin-glass at temperatures $T_1 = 1-\tau_1$ and $T_2 = 1-\tau_2$ to have a fixed overlap q_c was obtained:

$$\Delta F = \frac{12}{35}\,|q_c|^7 \Delta T^2. \tag{22}$$

As expected, this quantity turned out to be of the ninth order and it was argued in [5] that this result is most likely exact at the leading order.

2. Chaos with respect to random perturbations

In this section we consider the general problem of two constrained systems when the Hamiltonian of the second system is slightly changed by a perturbation uncorrelated with the original Hamiltonian. This includes two important cases:

(i) turning on of a small magnetic field which acts essentially as a random perturbation if there was no field in the original systems and
(ii) a small random perturbation in the couplings J_{ij} of the second system (bond chaos).

In these cases chaos can be assessed on the basis of a general principle. The argument was first proposed by G. Parisi for case (i) in the context of the phase state picture of the model (the so-called TAP approach [2]); here we will show that this argument can be formulated in the context of constrained systems and extended it to any random perturbation. We note that many properties of Parisi theory that are usually interpreted in terms of physical states can be phrased in terms of constrained systems and the latter framework is considered more suitable for a rigorous proof of them. Therefore we think it is important that this classical argument for chaos has a counterpart in the constrained systems context.

Parisi argued that the existence of two different magnetic susceptibilities in the spin-glass phase implies chaos with respect to the magnetic field. Indeed the equilibrium states in zero field have a susceptibility given by $\beta(1-q_{\rm EA})$; therefore, when a small field is turned on they develop a magnetization equal to $\beta(1-q_{\rm EA})h$. On the other hand the thermodynamic susceptibility of the spin-glass is $\beta(1-\overline{q})$ (where $\overline{q} = \int q(x)\mathrm{d}x$), therefore the equilibrium states in presence of a field h have a magnetization proportional tyo $\beta(1-\overline{q})h$ and cannot be the analytical continuation of the old equilibrium states. The argument relies on the possibility of expressing the susceptibility in terms of the overlaps; this is possible because the perturbation is uncorrelated with the original Hamiltonian and it can be extended to any random perturbation but not to the chaos in temperature problem.

In general it can be argued that if we add to a spin-glass Hamiltonian a random p-spin term of the form $-h_p \sum_{i_1<\cdots<i_p} \tilde{J}_{i_1\ldots i_p} s_{i_1}\ldots s_{i_p}$ where $\tilde{J}_{i_1\ldots i_p}$ are i.i.d. random variables of zero mean and variance $p!/(N^{p-1})$ uncorrelated with the original couplings J_{ij} the susceptibility of a state is $\beta(1-q_{\mathrm{EA}}^p)$ while the thermodynamic susceptibility is $\beta(1-\overline{q^p})$ and again Parisi's argument can be applied.

Now consider the free energy of two SK systems constrained at overlap q_c, *i.e.*, expression (16) with $\beta_1=\beta_2=\beta$. It is easy to see that the addition of a random p-spin interaction on the second system will modify the variational expression by a term $-h_p^2\beta^2(1-\int q_2^p(x)\mathrm{d}x)/2+o(h_p^2)$, therefore for small h_p the leading-order correction to the constrained free energy will be:

$$F(\beta,\beta,q_c,h_p)=F(\beta,\beta,q_c,0)-\frac{h_p^2\beta^2}{2}\left(1-\int q_2^p(x)\mathrm{d}x\right)+o(h_p^2) \tag{23}$$

where the correction term has to be evaluated using the $q_2(x)$ corresponding to $h_p=0$ which is given by expression (20) for $q_c \leq q_{\mathrm{EA}}$. Looking back at expression (20) one realizes that the smaller increase of $F(\beta,\beta,q_c,h_p)$ with h_p is for $q_c=0$ while a finite value of q_c yields a larger increase, therefore at small h_p we obtain the following expression of the free energy cost describing the large deviations of the $P(q)$:

$$P(q_c)\propto \exp\left[-\frac{Nh_p^2\beta^2}{4}\left(\int_0^{x_{\mathrm{Parisi}}(q_c)}(q_c^p-q_{\mathrm{Parisi}}^p(x))\mathrm{d}x\right)+o(h_p^2)N\right]. \tag{24}$$

At small finite values of q_c this leads to:

$$P(q_c)\propto \exp\left[-\frac{Nh_p^2\beta^2 p}{4\dot{q}(0)(p+1)}q_c^{p+1}+o(h_p^2)N+o(q_c^{p+1})N\right]. \tag{25}$$

As noted in [12] the above expression however is valid for $h_p \ll q_c \ll 1$, while the limit $q \ll 1$ at finite h_p can be studied at the level of Gaussian fluctuations around the $q_c=0$ solution. This problem has been recently addressed by Aspelmeier [12] in the context of the so-called truncated SK model. In the following we present a general treatment valid for the full model. This is possible because one can compute exactly the gaussian fluctuations of P_{ab} around $P_{ab}=0$ as was shown originally by Kondor in [9]. The bond chaos action for small n is [12]:

$$n\beta F(q_c)=N\left[\frac{1}{2}\beta^2\,\mathrm{Tr}\,Q^2+\frac{1}{2}\beta^2 B\,\mathrm{Tr}\,P^2-B\epsilon\,\mathrm{Tr}\,P+\frac{\epsilon^2 B}{2\beta^2}-\ln Z[\hat{Q}]+q_c\epsilon\right], \tag{26}$$

$$Z[\hat{Q}] = \sum_{\{S_a^1, S_a^2\}} \exp\left[\frac{1}{2}\beta^2 \sum_{a\neq b} Q_{ab} S_a^1 S_b^1 + \frac{1}{2}\beta^2 \sum_{a\neq b} Q_{ab} S_a^2 S_b^2 + \beta^2 \sum_{ab} P_{ab} S_a^1 S_b^2\right], \tag{27}$$

where again by S.P. we mean the value computed at the saddle point with respect to ϵ (as before actually we have a set of n terms $\{\epsilon_a\}$ but we assume that they are all equal to ϵ) and to the order parameter which is a $2n \times 2n$ matrix

$$\hat{Q} = \begin{pmatrix} Q & P \\ P^t & Q \end{pmatrix}.$$

The constant $B \geq 1$ is a measure of the bond perturbation and is equal to 1 if the second Hamiltonian is identical to the first. For $q_c = 0$ we have $\epsilon = 0$ and $P = 0$, now ϵ is coupled to the matrix $\hat{Q}$ only through the term $-B\epsilon \operatorname{Tr} P$, therefore we have to expand the free energy around the $P = 0$ solution for small P. The first non-trivial contributions are at second order, we have:

$$\begin{aligned}\frac{1}{\beta^2}\frac{\partial^2 n\beta F}{\partial P_{ab}\partial P_{mn}} &= B\delta_{am}\delta_{bn} - \beta^2\left(\langle S_a^1 S_b^2 S_m^1 S_n^2\rangle - \langle S_a^1 S_b^2\rangle\langle S_m^1 S_n^2\rangle\right) \\ &= B\delta_{ab,mn} - \beta^2 \tilde{Q}_{am}\tilde{Q}_{bn}\end{aligned} \tag{28}$$

where the matrix $\tilde{Q}$ has the same off diagonal elements as Q but all its diagonal coefficients are equal to 1, *i.e.*, $\tilde{Q} = I + Q$. The above contribution is the only relevant one, since it is easily seen that the coefficient of the mixed terms $P\delta Q$ vanishes. Therefore the P dependent part of the action (26) reads :

$$\frac{\beta^2 B}{2}\operatorname{Tr} P^2 - \frac{\beta^4}{2}\operatorname{Tr} P\tilde{Q}^2 P - B\epsilon \operatorname{Tr} P. \tag{29}$$

Going to the base where the general Parisi matrix is diagonal and taking the limit $n \to 0$ (see, *e.g.*, [4]), the previous expression becomes:

$$\frac{1}{2}\sum_a (\beta^2 B - \beta^4 \tilde{q}_a^2) p_a^2 - B\epsilon \sum_a p_a, \tag{30}$$

where a labels the n eigenvalues of the general Parisi hierarchical matrix and p_a represents the component of P in this base. The above expression can be easily extremized yielding:

$$p_a = \epsilon \frac{B}{\beta^2 B - \beta^4 \tilde{q}_a^2}, \tag{31}$$

thus the free energy cost at second order in q_c is:

$$\Delta F(q_c) = \frac{q_c^2 \beta}{2B^2}\left(\sum_a \frac{1}{B - \beta^2 \tilde{q}_a^2}\right)^{-1} + o(q_c^2). \tag{32}$$

The constant has to be evaluated taking the limit $n \to 0$ (see, *e.g.*, the appendix of [4]), and can be simplified using the fact that $\beta^2 \tilde{q}_0^2 = 1$ [7]. The result is:

$$\frac{1}{n}\sum_a \frac{1}{B - \beta^2 \tilde{q}_a^2} = \frac{1}{B-1} - \int_0^1 \frac{1}{x^2}\left(\frac{1}{B - \beta^2 \lambda_{\tilde{q}}^2(x)} - \frac{1}{B-1}\right) \mathrm{d}x. \tag{33}$$

The function $p(x)$ can be obtained as:

$$p(x) = \frac{\epsilon B}{\beta^2} \int_0^x \dot{q} \frac{2\beta^2 \lambda_{\tilde{q}}(y)}{(B - \beta^2 \lambda_{\tilde{q}}^2(y))^2} \mathrm{d}y, \tag{34}$$

where $\lambda_{\tilde{q}}(x)$ yields the eigenvalues of the Parisi matrix represented by $\tilde{q}(x)$: $\lambda_{\tilde{q}}(x) = 1 - xq(x) - \int_x^1 q(y)\mathrm{d}y$. For small $B - 1 = \Delta$ only the small x region is relevant, in this region the eigenvalue function behaves like $\lambda_{\tilde{q}}(x) = -x^2\dot{q}(0)/2 + o(x^2)$ where $\dot{q}(0)$ is the derivative of $q(x)$ in zero. Introducing the rescaled variable $z = \sqrt{\frac{T\dot{q}(0)}{\Delta}}\beta x$ the integrals in eqs. (33) and (34) can be computed explicitly and we have:

$$\Delta F(q_c) = \frac{q_c^2 \Delta^{3/2}}{\pi\sqrt{T\dot{q}(0)}} + o(q_c^2, \Delta^{3/2}) \tag{35}$$

$$p(x) = \frac{2q_c\sqrt{T\dot{q}(0)}}{\pi} f\left(\sqrt{\frac{T\dot{q}(0)}{\Delta}}\beta x\right) \tag{36}$$

where the function $f(x)$ is:

$$f(x) := \frac{x}{1+x^2} + \arctan x. \tag{37}$$

As expected the results obtained for the truncated model [12] are recovered near the critical temperature. It is interesting to see why the two arguments for chaos derived above fail in the case of chaos in temperature. In the case of finite q_c and small temperature difference this happens because the analog of expression (23) depends on the quantity $\int_0^1 q_2^2(x)\mathrm{d}x - p_d^2 + \int_0^1 p^2(x)\mathrm{d}x$. Plugging the solution (20) into this expression leads to an expression that does not depend on q_c, therefore chaos cannot be assessed at this level. In the case of small q_c and finite temperature difference the matrix of fluctuation (28) cannot be inverted because of the relationship $\beta^2\tilde{q}_0^2 = 1$ valid at all temperatures [9].

For recent mathematical work on chaos with respect to random perturbation see [16].

3. Conclusions and open problems

In the following we summarize the results obtained in the physics literature on the problem of chaos and discuss the main open problems.

For coupled systems with the same external parameters we know that there is no free energy cost when the constraint is inside the support of the overlap distribution. Indeed in this case we have solutions (expression (20)) of the variational equations that satisfy the constraint and have the same free energy of the unconstrained case. These solutions in general are constructed as permutations in the replica indices of the original single-system solution both in 1RSB and full-RSB models [4].

For coupled systems with the same external parameters, when the overlap is outside the support of the overlap distribution of the single system, we know that both for the SK model and for the p-spin model solutions of the variational equations have been found that display an increase in free energy [10]. Nevertheless it is not clear if these predictions are correct or they just give a bound on the true constrained free energy. We feel that the problem is particularly intriguing in the 1RSB case for values of the overlap between 0 and q_{EA}.

For coupled systems with different external parameters we can find models where there is no free energy cost in imposing a constraint $q_c \neq 0$ for some values of q_c. In particular we know that for the spherical $p > 2$ purely p-spin model there is no free energy cost in imposing that two copies at different temperatures $T_1 < T_c$ and $T_2 < T_c$ have an overlap $q_c = \sqrt{q_{\mathrm{EA}}(T_1) q_{\mathrm{EA}}(T_2)}$. We also know that for the spherical model with multi-p interactions that displays full RSB there is no free energy cost in imposing that two copies at different temperatures $T_1 < T_2 < T_c$ have an overlap q that can take any value between 0 and $q_{\mathrm{EA}}(T_2)$. In these cases however a detailed understanding of the solutions when the constraint is outside these non-chaotic values is still lacking. In the 1RSB case much as in the equal temperature case, the solutions can be computed quite easily and yield a positive free energy cost but it is not clear if this estimate is the correct one. In the full RSB case instead the study of the solutions for $q_c > q_{\mathrm{EA}}(T_2)$ has not been done at all.

The above results rely on the possibility of extremizing the free energy using the non-chaotic solutions described in the previous sections. In the SK model we know that this is impossible and that for any non-zero value of the constraint between two systems at different temperatures the free energy cost has to be positive. An estimate of the free energy cost was obtained using the non-chaotic equations as a variational ansatz but we still do not know the precise chaotic solutions possibly because of the presence of a non analytic term of the form ΔT^α with non integer α in the correct solutions. Despite these technical difficulties it is not even clear that in this case the ansatz will yield the correct value of the free energy. Furthermore, the non-chaotic solutions have the property that the overlaps of the two systems are essentially independent for values larger than the constraint because $p_d = p(1)$, while this is not the case in presence of chaos because $p_d \neq p(1)$ (see also the discussion in [7]), and we feel that this poses also conceptual problems to the interpretation of the solutions. The same conceptual problems are also present for different types of perturbations (*e.g.*, magnetic field and bond chaos) where, as we saw in the last section, chaos can be assessed easily on the basis of the random nature of the perturbation.

On the other hand another set of problems connected with the study of coupled systems with different external parameters has been pointed out by M. Talagrand [14]. Using Guerra's techniques one can show that the variational expression of the constraint free energy expressed in terms of Parisi matrices is a lower bound to the actual value. On the other hand the actual value of the constraint free energy must be larger than the sum of the free energy of the two systems separately, and if bound is sharp, as in the case of a single system, one should be able to show *a priori* that the maximum of the variational constrained free energy is larger or equal to the sum of the maxima of the variational expression of the two systems. The proof of this statement has eluded all efforts up to now. The importance of such a proof relies in the fact that it could possibly pave the way to prove that the resulting bound is sharp. Instead, if one finds that the maximum of the variational free energy is smaller than the sum of the free energies of the free systems, it should be concluded that the bound is not sharp and we should look for a better one. However, in all cases studied up to now the free energy cost was never negative and the question is whether the true value is larger than the one obtained; we feel that especially in the case of chaotic full-RSB systems the standard ansatz has some conceptual drawbacks but it is not at all clear how to amend them. The same kind of problems appears if one tries to prove ultrametricity considering three copies of the same systems with the same external fields and constraints that violate ultrametricity.

Finally let us mention that recent research on these problem has obtained a surprising result [15]. It is often assumed in the physics literature that the possible values of the overlap between two replicas are those for which the constrained free energy cost is zero. The spherical model with $p = 2$ provides an important counterexample which possibly makes things even more complicate. Indeed, while the support of the overlap of two copies of the same systems at the same temperature T is concentrated on $q_{\mathrm{EA}} = 1 - T$, it was shown in [15] that the free energy cost when constraining two copies at overlap q_c is zero for all $q_c < q_{\mathrm{EA}}$. Therefore it is not true in general that the support of the overlap distribution is given by all the values for which there is no extensive free energy cost. Actually the problem is present also considering the ultrametricity problem. Indeed it was shown in [15] that considering n constrained replicas at the same temperature there is no free energy cost provided the constraint matrix q_c^{ij} with $q_c^{ii} = 1$ is such that $Q_c - TI$ is nonnegative definite. We note that the problem is even more obnoxious. Indeed, even if the replicas have *different* temperatures T_i one can exhibit a RS bound that yields no free energy cost provided the matrix $q_c^{ij} - \delta_{ij} T_i$ is nonnegative definite.

References

[1] The problem of chaos has attracted much attention and has been studied by various approaches including: scaling arguments and real space renormalization group analysis A.J. Bray and M.A. Moore, Phys. Rev. Lett. 58, 57 (1987). D.S. Fisher and D.A. Huse, Phys. Rev. B 38, 386 (1988). J.R. Banavar and A.J. Bray, Phys. Rev. B 35, 8888 (1987). M. Nifle and H.J. Hilhorst, Phys. Rev. Lett 68, 2992 (1992); analytical and numerical studies on the mean field models, A. Billoire and B. Coluzzi Phys. Rev. E 67, 036108 (2003); A. Billoire and E. Marinari, cond-mat/020247; T. Rizzo and A. Crisanti, Phys. Rev. Lett. 90, 137201 (2003); numerical studies on finite-dimensional Edwards–Andersion models, F. Ritort, Phys. Rev. B 50, 6844 (1994). M. Ney-Nifle, Phys. Rev. B 57, 492 (1997). D.A. Huse and L-F. Ko, Phys. Rev. B 56, 14597 (1997). A. Billoire and E. Marinari, J. Phys. A 33, L265 (2000). T. Aspelmeier, A.J. Bray, and M.A. Moore, Phys. Rev. Lett. 89, 197202 (2002). M. Sasaki, K. Hukushima, H. Yoshino, H. Takayama, cond-mat0411138; analytical and numerical studies on elastic manifolds in random media, D.S. Fisher and D.A. Huse, Phys.

Rev. B 43, 10728 (1991). M. Sales and H. Yoshino Phs. Rev. E 65 066131 (2002). A. da Silveira and J.P. Bouchaud, Phys. Rev. Lett. 93, 015901 (2004). Pierre Le Doussal, cond-mat/0505679. It has been proposed that the chaos effect is a possible mechanism of the so-called rejuvenation effect found experimentally. See for example, P.E. Jonsson, R. Mathieu, P. Nordblad, H. Yoshino, H. Aruga Katori, A. Ito, Phys. Rev. B 70, 174402 (2004) and references there in. The problem of the temperature dependence of the Gibbs measure of the Random Energy Model has been also investigated both in the physics literature (M. Sales and J.-P. Bouchaud, Europhys. Lett. 56, 181 (2001)) and in the mathematical physics literature, I. Kurkova, J. Stat. Phys. 111, 35, (2003).

[2] M. Mézard, G. Parisi, and M.A. Virasoro, "Spin glass theory and beyond", World Scientific (Singapore 1987).

[3] David J. Earl and Michael W. Deem, Phys. Chem. Chem. Phys., 2005, 7, 3910.

[4] T. Rizzo, Europhys. Jour. B 29, 425 (2002).

[5] T. Rizzo and A. Crisanti, Phys. Rev. Lett. 90, 137201 (2003).

[6] H. Yoshino and T. Rizzo, Phys. Rev. B 73, 064416 (2006), Phys. Rev. B 77, 104429 (2008).

[7] T. Rizzo, J. Phys. A: Math. Gen. **34**, 5531–5549 (2001).

[8] S. Franz and M. Ney-Nifle, J. Phys. A: Math. Gen. **26**, L641 (1993).

[9] I. Kondor, J. Phys. A 22 (1989) L163.

[10] S. Franz, G. Parisi, M.A. Virasoro, J.Physique I **2** (1992) 1969.

[11] T. Nieuwenhuizen, Phys. Rev. Lett. **74** 4289 (1995).

[12] T. Aspelmeier, J. Phys. A: Math. Theor. 41 (2008) 205005, J. Stat. Mech. (2008) P04018, Phys. Rev. Lett. 100:117205, 2008.

[13] M. Talagrand, The Parisi formula. Annals of Mathematics 163, no 1, 2006, 221–263.

[14] M. Talagrand, Lecture Notes in Math., 1900, Springer, Berlin, 2007, 63–80.

[15] D. Panchenko and M. Talagrand, Ann. Probab. 35, 2007, no 6, 2321–2355.

[16] S. Chatterjeee, Chaos, concentration, and multiple valleys, arXiv:0810.4221 (2008).

Tommaso Rizzo
"E. Fermi" Center,
Via Panisperna 89 A
Compendio Viminale
I-00184, Roma, Italy

and

Dipartimento di Fisica
Università di Roma "La Sapienza"
P.le Aldo Moro 2
I-00185 Roma, Italy
e-mail: `tommaso.rizzo@inwind.it`

Progress in Probability, Vol. 62, 159–176

A non Gaussian Limit Law for the Covariances of Spins in a SK Model with an External Field

Albert Hanen

Abstract. We give an overview of the main steps of the proof of a non gaussian limit theorem for the covariance, for Gibbs' measure, of spins at two fixed sites in a SK model with an external field.

Mathematics Subject Classification (2000). 82D30, 60G15, 60F05.

Keywords. SK models, Gibbs' measure, exponential inequalities, C.L.T.

1. Introduction

Talagrand, in his book [5, Chapter 2], after Sherrington and Kirkpatrick's work [4], studied the following model:

Let N be a positive integer, β and $h \in \mathbb{R}^+$, $(g_{i,j})_{1\leq i<j}$ a set of i.i.d. Gaussian standard random variables, defining the disorder of the model. Let

$$\sigma = (\sigma_i)_{i=1}^N \in \{-1,+1\}^N \stackrel{\text{def}}{=} \Sigma_N.$$

The quantity σ_i is called the spin at site i of the configuration σ. We will also set

$$\rho = (\sigma_i)_{i=1}^{N-1} \in \Sigma_{N-1},\ \varepsilon = \sigma_N,\ \sigma = (\rho,\varepsilon).$$

Let us consider the Hamiltonian of the SK model with an external field h, i.e., the quantity

$$-H_N(\sigma,\beta,h) = \frac{\beta}{\sqrt{N}} \sum_{1\leq i<j\leq N} g_{i,j}\sigma_i\sigma_j + h\Big(\sum_{i\leq N}\sigma_i\Big). \tag{1}$$

The probability on Σ_N, denoted by $G_N(\beta,h)$ or G_N, associated to the weight function $\exp(-H_N(\sigma))$, $\sigma\in\Sigma_N$, is the Gibbs' measure of the model. If f is a numerical function defined on Σ_N^n, the integral of f with respect to the product measure G_N^n is denoted by $\langle f\rangle$, the expectation (with respect to the disorder of the model) of this integral is denoted by $\nu(f)$.

An element of Σ_N^n is a sequence of n configurations or replicas

$$(\sigma^1, \sigma^2, \dots, \sigma^n).$$

While G_N^n and ν are invariant by permutations of replicas, ν is also invariant by permutations of sites (this last property is called symmetry between sites).

This model has been extensively studied, specially with a high temperature condition such as β being sufficiently small ($\beta \leq \beta_0$). In particular, it is proved in [5, Theorem 2.4.10] that the square of the total variation distance between the law (for Gibbs' measure) of the sequence of spins at the first k sites and the product of their marginal laws has an expectation that decreases like $1/N$ and then the spins at a fixed set of sites are asymptotically independent in that sense.

It is then natural to study the covariance, for Gibbs' measure, denoted by $\gamma_{i,j}$, of spins at two fixed sites i and j, which must be small when taking expectation with respect to the disorder, and to look for a limit theorem when $N \to +\infty$.

Talagrand found ([5, Corollary 2.6.2], for β sufficiently small, the limit of the second-order moment of $\sqrt{N}\gamma_{i,j}$ if $N \to +\infty$.

In [2] the author computed, for each integer p, the limit of the p-order moment of this random variable, and discovered the unexpected fact that this random variable has a (generally) non Gaussian limit law.

Let

$$Y = \beta z \sqrt{q} + h, \tag{2}$$

where q satisfies the relation

$$\mathbb{E}(th^2(Y)) = q,$$

and where z is a standard Gaussian random variable. It is known ([5] Proposition 2.4.8) that q exists and is unique if $h > 0$. We will set

$$U = 1 - th^2(Y).$$

Let z_1 and z_2 be two independent standard Gaussian random variables, also independent of z, generating in the same way variables denoted by U_1 et U_2, which are then independent with the same law, that of U. In [2], we proved the following:

Theorem 1.1. *For β sufficiently small, the p-order moments of $\sqrt{N}\gamma_{i,j}$ converge, if $N \to +\infty$, towards those of the random variable:*

$$\frac{\beta}{\sqrt{1 - \beta^2\,\mathbb{E}(U^2)}}\, z U_1 U_2.$$

Remark 1.1. When $p = 2$, this theorem gives the limit found by Talagrand

$$\sigma^2 \stackrel{\text{def}}{=} \frac{\beta^2}{1 - \beta^2\,\mathbb{E}(U^2)}\,\mathbb{E}^2(U^2).$$

On the other hand, the limit law found is centered but non Gaussian, its moments of order $2l$ not being equal to $\sigma^{2l}\,\mathbb{E}(z^{2l})$. However, if $h = 0$ and $\beta < 1$, we have $q = 0$, $U = 1$. In this case the limit law is Gaussian.

The goal of the present paper is to give an overview of the main steps of the proof of Theorem 1.1, omitting many technicalities and secondary details. After a first study of the moments of p-order of the covariance, we will describe in Sections 3 and 4 some of the main tools used in [2]. We will then introduce a kind of heuristics and we will describe how it leads in a natural way to the computation of the limit in Theorem 1.1 for the second-order moment. The last part, more involved, will study in a similar spirit the general case of p-order moments.

We hope that this work would stimulate the reader to study this kind of problematic and would help the most courageous one to go inside the complete and sometimes delicate proofs contained in article [2] and omitted here.

2. Moments of p-order of the covariance: a first study

We know that the covariance of spins at sites i and j is given by

$$\gamma_{i,j} = \langle \sigma_i \sigma_j \rangle - \langle \sigma_i \rangle \langle \sigma_j \rangle.$$

Let $\tilde{\sigma} = \sigma^1 - \sigma^2$ be the difference of two replicas σ^1 and σ^2 ($\tilde{\sigma}_N = \varepsilon^1 - \varepsilon^2 = \tilde{\varepsilon}$).

Definition 2.1. Let us call symmetrized covariance of spins at sites i and j the quantity

$$\tilde{\gamma}_{i,j} = \langle \tilde{\sigma}_i \tilde{\sigma}_j \rangle.$$

It is easily seen that

$$\gamma_{i,j} = \frac{1}{2} \tilde{\gamma}_{i,j} \,. \tag{3}$$

Then it will be sufficient to study the p-order moments of $\tilde{\gamma}_{i,j}$, which are independent of the choice of i and j.

Let us consider a sequence of $2p$ replicas

$$(\sigma^1, \ldots, \sigma^{2r-1},\, \sigma^{2r}, \ldots,\, \sigma^{2p}),$$

and let us set

$$\tilde{\sigma}^r = \sigma^{2r-1} - \sigma^{2r}, \quad \text{for all } r \in [1,\, 2,\, \ldots,\, p], \tag{4}$$

$$\tilde{\sigma}_i^{\otimes p} = \prod_{r=1}^{r=p} \tilde{\sigma}_i^r, \tag{5}$$

$$\tilde{\sigma}_N^{\otimes p} = \tilde{\varepsilon}^{\otimes p}. \tag{6}$$

Using replicas and taking $i = 1$, $j = N$, we get, using also symmetry between sites:

Theorem 2.1. *If* $f^- = \tilde{\sigma}_1^{\otimes p}$, *then*

$$\mathbb{E}(\tilde{\gamma}_{i,j}^p) = \mathbb{E}(\langle \tilde{\sigma}_1 \tilde{\sigma}_N \rangle^p) = \nu(f^- \tilde{\varepsilon}^{\otimes p}). \tag{7}$$

So we have to study an integral, with respect to ν, of a product of two functions depending respectively only of the spins at the first and at the last sites. To study that kind of integral, it is natural to use the "smart path" method introduced by Talagrand ([5, Chapter 2]).

3. The smart path method

Talagrand introduced a continuous family of Hamiltonians

$$- H_{N,t}, \quad t \in [0,1],$$
$$\text{with } - H_{N,1} = -H_N.$$

In the same way as in the case $t = 1$, this family induces Gibbs' measures $G_{N,t}$ and integrals $\langle f \rangle_t$ having expectations $\nu_t(f)$. The disorder of $-H_{N,t}$ depends on the disorder of $-H_N$ and on an independent Gaussian standard random variable z.

3.1. The case $t = 0$

Let us consider the variable Y associated to z defined in (2). The spin $\sigma_N \stackrel{\text{def}}{=} \varepsilon$ verifies $\langle \varepsilon \rangle_0 = th(Y)$ and is independent of the spins at the $N-1$ first sites, which have the law $G_{N-1}(\beta\sqrt{\frac{N-1}{N}}, h)$. This law induces integrals denoted by $\langle \cdot \rangle_-$ having expectation $\nu_-(\cdot)$.

Then if a numerical function f of n replicas verifies $f = f^- h_0$, where f^- only depends on the spins of these replicas at the first $N-1$ sites and h_0 only depends on their spins at site N, we have the relations:

$$\langle f \rangle_0 = \langle f^- \rangle_0 \langle h_0 \rangle_0 = \langle f^- \rangle_- \langle h_0 \rangle_0, \tag{8}$$
$$\nu_0(f) = \nu_0(f^-)\nu_0(h_0) = \nu_-(f^-)\nu_0(h_0)\,. \tag{9}$$

3.2. Derivatives of $\nu_t(f)$

Definition 3.1. Let us call, for u and $v \in \mathbb{R}^N$,

$$R(u,v) = \frac{1}{N}\sum u_i v_i \stackrel{\text{def}}{=} R^-(u,v) + \frac{u_N v_N}{N}\,. \tag{10}$$

- When u and v are in Σ_N, R (resp. R^-) is called the "overlap" (resp. the "M-overlap") of these configurations.
- When u and v are differences of elements of Σ_N, R (resp. R^-) is called a "symmetrized overlap" (resp. a "M-symmetrized overlap").

Let us introduce some objects associated to overlaps. They will be used to compute the derivatives of $\nu_t(f)$. If $J = \{r,s\}, r < s$ is a pair of integers indexing replicas σ^r and σ^s, we set

$$R_J = R(\sigma^r, \sigma^s) = R_{r,s},$$
$$\dot{R}_J = R_{r,s} - q \stackrel{\text{def}}{=} \dot{R}_J^- + \frac{\varepsilon^r \varepsilon^s}{N} \stackrel{\text{def}}{=} \dot{R}_J^- + \frac{\varepsilon^J}{N}\,.$$

The quantity $\dot{R}_J$ (respectively $\dot{R}_J^-$) is called a "recentered overlap" (respectively a "recentered M-overlap").

Let $l \geq 1$ be an integer and let us introduce also:

$$J_i = \{r_i, s_i\} \text{ a pair of integers} \in [1, 2, \ldots, n+2l], \quad \text{defined for each integer } i \leq l, \tag{11}$$

$$\hat{J} = (J_1, J_2, \ldots, J_i, \ldots J_l), \tag{12}$$

$$c_{\hat{J}}(n) \text{ some constant } (= 1 \text{ if for all } i, J_i \subset [1, 2, \ldots, n]), \tag{13}$$

$$T_{\hat{J},t}(f) = \nu_t\Big(f \prod_{i=1}^{i=l} \varepsilon^{J_i} \dot{R}^-_{J_i}\Big). \tag{14}$$

Theorem 3.1. *Let f be a numerical function depending on n replicas. Then $\nu_t(f)$ has derivatives of every order l, given by*

$$\nu_t^{(l)}(f) = \beta^{2l} \sum_{\hat{J}} c_{\hat{J}}(n)\nu_t\left(f \prod_{i=1}^{i=l}\left(\varepsilon^{J_i}\dot{R}^-_{J_i}\right)\right), \tag{15}$$

$$= \beta^{2l} \sum_{\hat{J}} c_{\hat{J}}(n) T_{\hat{J},t}(f). \tag{16}$$

This theorem has been proved in [5, Proposition 2.4.5] for $l = 1$ and in [2, Theorem 2.3] for the general case.

Remark 3.1. When $f = f^- h_0$, such as in equations (8) (or (7)), and $t = 0$, we have

$$T_{\hat{J},0}(f) = \nu_0\left(h_0 \prod_{i=1}^{i=l}\left(\varepsilon^{J_i}\right)\right)\nu_-\left(f^- \prod_{i=1}^{i=l} \dot{R}^-_{J_i}\right) \stackrel{\text{def}}{=} T_{1,\hat{J}}(h_0) T_{2,\hat{J}}(f^-). \tag{17}$$

To pursue the evaluation of the p-order moment of the symmetrized covariance, it is natural, considering equation (7) and Theorem 3.1, to try a Taylor expansion of $\nu_t(f^- \tilde{\varepsilon}^{\otimes p})$, with $f^- = \tilde{\sigma}_1^{\otimes p}$, for $t = 1$, at $t = 0$, up to some order d_0.

It is then necessary to specify what is meant here by asymptotical evaluation and to recall some known results.

4. Asymptotic evaluations

It has been proved ([5, Proposition 2.4.6] that there exists a number, depending only on β and n, denoted by $K_1(n, \beta)$, such that, f being a positive numerical function of n replicas, we have:

$$\nu_t(f) \leq K_1(n, \beta)\nu(f). \tag{18}$$

We may omit the dependence on β if $\beta \leq \beta_0$.

It also has been proved in [5, Section 2.5] that there exists a number β_0 such that the recentered overlaps $\dot{R}_{1,2} = R(\sigma^1, \sigma^2) - q$ and also the recentered M-overlaps $\dot{R}^-_{1,2} = R^-(\sigma^1, \sigma^2) - q$ have absolute moments of order j bounded by

$(L(j+1)/N)^{j/2}$, for some number L, when $\beta \leq \beta_0$. These inequalities are also called "exponential inequalities".

Definition 4.1. A numerical expression $F(N,\beta,\theta)$, defined for $\beta \leq \beta_0$, is said to verify the relation

$$F(N,\beta,\theta) = O(k),$$

if there exists a constant K, not depending on β and on N, but possibly depending on the parameter θ, such that

$$|F(N,\beta,\theta)| \leq \frac{K}{N^{\frac{k}{2}}}.$$

Using this notation we get the following weaker form of exponential inequalities:

$$\nu(|R_{1,2} - q|^j) = O(j). \tag{19}$$

We have the same relation if we replace R by R^-. We deduce from these relations the following

Theorem 4.1. *Let f be a numerical function of n replicas. Then we have*

$$\nu_t^{(l)}(f) = O(l)\nu^{1/2}(f^2). \tag{20}$$

The proof uses the exponential inequalities as given in equation (19), Hölder's inequality and Schwarz inequality applied to each term $T_{\hat{J},t}(f)$ in equation (15), and relation (18).

Remark 4.1. Let us suppose that $f = f^- \tilde{\varepsilon}^{\otimes p}$, where f^- is bounded (e.g., if $f^- = \tilde{\sigma}_1^{\otimes p}$). If $\beta \leq \beta_0$, the Taylor expansion of $\nu_t(f)$ for $t=1$ at $t=0$ up to the order p has a remainder term which is an $O(p+1)$. If we want to find the limit of $\sqrt{N}^p \nu_1(f)$ when $N \to +\infty$, it will then be sufficient to find the limit of

$$\sqrt{N}^p \nu_0^{(l)}(f), \text{ for all } l \in [0,p].$$

5. The analogy rule

5.1. Some heuristics

Let us give now some definition:

Definition 5.1. Let g and g^- be two numerical functions of n replicas, such that g^- depends only on their spins at the $N-1$ first sites, while g may depend on the spins at all N sites.

If there exists a function $C(\beta,h)$ (possibly $=0$), not depending on N, such that for some integer k we have:

$$\nu(g) \text{ (and also } \nu_-(g^-)) = \frac{C(\beta,h)}{\sqrt{N}^k} + O(k+1),$$

we will say that $\nu(g)$ (and also $\nu_-(g^-)$) have the same asymptotic behavior at order k, or equivalently that the *analogy rule* is valid for the pair (g,g^-) at this same order.

Remark 5.1. In this case, we have:

$$\begin{aligned}\nu(g) \text{ (and also } \nu_-(g^-)) &= O(k) = O(r) \text{ if } r \leq k,\\ &= O(k+1) \qquad \text{if } C(\beta,h) = 0.\end{aligned}$$

Then the *analogy rule* is also valid for the pair (g, g^-) at any order $r \leq k$.

If the last condition $C(\beta, h) = 0$ was not allowed in the definition of the *analogy rule* given above, k would be unique.

We will use this rule for particular pairs (g, g^-) such that g is some "algebraic analog" of g^- for N sites, often built by replacing each M-recentered overlap $\dot{R}_J^-$ appearing in g^- by $\dot{R}_J$ (or each symmetrized M-overlap appearing in g^- by the corresponding one for N sites).

If we want to study $\nu_-(g^-)$, it will often be simpler to study instead $\nu(g)$. The asymptotic study of $\nu(g)$ will generally give, often with difficulty, its asymptotic behavior at a convenient order k. As it seems in many cases difficult to prove that $\nu_-(g^-)$ has the same asymptotic behavior at order k (or equivalently that $\nu(g) - \nu_-(g^-) = O(k+1)$), we will assume *here* without proofs that the *analogy rule* is valid for each of these particular pairs (g, g^-) at an order k (depending on g) given by the asymptotic study of $\nu(g)$: it's our *heuristics*.

These assumptions will allow us to focus on the logic of some important computations made in [2], where complete and rigorous proofs are given, including direct or indirect proofs of the assumptions mentioned above.

However, let us give some examples where the study of the difference

$$\nu(g) - \nu_-(g^-)$$

is easy.

Let us suppose that $\nu^{1/2}(g^2) = O(k)$ and that $\nu(g)$ has a known asymptotic behavior at order k. We may write the following relation:

$$\nu(g) - \nu_-(g^-) = \nu(g) - \nu_0(g^-) = \nu(g) - \nu_0(g) + \nu_0(g - g^-). \tag{21}$$

We know, using Taylor's formula at order $l = 0$, that the difference

$$\nu(g) - \nu_0(g) = O(1)\nu^{1/2}(g^2) = O(k+1).$$

If the second term of the last equality in (21) is also evaluated as $O(k+1)$, then $\nu(g)$ and $\nu_-(g^-)$ have the same asymptotic behavior at order k.

Using exponential inequalities (19) and [5, Lemma 2.5.4] this method works for the pair

$$g^- = \dot{R}_J^-, \quad g = \dot{R}_J, \quad \text{with } k = 1,\ C(\beta, h) = 0.$$

It also works for the pair

$$g^- = (\dot{R}_J^-)^2, \quad g = \dot{R}_J^2, \quad \text{with } k = 2.$$

In this case, $C(\beta, h)$ is a known function given in [5, Corollary 2.6.10].

This direct method described above fails, e.g., if $\nu(g)$ has an asymptotic behavior at some order k such that we are only able to prove that

$$\nu^{1/2}(g^2) = O(r), \text{ with } r < k.$$

As a first example, let us consider, using relation (5), the case, found in relations (32) and (33) given in the next subsection, of the pair

$$g^- = \tilde{\sigma}_1^1 \tilde{\sigma}_1^2 R^-(\tilde{\sigma}^1, \tilde{\sigma}^2),$$
$$g = \tilde{\sigma}_1^1 \tilde{\sigma}_1^2 R(\tilde{\sigma}^1, \tilde{\sigma}^2).$$

It is easy to see that:

$$g - g^- = \frac{1}{N} \tilde{\sigma}_1^1 \tilde{\sigma}_1^2 \tilde{\varepsilon}^1 \tilde{\varepsilon}^2.$$

We have $\nu_0(\tilde{\varepsilon}^1 \tilde{\varepsilon}^2) = 0$ so that, using relation (9), we have $\nu_0(g - g^-) = 0$. In the next subsection we will find the asymptotic behavior at order $k = 2$ of $\nu(g)$. We then have to prove that

$$\nu(g) - \nu_0(g) = O(3). \tag{22}$$

However, we are only able, using the majoration of the terms $\tilde{\sigma}_1^1 \tilde{\sigma}_1^2$ and exponential inequalities (19) to prove that $\nu^{1/2}(g^2) = O(1)$, so the direct method fails. Proving relation (22) would need a Taylor expansion at order $l = 1$ of $\nu(g)$, i.e., a precise computation of $\nu_0'(g)$ and a proof that $\nu_0'(g) = O(3)$.

In this case (as in other more complicated cases where symmetrized M-overlaps appear), it is more convenient, using the known asymptotic behavior of $\nu(g)$ for $N-1$ (instead of N) sites and for $\beta\sqrt{\frac{N-1}{N}}$ instead of β, to directly study $\nu_-(g^-)$.

Let us give another example and let us set, using relation (5):

$$h_2^- = \prod_{i=1}^{2} (\dot{R}_{J_i}^-), \quad h_2 = \prod_{i=1}^{2} (\dot{R}_{J_i}),$$
$$g^- = \tilde{\sigma}_1^{\otimes 3} h_2^-, \qquad g = \tilde{\sigma}_1^{\otimes 3} h_2.$$

We will prove below (see the proof of Theorem 7.1 when $p = 3$), that $\nu(g) = O(4)$, so it has an asymptotic behavior at order $k = 3$ in our sense. Using that the quantities $\tilde{\sigma}_1^{\otimes 3}$ are bounded, we may only prove that $\nu^{1/2}(g^2) = O(2) (\neq O(3))$, and we are unable without further investigation to prove or disprove the relation $\nu_0(g - g^-) = O(4)$, so we can't prove by this method that $\nu_-(g^-) = O(4)$.

We then would also have to directly study $\nu_-(g^-)$ and to adapt to this case the proof that $\nu(g) = O(4)$.

We will study in the next subsection how the *analogy rule* may work in the case $p = 2$ and gives Theorem 1.1 in this case.

5.2. The case $p = 2$

Let us study the case of the second-order moment of the symmetrized covariance, given by the relation (7). By taking four replicas we get:

$$\mathbb{E}(\tilde{\gamma}_{i,j}^2) = \nu(f^- h_0) \quad (\text{where } f^- = \tilde{\sigma}_1^{\otimes 2},\ h_0 = \tilde{\varepsilon}^{\otimes 2}), \tag{23}$$
$$\stackrel{\text{def}}{=} \nu(f), \tag{24}$$
$$= \nu((\sigma_1^1 - \sigma_1^2)(\sigma_1^3 - \sigma_1^4)(\varepsilon^1 - \varepsilon^2)(\varepsilon^3 - \varepsilon^4)). \tag{25}$$

Using now Remark 4.1, we only have to study

$$\nu_0(f),\ \nu_0'(f),\ \nu_0''(f).$$

We will find that $\nu_0(f) = 0$ and that, when $N \to +\infty$,

$$N\nu_0'(f) \xrightarrow[N\to\infty]{} 4\frac{\beta^2 E^2(U^2)}{(1-\beta^2\,\mathbb{E}(U^2))}, \tag{26}$$

$$N\nu_0''(f) \xrightarrow[N\to\infty]{} 0. \tag{27}$$

So $N\,\mathbb{E}(\tilde{\gamma}_{i,j}^2)$ and $N\nu_0'(f)$ have the same limit when $N \to +\infty$. Applying this result to

$$N\,\mathbb{E}(\gamma_{i,j}^2) = \frac{1}{4}N\,\mathbb{E}(\tilde{\gamma}_{i,j}^2),$$

we get Theorem 1.1 for $p = 2$.

Let us begin to prove the results previously announced. It is clear that, using independence of replicas under $G_{N,0}$ and the fact that, for all j,

$$\langle \varepsilon^j \rangle_0 = th(Y),$$

we have

$$\langle h_0 \rangle_0 = \langle (\varepsilon^1 - \varepsilon^2) \rangle_0 \langle (\varepsilon^3 - \varepsilon^4) \rangle_0 = 0.$$

Then, using relations (9), we have

$$\nu_0(f) = \nu_0(h_0)\nu_-(f^-) = 0.$$

Let us now study $\nu_0'(f)$. Using equations (15) and (17) with $l = 1$ and $n = 4$, we get:

$$\nu_0'(f) = \beta^2 \sum_{J=(r,s)\subset[1,\dots,6]} c_J(4)\nu_0((\varepsilon^1-\varepsilon^2)(\varepsilon^3-\varepsilon^4)\varepsilon^r\varepsilon^s)\nu_-(f^-\dot{R}_J^-). \tag{28}$$

It is easy to prove that we have

$$\langle (\varepsilon^1-\varepsilon^2)(\varepsilon^3-\varepsilon^4)\varepsilon^r\varepsilon^s \rangle_0 = 0, \text{ except if } r \in \{1,2\},\ s \in \{3,4\}.$$

In this last case, we will find:

$$\langle (\varepsilon^1-\varepsilon^2)(\varepsilon^3-\varepsilon^4)\varepsilon^r\varepsilon^s \rangle_0 = (-1)^{r+s}U^2.$$

We also have in this case $c_J(4) = 1$, and

$$\nu_0'(f) = \beta^2\,\mathbb{E}(U^2) \sum_{\substack{r\in\{1,2\}\\ s\in\{3,4\}}} \nu_-(f^-(-1)^{r+s}\dot{R}_{r,s}^-). \tag{29}$$

If we set $\tilde{\sigma}^1 = \sigma^1 - \sigma^2$, $\tilde{\sigma}^2 = \sigma^3 - \sigma^4$, we get:

$$\nu_0'(f) = \beta^2\,\mathbb{E}(U^2)\nu_-(\tilde{\sigma}_1^1\tilde{\sigma}_1^2 R^-(\tilde{\sigma}^1,\tilde{\sigma}^2)). \tag{30}$$

Let us try to apply the analogy rule described in Definition 5.1 to the pair

$$g^- = \tilde{\sigma}_1^1\tilde{\sigma}_1^2 R^-(\tilde{\sigma}^1,\tilde{\sigma}^2), \quad g = \tilde{\sigma}_1^1\tilde{\sigma}_1^2 R(\tilde{\sigma}^1,\tilde{\sigma}^2). \tag{31}$$

We then have to study the asymptotic behavior of $\nu(g)$. Using first symmetry between sites, then taking an arithmetical mean, we have, for all $i \in [1,\ldots, N]$,

$$\nu(g) = \nu(\tilde{\sigma}_i^1\tilde{\sigma}_i^2 R(\tilde{\sigma}^1,\tilde{\sigma}^2)) = \nu(R^2(\tilde{\sigma}^1,\tilde{\sigma}^2)). \tag{32}$$

It has been proved in [5, Proposition 2.6.1] that if $\beta \le \beta_0$, we have:

$$\nu(R^2(\tilde{\sigma}^1,\tilde{\sigma}^2)) = 4\frac{\mathbb{E}(U^2)}{N(1-\beta^2\,\mathbb{E}(U^2))} + O(3). \tag{33}$$

We then have, using the analogy rule, assumed to be valid at the order $k = 2$ found above and equation (30):

$$\nu_0'(f) = 4\frac{\beta^2 E^2(U^2)}{N(1-\beta^2\,\mathbb{E}(U^2))} + O(3). \tag{34}$$

We easily deduce from (34) the result announced in (26).

Let us now study $\nu_0''(f)$. Using equations (15) and (17) with $l = 2, n = 4$, and setting, as usual $\hat{J} = (J_1, J_2)$, where J_i is a pair of integers of the interval $[1,\ldots, 8]$, it is easily found that each term

$$T_{1,\hat{J}}(h_0) = T_{1,\hat{J}}(\tilde{\varepsilon}^1\tilde{\varepsilon}^2) = \nu_0(\tilde{\varepsilon}^1\tilde{\varepsilon}^2\varepsilon^{J_1}\varepsilon^{J_2})$$

is bounded by 4.

Then let us study the terms

$$T_{2,\hat{J}}(f^-) = \nu_-(\tilde{\sigma}_1^1\tilde{\sigma}_1^2\dot{R}_{J_1}^-\dot{R}_{J_2}^-) \stackrel{\text{def}}{=} \nu_-(g^-).$$

Let us take $g = \tilde{\sigma}_1^1\tilde{\sigma}_1^2\dot{R}_{J_1}\dot{R}_{J_2}$ as an analog of g^- for N sites and let us try to apply the analogy rule to the pair (g, g^-). We have

$$\begin{aligned}
\nu(g) &= \nu(\tilde{\sigma}_1^1\tilde{\sigma}_1^2\dot{R}_{J_1}\dot{R}_{J_2}),\\
&= \nu(\tilde{\sigma}_i^1\tilde{\sigma}_i^2\dot{R}_{J_1}\dot{R}_{J_2}), \quad \text{for all } i \in [1,\ldots, N],\\
&= \nu(\tilde{\varepsilon}^1\tilde{\varepsilon}^2\dot{R}_{J_1}\dot{R}_{J_2}),\\
&= \nu\Big(\tilde{\varepsilon}^1\tilde{\varepsilon}^2\Big(\dot{R}_{J_1}^- + \frac{\varepsilon^{J_1}}{N}\Big)\Big(\dot{R}_{J_2}^- + \frac{\varepsilon^{J_2}}{N}\Big)\Big),\\
&= \nu(\tilde{\varepsilon}^1\tilde{\varepsilon}^2\dot{R}_{J_1}^-\dot{R}_{J_2}^-) + \frac{\nu(\tilde{\varepsilon}^1\tilde{\varepsilon}^2(\dot{R}_{J_1}^-\varepsilon^{J_2} + \dot{R}_{J_2}^-\varepsilon^{J_1}))}{N} + \frac{\nu(\tilde{\varepsilon}^1\tilde{\varepsilon}^2\varepsilon^{J_1}\varepsilon^{J_2})}{N^2}.
\end{aligned}$$

Let us evaluate each of the three terms obtained.

- The last term is bounded by $\frac{4}{N^2}$ and then is $O(4)$.
- The second term is bounded by $\frac{4}{N}(\nu(|\dot{R}_{J_1}|) + \nu(|\dot{R}_{J_2}|))$. Using exponential inequalities (19) with $j = 1$, we get that this second term is $O(3)$.
- The first term is such that

  $$\nu_0(\tilde{\varepsilon}^1\tilde{\varepsilon}^2\dot{R}_{J_1}^-\dot{R}_{J_2}^-) = \nu_0(\tilde{\varepsilon}^1\tilde{\varepsilon}^2)\nu_-(\dot{R}_{J_1}^-\dot{R}_{J_2}^-) = 0.$$

  This term is then, using [5, equation 2.248], an $O(1)\nu^{1/2}(\dot{R}_{J_1}^-\dot{R}_{J_2}^-)^2$. Using Schwarz and exponential inequalities, we find that the first term is also $O(3)$.

We then have

$$\nu(g) = O(3).$$

We may now apply the analogy rule, supposed to be valid (in our sense) at the order 2 found above, to the pair (g, g^-). We get that $T_{2,\hat{J}}(f^-) = O(3)$ and we then have

$$\nu_0''(f) = O(3). \tag{35}$$

Relation (27) is then proved.

6. Back to the general case

Going back to the general case for the p-order, let us study, according to Remark 4.1, the derivatives $\nu_0^{(l)}(f)$ when $f = f^- \tilde{\varepsilon}^{\otimes p}$, for any given l. This study will allow us to prove Theorem 1.1.

Equations (16) and (17) lead us to study the quantities $T_{1,\hat{J}}(\tilde{\varepsilon}^{\otimes p})$ and $T_{2,\hat{J}}(\tilde{\sigma}_1^{\otimes p})$ for any given sequence $\hat{J} = (J_1, \ldots, J_i, \ldots, J_l)$.

6.1. An overview

Let us briefly describe the contents of the next subsection and also of the following sections:

Subsection 6.2 is devoted to the study of $T_{1,\hat{J}}(\tilde{\varepsilon}^{\otimes p})$, for which, using combinatorial arguments, we give it's exact value.

Section 7 studies the derivatives $\nu_0^{(l)}(f)$ when $2l \neq p$. We mainly get the following:

$$\nu_0^{(l)}(\tilde{\varepsilon}^{\otimes p} f^-) = \begin{cases} 0 & \text{if } \ 2l < p\,, \\ O(p+1) & \text{if } \ 2l > p\,. \end{cases} \tag{36}$$

To get the last result above, we will have to study the quantities $T_{2,\hat{J}}(\tilde{\sigma}_1^{\otimes p})$.

So, as in the case $p = 2$, if we want to study the limit, when $N \to +\infty$, of each quantity $(\sqrt{N})^p \nu_0^{(l)}(f)$, this limit is 0 in each case, except when p is even ($p = 2l_0$) and $l = l_0$.

Section 8 finds the limit in this case and proves Theorem 1.1.

6.2. A combinatorial result

As previously announced, let us study the quantities $T_{1,\hat{J}}(\tilde{\varepsilon}^{\otimes p})$, for any p-order. We know that we have:

$$T_{1,\hat{J}}(\tilde{\varepsilon}^{\otimes p}) = \mathbb{E}\Big[\Big\langle \tilde{\varepsilon}^{\otimes p}\big(\prod_{i=1}^{i=l} \varepsilon^{J_i}\big)\Big\rangle_0\Big].$$

The symmetric difference $A \triangle B$ of two sets A and B is associative and we have the relation:

$$\prod_{i=1}^{i=l} \varepsilon^{J_i} = \varepsilon^{J_1 \triangle J_2 \triangle \cdots \triangle J_l} \stackrel{\text{def}}{=} \varepsilon^{B'}.$$

Definition 6.1. Let us set

- $\hat{B} = B' \cap [1, \ldots, 2p]$,
- $C = B' \cap [1, \ldots, 2p]^c$.
- $r(B) = \sum_{x \in B}(x+1)$.

Let us introduce now a class of subsets of the interval $[1, \ldots, 2p]$.

Definition 6.2. We denote by $\mathcal{C}$ the class of subsets $B \subset [1, \ldots, 2p]$ such that

$$|B| = p \text{ and } |B \cap \{2r-1, 2r\}| = 1 \text{ for all } r \in [1, \ldots, p]. \tag{37}$$

In this case, we have

$$B = (a_1, a_2, \ldots, a_r, \ldots, a_p),$$

where $a_r \in \{2r-1, 2r\}$ for all $r \in [1, 2, \ldots, p]$.

We then have

$$\left\langle \tilde{\varepsilon}^{\otimes p} \Big(\prod_{i=1}^{i=l} \varepsilon^{J_i} \Big) \right\rangle_0 = \langle \tilde{\varepsilon}^{\otimes p} \varepsilon^{B'} \rangle_0 = \langle \tilde{\varepsilon}^{\otimes p} \varepsilon^{\hat{B}} \rangle_0 \langle \varepsilon^C \rangle_0 = \langle \tilde{\varepsilon}^{\otimes p} \varepsilon^{\hat{B}} \rangle_0 th(Y)^{|C|}.$$

Using only combinatorial arguments, we may prove ([2, Proposition 3.6]) the following proposition, generalizing some results obtained when $p = 2$:

Proposition 6.1. *We have*

$$\langle \tilde{\varepsilon}^{\otimes p} \varepsilon^{\hat{B}} \rangle_0 = \begin{cases} 0 & \text{if } \hat{B} \notin \mathcal{C}, \\ (-1)^{r(\hat{B})} U^p & \text{if } \hat{B} \in \mathcal{C}. \end{cases} \tag{38}$$

Taking expectations, we get the following:

Corollary 6.1. *We have*

$$T_{1,\hat{J}}(\tilde{\varepsilon}^{\otimes p}) = \begin{cases} 0 & \text{if } \hat{B} \notin \mathcal{C}, \\ (-1)^{r(\hat{B})} \mathbb{E}(U^p th(Y)^{|C|}) & \text{if } \hat{B} \in \mathcal{C}. \end{cases} \tag{39}$$

7. Study of the derivatives $\nu_0^{(l)}(\tilde{\varepsilon}^{\otimes p} f^-)$ when $2l \neq p$

This section is divided in two subsections: in the first one, we study the cases $2l < p$, in the second one the cases $2l > p$.

7.1. The cases $2l < p$

Using our previous results, we are now able to prove the following:

Theorem 7.1. *When $2l < p$, we have $\nu_0^{(l)}(\tilde{\varepsilon}^{\otimes p} f^-) = 0$.*

Proof. In this case, for any sequence $\hat{J}$ of length l, we have

$$\mid \hat{B} \mid \leq \mid B' \mid \leq 2l < p, \text{ thus } \hat{B} \notin \mathcal{C}.$$

Using the first half of equation (39) and equation (17), we get:

$$T_{1,\hat{J}}(\tilde{\varepsilon}^{\otimes p}) = 0 \text{ and } T_{\hat{J},0}(f) = 0 \text{ for all } \hat{J}.$$

Using now equation (16) for $t = 0$, we get our theorem. □

Corollary 7.1. *When $p = 2l_0$ or $2l_0 - 1$, we have the relation:*

$$\nu(\tilde{\varepsilon}^{\otimes p} f^-) = O(l_0)\nu^{\frac{1}{2}}((f^-)^2).$$

Proof. When taking a Taylor expansion up to the order $l_0 - 1$, the terms of this expansion vanish, and the remainder term, using relation (20), is $O(l_0)\nu^{\frac{1}{2}}((f^-)^2)$. □

7.2. The cases $2l > p$

Let us study now the derivatives $\nu_0^{(l)}(\tilde{\varepsilon}^{\otimes p} f^-)$ when $2l > p$ and $f^- = \tilde{\sigma}_1^{\otimes p}$.

We will study for these cases the term $T_{2,\hat{J}}(\tilde{\sigma}_1^{\otimes p})$ appearing in the equation (17). Let us set:

$$h_l^- = \prod_{i=1}^{i=l}(\dot{R}_{J_i}^-), \quad h_l = \prod_{i=1}^{i=l}(\dot{R}_{J_i}).$$

Let us consider the following pair:

$$g^- = \tilde{\sigma}_1^{\otimes p} h_l^-, \quad g = \tilde{\sigma}_1^{\otimes p} h_l.$$

We know that we have

$$T_{2,\hat{J}}(\tilde{\sigma}_1^{\otimes p}) = \nu_-(g^-).$$

Let us try to apply the analogy rule to the pair (g, g^-) defined above, i.e., let us study the asymptotic behavior of $\nu(g)$. Using symmetry between sites, we have:

$$\begin{aligned}\nu(g) &= \nu(\tilde{\sigma}_i^{\otimes p} h_l), \text{ for all } i \in [1, \ldots, N], \\ &= \nu(\tilde{\varepsilon}^{\otimes p} h_l), \\ &= \nu\Big(\tilde{\varepsilon}^{\otimes p} \prod_{i=1}^{i=l}\Big(\dot{R}_{J_i}^- + \frac{\varepsilon^{J_i}}{N}\Big)\Big).\end{aligned}$$

By expanding the product, we get a sum of terms of the kind

$$S_k \stackrel{\text{def}}{=} \frac{\nu((\prod_{i=1}^{i=k} \varepsilon^{J_i})\tilde{\varepsilon}^{\otimes p} h_{l-k}^-)}{N^k}, \quad k \in [0, \ldots, l].$$

Let us set, as in Subsection 6.2:

$$\prod_{i=1}^{i=k}(\varepsilon^{J_i}) = \varepsilon^{B_k'}, \text{ where } |B_k'| \leq 2k.$$

- If $2k < p$, we could prove in [2, Section 3] analogues of Theorem 7.1 and Corollary 7.1, with $f^- = h^-_{l-k}$, replacing the condition $2l < p$ by $2l < p - 2k$ and l_0 by $l_0 - k$. So we get, for $2k < p$:

$$S_k = O(2k)O(l_0 - k)O(l - k) = O(l_0 + l).$$

- If $2k \geq p$, we also can get the same result. If $l_0 \geq \frac{p}{2}$ and $l > \frac{p}{2}$, we have $l_0 + l \geq p + 1$. Then, for all $k \in [0, \dots, l]$ we have

$$S_k = O(l_0 + l) = O(p + 1).$$

We then have

$$\nu(g) = O(p + 1).$$

We may then apply the analogy rule, supposed to be valid at order p, to the pair (g, g^-).

Using also the fact that $|T_{1,\hat{\jmath}}(\tilde{\varepsilon}^{\otimes p})| \leq 2^p$ and equations (16) and (17), we finally get the following result ([2, Theorem 4.4]), generalizing relation (35) obtained when $p = 2 = l$:

Proposition 7.1. *If $2l > p$, we have:*

$$\nu_0^{(l)}(\tilde{\varepsilon}^{\otimes p}\tilde{\sigma}_1^{\otimes p}) = O(p + 1). \tag{40}$$

We may deduce from this proposition the following

Theorem 7.2. *When p is odd,*

$$\sqrt{N}^p E(\gamma_{i,j}^p) \xrightarrow[N \to \infty]{} 0. \tag{41}$$

Proof. We know that:

$$\sqrt{N}^p E(\gamma_{i,j}^p) = \sqrt{N}^p \frac{1}{2^p} \nu(\tilde{\varepsilon}^{\otimes p}\tilde{\sigma}_1^{\otimes p}).$$

Using Taylor's formula up to the order p to calculate $\nu(\tilde{\varepsilon}^{\otimes p}\tilde{\sigma}_1^{\otimes p})$, we know, using Theorem 7.1 and equation (40), that the l-order derivatives at $t = 0$ of $\nu_t(\tilde{\varepsilon}^{\otimes p}\tilde{\sigma}_1^{\otimes p})$ for $t = 1$ satisfy the following relation:

$$\nu_0^{(l)}(\tilde{\varepsilon}^{\otimes p}\tilde{\sigma}_1^{\otimes p}) = \begin{cases} 0 & \text{if } 2l < p\,, \\ O(p+1) & \text{if } 2l > p\,. \end{cases} \tag{42}$$

We can't have $2l = p$ if p is odd.

As the remainder term is also an $O(p + 1)$, knowing that

$$\sqrt{N}^p O(p + 1)) \xrightarrow[N \to \infty]{} 0,$$

we get the result. □

8. The case $p = 2l_0$

Using the previous results, we see that the derivatives

$$\nu_0^{(l)}(\tilde{\varepsilon}^{\otimes 2l_0}\tilde{\sigma}_1^{\otimes 2l_0}) = \begin{cases} 0 & \text{if } l < l_0, \\ O(2l_0+1) & \text{if } l > l_0. \end{cases}$$

Using Remark 4.1, we then see that the two expressions

$$\sqrt{N}^{2l_0}\nu(\tilde{\varepsilon}^{\otimes 2l_0}\tilde{\sigma}_1^{\otimes 2l_0}) \text{ and } \frac{1}{l_0!}\sqrt{N}^{2l_0}\nu_0^{(l_0)}(\tilde{\varepsilon}^{\otimes 2l_0}\tilde{\sigma}_1^{\otimes 2l_0})$$

have the same limit, if any.

In the first part of this section we give an explicit computation, in equation (45), of the derivative $\nu_0^{(l_0)}(\tilde{\varepsilon}^{\otimes 2l_0}\tilde{\sigma}_1^{\otimes 2l_0})$. In the second part of this section we find its asymptotic behavior at the order $2l_0$, given in relation (52), and prove Theorem 1.1 in the general case.

According to equation (16) used for $t = 0$, we have to study the products

$$c_{\hat{J}}(4l_0)T_{1,\hat{J}}(\tilde{\varepsilon}^{\otimes 2l_0})T_{2,\hat{J}}(\tilde{\sigma}_1^{\otimes 2l_0}),$$

when $\hat{J}$ is a sequence of l_0 pairs J_i.

Using Definition 6.1 and relation (38), we may prove by combinatorial means the following:

Proposition 8.1. *If $\hat{J} = (J_1, J_2, \ldots, J_{l_0})$ is such that $T_{1,\hat{J}}(\tilde{\varepsilon}^{\otimes 2l_0}) \neq 0$, there exists a subset*

$$\hat{B} = (a_1, a_2, \ldots, a_{2l_0}) \in \mathcal{C}$$

such that the J_i are an ordered partition of $\hat{B}$. In this case, there exists an ordered partition, denoted by π_{l_0}, of the interval $[1, 2, \ldots, 2l_0]$ in l_0 pairs $\{r_i, s_i\}$ such that, for all $i \in [1, 2, \ldots, l_0]$,

$$J_i = \{a_{r_i}, a_{s_i}\}$$

and we have

$$T_{1,\hat{J}}(\tilde{\varepsilon}^{\otimes 2l_0}) = (-1)^{\sum_i (a_{r_i}+a_{s_i})}\,\mathbb{E}(U^{2l_0}). \tag{43}$$

We also have the pairs $J_i \subset [1, \ldots, 4l_0]$, thus the constants $c_{\hat{J}}(4l_0) = 1$. So for such a $\hat{J}$, we get:

$$c_{\hat{J}}(4l_0)T_{1,\hat{J}}(\tilde{\varepsilon}^{\otimes 2l_0})T_{2,\hat{J}}(\tilde{\sigma}_1^{\otimes 2l_0}) = (-1)^{\sum_i (a_{r_i}+a_{s_i})}\,\mathbb{E}(U^{2l_0})\nu_-\Big(\tilde{\sigma}_1^{\otimes 2l_0}\prod_{i=1}^{i=l_0} R^-_{a_{r_i},a_{s_i}}\Big).$$

Summing now the terms used above first over the J_i associated to a same partition π_{l_0}, and using (16) for $t = 0$, we get the following result:

Proposition 8.2. *We have*

$$\nu_0^{(l_0)}(\tilde{\varepsilon}^{\otimes 2l_0}\tilde{\sigma}_1^{\otimes 2l_0}) = \beta^{2l_0}\,\mathbb{E}(U^{2l_0})\sum_{\pi_{l_0}}\nu_-\Big(\tilde{\sigma}_1^{\otimes 2l_0}\prod_{i=1}^{i=l_0}R^-(\tilde{\sigma}_{r_i},\tilde{\sigma}_{s_i})\Big)$$
$$\stackrel{\text{def}}{=}\beta^{2l_0}\,\mathbb{E}(U^{2l_0})\sum_{\pi_{l_0}}\nu_-(\tilde{\sigma}_1^{\otimes 2l_0}R^-_{\pi_{l_0}}).$$

Remark 8.1. We get the same result replacing $\tilde{\sigma}_1^{\otimes 2l_0}$ by any function f^-.

However, when $f^- = \tilde{\sigma}_1^{\otimes 2l_0}$, we get the following more precise result:

Proposition 8.3. *Let $\pi_{l_0,0}$ be the "canonical" partition such that*

$$\forall i \le l_0, \quad r_i = 2i-1, \quad s_i = 2i. \tag{44}$$

We have:

$$\nu_0^{(l_0)}(\tilde{\varepsilon}^{\otimes 2l_0}\tilde{\sigma}_1^{\otimes 2l_0}) = \beta^{2l_0}\,\mathbb{E}(U^{2l_0})\frac{(2l_0)!}{2^{l_0}}\nu_-(\tilde{\sigma}_1^{\otimes 2l_0}R^-_{\pi_{l_0,0}}). \tag{45}$$

Remark 8.2. We see that (45) generalizes (30) obtained when $p = 2$. We also have:

$$\frac{(2l_0)!}{2^{l_0}} = l_0!E(z^{2l_0})$$

where z is a standard Gaussian random variable.

Let us quickly describe now the main steps of [2, Section 5]. Equation (45) leads us, trying to use the analogy rule for the pair

$$(g^- = \tilde{\sigma}_1^{\otimes 2l_0}R^-_{\pi_{l_0,0}},\ g = \tilde{\sigma}_1^{\otimes 2l_0}R_{\pi_{l_0,0}}), \tag{46}$$

to evaluate the asymptotic behavior of $\nu(g)$.

Using symmetry between replicas and obvious notations, we get:

$$\nu(\tilde{\sigma}_1^{\otimes 2l_0}R_{\pi_{l_0,0}}) = \nu(\tilde{\sigma}_i^{\otimes 2l_0}R_{\pi_{l_0,0}}),$$
$$= \nu(\tilde{\varepsilon}^{\otimes 2l_0}R_{\pi_{l_0,0}}),$$

where

$$R_{\pi_{l_0,0}} = \prod_{r=1}^{r=l_0}\Big(R^-(\tilde{\sigma}^{2r-1},\tilde{\sigma}^{2r}) + \frac{\tilde{\varepsilon}^{2r-1}\tilde{\varepsilon}^{2r}}{N}\Big).$$

By expanding this product and using an extension of Proposition 8.2, we finally may get the following

Proposition 8.4 ([2, Proposition 5.4]). *If $\hat{\pi}_l$ denotes the partition π_l of the interval $[1, 2, \ldots, l]$ with no order between its atoms and if*

$$S_{0,k} = \sum_{\hat{\pi}_{l_0-k}}\nu_-(R^-_{\pi_{l_0-k,0}}R^-_{\hat{\pi}_{l_0-k}}), \tag{47}$$

we have:

$$\nu(\tilde{\sigma}_1^{\otimes 2l_0} R_{\pi_{l_0,0}}) = \mathbb{E}(U^{2l_0})\Big[\sum_{k=0}^{k=l_0} C_{l_0}^k \Big(\frac{4}{N}\Big)^k (\beta^2)^{l_0-k} S_{0,k}\Big] + O(2l_0+1). \tag{48}$$

Let us try now to apply the analogy rule to each term in the sum defining $S_{0,k}$. We get first the following result:

Proposition 8.5. *Let us set* $l' = l_0 - k$.

$$\nu(R_{\hat{\pi}_{l'}} R_{\pi_{l',0}}) = \begin{cases} O(2l'+1) & \text{if } \hat{\pi}_{l'} \neq \pi_{l',0}\,, \\ \Big(\frac{4\,\mathbb{E}(U^2)}{N(1-\beta^2\,\mathbb{E}(U^2))}\Big)^{l'} + O(2l'+1) & \text{if } \hat{\pi}_{l'} = \pi_{l',0}\,. \end{cases} \tag{49}$$

This proposition is proved in [2, end of Section 5] and needs a central limit theorem proved by Talagrand for the overlaps ([5, Section 2.7]).

Using now the analogy rule, assumed to be valid at order $k = 2l'$, for each pair $(R_{\hat{\pi}_{l'}}, R_{\pi_{l',0}}, R^-_{\hat{\pi}_{l'}}, R^-_{\pi_{l',0}})$, we get:

$$S_{0,k} = \Big(\frac{4\,\mathbb{E}(U^2)}{N(1-\beta^2\,\mathbb{E}(U^2))}\Big)^{l_0-k} + O(2(l_0-k)+1). \tag{50}$$

Reporting this evaluation in equation (48), we get:

$$\nu(\tilde{\sigma}_1^{\otimes 2l_0} R_{\pi_{l_0,0}}) = \mathbb{E}(U^{2l_0})\Big(\frac{4}{N(1-\beta^2\,\mathbb{E}(U^2))}\Big)^{l_0} + O(2l_0+1). \tag{51}$$

We may now use the analogy rule, assumed to be valid at the order $k = 2l_0$, for the pair (g, g^-) defined in equation (46). Using equation (45), we get the following generalization of (34):

$$\nu_0^{(l_0)}(\tilde{\varepsilon}^{\otimes 2l_0}\tilde{\sigma}_1^{\otimes 2l_0}) = \beta^{2l_0}\frac{(2l_0)!}{2^{l_0}} E^2(U^{2l_0})\Big(\frac{4}{N(1-\beta^2\,\mathbb{E}(U^2))}\Big)^{l_0} + O(2l_0+1). \tag{52}$$

We then have, using Remark 8.2, relations (7), (42), (52):

$$E(\tilde{\gamma}_{i,j}^{2l_0}) = \Big(\frac{4\beta^2}{N}\Big)^{l_0} E^2(U^{2l_0})E(z^{2l_0})\Big(\frac{1}{(1-\beta^2\,\mathbb{E}(U^2))}\Big)^{l_0} + O(2l_0+1), \tag{53}$$

$$E(\gamma_{i,j}^{2l_0}) = \Big(\frac{\beta^2}{N}\Big)^{l_0} E^2(U^{2l_0})E(z^{2l_0})\Big(\frac{1}{(1-\beta^2\,\mathbb{E}(U^2))}\Big)^{l_0} + O(2l_0+1). \tag{54}$$

This is the result previously announced in Theorem 1.1 when the order p of the moments is even.

9. Conclusion and perspectives

This result has been proved with some hypothesis (β sufficiently small) implying exponential inequalities and central limit theorems for overlaps; it could be natural to study our problem under more general hypothesis having the same implications. It is also possible to look, if β is sufficiently small, for other limit theorems, e.g., for the arithmetic mean of spins over all sites ([3]), or for an extension of our results in the case of p-spins as studied in [1].

Acknowledgement

The author is very grateful to Michel Talagrand for his constant interest in this work.

References

[1] Bardina, X., Márquez-Carreras, D., Rovira, C., Tindel, S.: The p-spin interaction model with external field. Potential Anal. **21**(4), 311–362 (2004).

[2] Hanen, A.: Un théorème limite pour les covariances des spins dans le modèle de Sherrington-Kirkpatrick avec champ externe. Ann. Probab. **35**(1), 141–179 (2007).

[3] Hanen, A.: A central limit theorem for the mean magnetistion in the Sherrington–Kirkpatrick model with an external field (2008). In preparation.

[4] Sherrington, D., Kirkpatrick, S.: Solvable model of a spin-glass. Phys. Rev. Lett. **35**(26), 1792–1796 (1975).

[5] Talagrand, M.: Spin glasses: a challenge for mathematicians, *Ergebnisse der Mathematik und ihrer Grenzgebiete. 3. Folge.*, vol. 46. Springer-Verlag, Berlin (2003). Cavity and mean field models.

Albert Hanen
University Paris X Nanterre
200 avenue de la République
F-92000 Nanterre, France
e-mail: ahanen3752@aol.com

Progress in Probability, Vol. 62, 177–201

A Limit Theorem for Mean Magnetisation in the Sherrington–Kirkpatrick Model with an External Field

Albert Hanen

Abstract. Some properties of mean magnetization (i.e., mean spin) are established in the case of a SK model with an external field, and a central limit theorem is proved.

Mathematics Subject Classification (2000). 82D30, 60G15, 60F05.

Keywords. SK models, Gibbs' measure, exponential inequalities, mean magnetization, C.L.T.

1. Introduction

Talagrand, in Chapter 2 of his book on spinglasses [5], studied the following model, introduced by Sherrington and Kirkpatrick [4].

Let β and $h \in \mathbb{R}^+$, $(g_{i,j})_{1\leq i<j}$ a set of i.i.d. Gaussian standard random variables, defining the disorder of the model, N be a positive integer. Let

$$\sigma = (\sigma_i)_{i=1}^N \in \{-1,+1\}^N \stackrel{\text{def}}{=} \Sigma_N \,.$$

The quantity σ_i is called the spin at site i of the configuration σ.

Let us consider the Hamiltonian of the SK model with a constant external field h, i.e., the quantity

$$-H_N(\sigma,\beta,h) = \frac{\beta}{\sqrt{N}} \sum_{1\leq i<j\leq N} g_{i,j}\sigma_i\sigma_j + h\Big(\sum_{i\leq N} \sigma_i\Big)\,. \tag{1}$$

The Gibbs' measure of the model is the probability on Σ_N, denoted by $G_N(\beta,h)$ or G_N, associated to the weight function $\exp(-H_N(\sigma))$, $\sigma \in \Sigma_N$. If f is a numerical function defined on Σ_N^n, the integral of f with respect to the product measure G_N^n is denoted by $\langle f\rangle$, the expectation (with respect to the disorder of the model) of this integral is denoted by $\nu(f)$. One may view ν, for each fixed n, as the expectation of the product measure G_N^n.

Let σ^1, σ^2 be two configurations or replicas. Talagrand extensively studied their overlap defined by

$$R(\sigma^1, \sigma^2) \stackrel{\text{def}}{=} \frac{1}{N} \sum_{i=1}^{N} \sigma_i^1 \sigma_i^2 = R_{1,2} .$$

Let

$$Y = \beta z \sqrt{q} + h, \text{ where } q \text{ satisfies the relation } \mathbb{E}(th^2(Y)) = q , \tag{2}$$

where z is a standard Gaussian random variable. It is known ([5, Proposition 2.4.8]) that q exists and is unique if $h > 0$. We will set also

$$\mathbb{E}(th^k(Y)) = q_k .$$

Let us then denote by $\dot{R}_{1,2}$ the "recentered" overlap $R_{1,2} - q_2$. The k-order moments, with respect to ν, of these recentered overlaps have been studied at sufficiently high temperature (i.e., for β sufficiently small) and these studies lead to prove Central Limit Theorems for overlaps ([5, Section 2.7]).

A non Gaussian limit theorem for the covariance (with respect to Gibbs' measure) of the spins at two sites is given in [2] and in [3] (where a detailed proof and a more complete bibliography are given).

We will be interested here in proving a Gaussian limit theorem for the mean spin:

$$\bar{\sigma} = \frac{1}{N} \sum_{i=1}^{i=N} \sigma_i . \tag{3}$$

Such a result will be a partial answer to some questions of Talagrand in his book [5, end of Section 2.7].

We will proceed in a manner similar to the study of overlaps. We will study the "recentered magnetization" denoted by $\dot{f}$, where

$$\dot{f}(\sigma) = \bar{\sigma} - q_1 . \tag{4}$$

We will prove mainly the following two theorems:

Theorem 1.1. *For β sufficiently small, there exists a constant V such that*

$$N\nu(\dot{f}^2) \xrightarrow[N \to +\infty]{} V. \tag{5}$$

Remark 1.1. The "constant" V above is in fact a complicated but explicit function, not depending on N, of the parameters β and h of the model, which is given below in (55), or equivalently in (100).

Definition 1.1. Let us set

$$X_N = \frac{\dot{f}}{(\frac{V}{N})^{\frac{1}{2}}} .$$

Theorem 1.2. *The k-order moment relative to the probability ν of the random variable X_N converges, when $N \to +\infty$, towards the k-order moment of a standard Gaussian variable z.*

The main method used to obtain these results, as in the case of overlaps, is the "Smart Path" method introduced by Talagrand. Some properties of this method are recalled in Section 2 which ends with the notation $O(k)$ used in [5]. Section 3 recalls the exponential inequalities for overlaps, i.e., some estimates of the $2k$-order moments of the recentered overlaps, and gives some useful properties of products of recentered overlaps. Section 4 studies the asymptotic behavior of $\nu(\dot{f})$ when $N \to +\infty$. Section 5 first gives an asymptotic bound of the second-order moment of $\dot{f}$ with respect to ν, then proves Theorem 1.1 and computes explicitly the constant V. The next section proves an exponential inequality for the "recentered magnetization" and the last section sets an induction relationship between the k-order and the $k-2$-order moments of $\dot{f}$, allowing us to give another (equivalent) formula for V and to prove Theorem 1.2.

2. The smart path method

To study the spinglass model defined by the Hamiltonian (1), Talagrand introduced a continuous family of Hamiltonians

$$-H_{N,t}, \quad t \in [0,1], \text{ with } -H_{N,1} = -H_N .$$

In the same way as in the case $t = 1$, this family induces Gibbs' measures $G_{N,t}$, integrals $\langle f \rangle_t$ having expectations $\nu_t(f)$. The disorder of $-H_{N,t}$ depends on the disorder of $-H_N$ and on an independent Gaussian standard random variable z.

In the rest of this section, we will first give some properties of the Gibbs' measure $G_{N,0}$, and of the integral $\nu_0(f)$. Then we will give a formula allowing to compute the derivatives $\nu_t'(f)$ and we will also give some useful bounds for $\nu_t(f)$. In the last subsection we will recall the notation $O(k)$.

2.1. The case $t = 0$

Let us consider the variable Y associated to z defined in the equation (2). The spin $\sigma_N \stackrel{\text{def}}{=} \varepsilon$ satisfies $\langle \varepsilon \rangle_0 = th(Y)$ and is independent of the spins at the $N-1$ first sites, which have law $G_{N-1}(\beta\sqrt{(N-1)/N}, h)$. This law induces integrals denoted by $\langle . \rangle_-$ having expectation $\nu_-(\cdot)$. Then if a numerical function f of n replicas satisfies $f = f^- h_0$, where f^- only depends on the spins of these replicas at the first $N-1$ sites and h_0 only depends on their spins at site N, we have the relations:

$$\langle f \rangle_0 = \langle f^- \rangle_0 \langle h_0 \rangle_0 = \langle f^- \rangle_- \langle h_0 \rangle_0 , \tag{6}$$

$$\nu_0(f) = \nu_0(f^-)\nu_0(h_0) = \nu_-(f^-)\nu_0(h_0) . \tag{7}$$

2.2. Derivatives of $\nu_t(f)$

Let us give some properties of $\nu_t(f)$, f being here a numerical function of n replicas. Let us introduce first some objects associated to overlaps, setting

$$R^-(\sigma^1, \sigma^2) \stackrel{\text{def}}{=} \frac{1}{N} \sum_{i=1}^{i=N-1} \sigma_i^1 \sigma_i^2 = R_{1,2}^- .$$

If $J = (r,s)$, $r < s$ is an ordered pair of integers indexing replicas σ^r and σ^s, we can also set:

$$\varepsilon^r = \sigma^r_N, \varepsilon^s = \sigma^s_N,$$

$$R_J = R(\sigma^r, \sigma^s) = R_{r,s},\ \dot{R}_J = R_{r,s} - q_2 \stackrel{\text{def}}{=} \dot{R}^-_J + \frac{\varepsilon^r \varepsilon^s}{N} \stackrel{\text{def}}{=} \dot{R}_J + \frac{\varepsilon^J}{N}.$$

$\dot{R}_J$ is called the recentered overlap associated to J, (8)

$\dot{R}^-_J$ is similarly called the recentered M-overlap associated to J. (9)

Let us set

$$\bar{R}_n = \sum_{1 \le r < s \le n} \varepsilon^r \varepsilon^s \dot{R}_{r,s} - n \sum_{r=1}^{r=n} \varepsilon^r \varepsilon^{n+1} \dot{R}_{r,n+1} + \frac{n(n+1)}{2} \varepsilon^{n+1} \varepsilon^{n+2} \dot{R}_{n+1,n+2}.$$

Let us also denote by $\bar{R}^-_n$ the quantity obtained by replacing the terms $\dot{R}_J$ by $\dot{R}^-_J$ in the definition of $\bar{R}_n$. Then we have, for all functions f defined on Σ^n_N, the relation:

$$\nu'_t(f) = \beta^2 \nu_t(f \bar{R}_n) = \beta^2 \nu_t(f \bar{R}^-_n). \tag{10}$$

This relation is an other form of [5, Proposition 2.4.5].

These properties may allow us to evaluate $\nu(f)$ through a Taylor expansion of $\nu_t(f)$ at $t = 0$, $\nu_0(f)$ (or $\nu'_0(f)$) being generally easier to compute. As it often works, the construction of $\nu_t(f)$ is called the method of "Smart Path".

Remark 2.1. We see that in equation (10) the derivative $\nu'_t(f)$ appears to be a known linear combination of expressions such as $\nu_t(f\dot{R}_J)$ or $\nu_t(f\dot{R}^-_J)$; these expressions introducing the overlaps and their connections with f will play an important role in the sequel, by trying either to compute them or to bound them using an Hölder's type inequality. For example, the computation of $\nu(f\dot{R}_J)$ when f is the recentered magnetization defined above in (4), will allow us to get expressions depending only on overlaps for which we may use their known asymptotic evaluations. This idea will be an important key to prove Theorem 1.1.

2.3. Some estimates of $\nu_t(f)$

Let us give some other useful properties.

- It is proved in [5, Proposition 2.4.6] that there exists a number K_0, depending only on n and β, such that, if f is a non negative function on Σ^n_N, we have:

$$\nu_t(f) \le K_0(n,\beta)\nu(f). \tag{11}$$

  If $\beta \le \beta_0$, we may omit dependence on β and replace $K_0(n,\beta)$ by $K_0(n)$.
- It is also proved in [5, Proposition 2.4.7] that there exists a number K_1, depending only on n and β, such that, if f is a numerical function on Σ^n_N, and if the numbers τ_1, τ_2 satisfy $1/\tau_1 + 1/\tau_2 = 1$, we have:

$$|\nu(f) - \nu_0(f)| \le K_1(n,\beta)\nu^{1/\tau_1}(|f|^{\tau_1})\nu^{1/\tau_2}(|\dot{R}_{1,2}|^{\tau_2}). \tag{12}$$

  As above, we may omit dependence on β if $\beta \le \beta_0$ and replace $K_1(n,\beta)$ by $K_1(n)$.

2.4. The notation $O(k)$

We will give this useful notation, as used by Talagrand in his book [5].

Definition 2.1. A numerical expression $F(N,\beta,\theta)$, defined for $\beta \leq \beta_0$, is said to verify the relation

$$F(N,\beta,\theta) = O(k)\,,$$

if there exists a constant K, not depending on β nor on N, but possibly depending on the parameter θ, such that

$$|F(N,\beta,\theta)| \leq \frac{K}{N^{\frac{k}{2}}}\,.$$

We will often use the following notation:

Definition 2.2. If a and b are two numerical quantities such that

$$a - b = O(k+1),$$

we will denote this relation as

$$a \stackrel{k+1}{\sim} b\,.$$

Remark 2.2. We may also say in this case that a and b are asymptotically equivalent at order $k+1$.

3. Exponential inequalities and applications

In this section, we recall first "exponential" inequalities for overlaps, valid for β sufficiently small, that we use to evaluate remainder terms in some Taylor expansion of $\nu_t(f)$. The last part of this section studies products, denoted by h_1^l, of l recentered overlaps and products, denoted by $(h^-)_1^l$, of their associated recentered M-overlaps and mainly proves the following:

$$\nu(h_1^l) \stackrel{l+1}{\sim} \nu_0((h^-)_1^l). \tag{13}$$

3.1. Exponential inequalities for overlaps

For β sufficiently small ($\beta \leq \beta_0$), Talagrand ([5, Section 2.5]) proved some exponential inequalities for recentered overlaps, such as:

$$\nu(\dot{R}_{1,2}^{2k}) \leq \left(\frac{\hat{L}k}{N}\right)^k, \quad \hat{L} \text{ being some constant.} \tag{14}$$

These inequalities imply:

$$\nu(|R_{1,2} - q_2|^j) = O(j)\,. \tag{15}$$

We have the same last inequalities if we replace R by R^-.

3.2. First- and second-order Taylor formulas for $\nu_t(f)$

Let us denote by $[g_i]^*$ some finite linear combination of quantities g_i having the same structure. Using (11) and (10), we can see that

$$|\nu_t'(f)| \leq [\nu(|fR_{J_i}|)]^*, \text{ and also} \tag{16}$$

$$|\nu_t'(f)| \leq [\nu(|fR^-_{J_i}|)]^*. \tag{17}$$

We then have ([5, Proposition 2.5.3]), for $\beta \leq \beta_0$:

$$\nu(f) = \nu_0(f) + O(1)\nu^{\frac{1}{2}}(f^2), \tag{18}$$

$$\nu(f) = \nu_0(f) + \nu_0'(f) + O(2)\nu^{\frac{1}{2}}(f^2). \tag{19}$$

Remark 3.1. In all the sequel of this paper, we will suppose $\beta \leq \beta_0$, where β_0 is chosen sufficiently small to have the exponential inequalities (14), (15) and the C.L.T. for the overlaps.

3.3. Products of l recentered overlaps

We will prove now a technical lemma which will be needed in the sequel. Let

$$J_i = \{r_i, s_i\}, \quad (i \in [1, \ldots, l])$$

be a sequence of l ordered pairs of positive integers, each pair indexing replicas $(\sigma^{r_i}, \sigma^{s_i})$. Let

$$h_1^l = \prod_{i=1}^{l} \dot{R}_{J_i}; \quad (h^-)_1^l = \prod_{i=1}^{l} \dot{R}^-_{J_i}.$$

Lemma 3.1. *We have the following relations:*

$$\nu(|h_1^l|) = O(l) = \nu(|(h^-)_1^l|), \tag{20}$$

$$\nu(h_1^l) = \nu_0((h^-)_1^l) + O(l+1). \tag{21}$$

Remark 3.2. Using Definition 2.2, we may write also (21) as (13).

Proof. We know (using (15)) that

$$\nu(|(\dot{R}_{J_i})^k|) = O(k) = \nu(|(\dot{R}^-_{J_i})^k|). \tag{22}$$

These inequalities and Hölder's inequalities imply the first part of Lemma 3.1.

To prove the second part, we shall prove first that, for all t,

$$\nu_t(h_1^l) \overset{l+1}{\sim} \nu_t((h^-)_1^l). \tag{23}$$

We have:

$$\begin{aligned} h_1^l - (h^-)_1^l &= \sum_{r=0}^{l-1}((h^-)_1^r h_{r+1}^l - (h^-)_1^{r+1} h_{r+2}^l) \\ &= \sum_{r=0}^{l-1} \frac{\varepsilon^{J_{r+1}}}{N}(h^-)_1^r h_{r+2}^l. \end{aligned}$$

Thus we have:

$$|h_1^l - (h^-)_1^l| \le \frac{1}{N}\sum_{r=0}^{l-1} |(h^-)_1^r h_{r+2}^l| .$$

Exponential and Hölder's inequalities prove that $\nu(|(h^-)_1^r h_{r+2}^l|) = O(l-1)$. Thus, n being the total number of replicas involved, using also (11), we have:

$$\begin{aligned} |\nu_t(h_1^l) - \nu_t((h^-)_1^l)| &\le \frac{1}{N}\sum_{r=0}^{l-1} \nu_t(|(h^-)_1^r h_{r+2}^l|) , \\ &\le \frac{K_0(n)}{N}\sum_{r=0}^{l-1} \nu(|(h^-)_1^r h_{r+2}^l|) , \\ &= O(l+1) . \end{aligned}$$

Thus (23) is proved.

We also have, using Taylor's formula and the equation (16) with $f = h_1^l$:

$$\nu(h_1^l) = \nu_0(h_1^l) + \nu_t'(h_1^l) ,$$

where we have

$$|\nu_t'(h_1^l)| \le [\nu(|h_1^l \dot{R}_{J_i}|)]^* .$$

As each term $\nu(|h_1^l \dot{R}_{J_i}|)$ is of the type

$$\nu(|h_1^{l+1}|) = O(l+1), \text{ using (20)},$$

we then have

$$\nu_t'(h_1^l) = O(l+1) \text{ and } \nu(h_1^l) \overset{l+1}{\sim} \nu_0(h_1^l) .$$

Thus, using (23) for $t=0$, we obtain (21):

$$\nu(h_1^l) \overset{l+1}{\sim} \nu_0((h^-)_1^l) .$$

□

4. The mean spin: first-order properties

In this section we begin our study of the recentered magnetization. We will prove mainly that, as in the cases of overlaps, $\nu(\dot{f}) = O(2)$. We will get this result as a corollary of Proposition 4.1 below, used several times in the sequel.

Let $\dot{f}$ be the recentered magnetization. We have, by symmetry between sites, the relation:

$$\nu(\dot{f}) = \nu((\varepsilon^1 - q_1)) .$$

Let us give some definitions and let us set:

Definition 4.1. We define α_1 and α_2 by

$$\begin{aligned} \alpha_1 &= \nu_0(((\varepsilon^1 - q_1)\varepsilon^1\varepsilon^r)(r>1) = q_1(1-q_2) , \\ \alpha_2 &= \nu_0(((\varepsilon^1 - q_1)\varepsilon^r\varepsilon^s)(1<r<s) = q_3 - q_1 q_2 . \end{aligned}$$

Definition 4.2. We define $\hat{R}_n^-$ by

$$\hat{R}_n^- = \alpha_1\Big(\sum_{r=2}^{n} \dot{R}_{1,r}^- - n\dot{R}_{1,n+1}^-\Big)$$
$$+ \alpha_2\Big(\sum_{2\leq r<s\leq n} \dot{R}_{r,s}^- - n\sum_{r=2}^{n} \dot{R}_{r,n+1}^- + \frac{n(n+1)}{2}\dot{R}_{n+1,n+2}^-\Big)\,. \quad (24)$$

We have the following

Proposition 4.1. *Let g^- be a function with real values, such that*

$$\hat{g} \stackrel{\text{def}}{=} (\varepsilon^1 - q_1)g^-$$

depends on n replicas denoted by $\sigma^1, \sigma^2, \ldots, \sigma^n$ and g^- does not depend on the spins of these replicas at the last site N. We have, using Taylor's formula:

$$\nu_0((\varepsilon^1 - q_1)g^-) = \nu_0(\hat{g}) = 0\,, \quad (25)$$
$$\nu((\varepsilon^1 - q_1)g^-) = \nu(\hat{g}) = \beta^2\nu_0(g^-\hat{R}_n^-) + O(2)\nu^{\frac{1}{2}}((g^-)^2)\,. \quad (26)$$

Remark 4.1. The derivative $\nu_0'(\hat{g})$, which is the first term of the right-hand side in equation (26) above, is a linear combination of expressions $\nu_0(g^-\dot{R}_J^-)$ that we will study in the sequel for certain g^-.

Before proving Proposition 4.1, let us first prove the following

Corollary 4.1. *We have the property:*

$$\nu(\dot{f}) = O(2)\,. \quad (27)$$

Proof. As $\nu(\dot{f}) = \nu(\varepsilon^1 - q_1)$, we can apply Proposition 4.1 with $g^- = 1$, $n = 1$. We then have:

$$\nu(\dot{f}) = \beta^2\nu_0(\hat{R}_1^-) + O(2)\,,$$
$$= \beta^2(-\alpha_1\nu_0(\dot{R}_{1,2}^-) + \alpha_2\nu_0(\dot{R}_{2,3}^-)) + O(2)\,.$$

We know, using (21) with $l = 1$, that

$$\nu_0(\dot{R}_{1,2}^-) \stackrel{2}{\sim} \nu(\dot{R}_{1,2}) \text{ and } \nu_0(\dot{R}_{2,3}^-) \stackrel{2}{\sim} \nu(\dot{R}_{2,3})\,.$$

As we know, using [5, equation (2.215)], that $\nu(\dot{R}_J) = O(2)$ for all J, the corollary is proved. □

Proof of Proposition 4.1. We know ([5, Lemma 2.4.4]) that:

$$\langle\varepsilon^1\rangle_0 = th(Y),$$

so that

$$\nu_0(\varepsilon^1 - q_1) = 0\,,$$

and

$$\nu_0((\varepsilon^1 - q_1)g^-) = \nu_0(\varepsilon^1 - q_1)\nu_0(g^-) = 0\,.$$

The last equality above is (25).

Using also the fact that

$$|\varepsilon^1 - q_1| \leq 2\,,$$

we then have, by Taylor's formula and (19):

$$\nu((\varepsilon^1 - q_1)g^-) = \nu_0'((\varepsilon^1 - q_1)g^-) + O(2)\nu^{\frac{1}{2}}((g^-)^2)\,.$$

Using (10) allows us to calculate $\nu_0'((\varepsilon^1 - q_1)g^-)$. We have:

$$\begin{aligned}
\nu_0'((\varepsilon^1 - q_1)g^-) &= \beta^2\nu_0((\varepsilon^1 - q_1)g^-\bar{R}_n^-)\,,\\
&= \beta^2\nu_0((\varepsilon^1 - q_1)g^-)\left[\sum_{r=2}^{n}\varepsilon^1\varepsilon^r\dot{R}_{1,r}^-\right]\\
&\quad + \beta^2\nu_0((\varepsilon^1 - q_1)g^-)\left[\sum_{2\leq r<s\leq n}\varepsilon^r\varepsilon^s\dot{R}_{r,s}^-\right]\\
&\quad - n\beta^2\nu_0\left((\varepsilon^1 \quad q_1)\,g^-\left[\sum_{r=2}^{n}\varepsilon^r\varepsilon^{n+1}\dot{R}_{r,n+1}^-\right]\right)\\
&\quad - n\beta^2\nu_0((\varepsilon^1 - q_1)g^-\varepsilon^1\varepsilon^{n+1}\dot{R}_{1,n+1}^-)\\
&\quad + \frac{n(n+1)}{2}\beta^2\nu_0((\varepsilon^1 - q_1)g^-\varepsilon^{n+1}\varepsilon^{n+2}\dot{R}_{n+1,n+2}^-)\,.
\end{aligned}$$

We know that we have

$$\left\langle\prod_{r\in K}\varepsilon^r\right\rangle_0 = (th(Y))^{|K|}\,,$$

and then

$$\nu_0\left(\prod_{r\in K}\varepsilon^r\right) = \mathbb{E}\big((th(Y))^{|K|}\big) = q_{|K|}\,,$$

where K is a finite set of integers indexing $|K|$ distinct configurations σ^r whose spins at site N are denoted by ε^r.

We then have, using (7), Definitions 4.1 and 4.2, the relation

$$\nu_0'((\varepsilon^1 - q_1)g^-) = \beta^2\nu_0(g^-\hat{R}_n^-)\,. \tag{28}$$

Proposition 4.1 is proved. □

5. Study of the second-order moment $\nu(\dot{f}^2)$

In this section, we will first prove the following.

Theorem 5.1. *We have the relation:*

$$\nu(\dot{f}^2) = O(2)\,. \tag{29}$$

Then we will make this relation more precise by proving Theorem 1.1. In the last subsection, the constant V appearing in Theorem 1.1 will be computed as an explicit function of the parameters of the model.

5.1. Estimation of $\nu(\dot{f}^2)$

To prove our first result, we will try to follow the main steps of the corresponding result for the overlaps ([5, Theorem 2.4.2]), but instead of finding at the end a direct inequality, we will get an inequality for the value at $x = \nu^{1/2}(\dot{f}^2)$ of a second-degree polynomial and we will have to study its roots to get our result. As an important consequence we will also find, in Corollary 5.1 below, that

$$\nu(\dot{f}\dot{R}_J) \overset{3}{\sim} \nu_0(\dot{f}^- \dot{R}_J^-). \tag{30}$$

Proof of Theorem 5.1. We have, by symmetry among sites:

$$\nu(\dot{f}^2) = \nu((\varepsilon^1 - q_1)\dot{f}) \overset{\text{def}}{=} \nu(g). \tag{31}$$

We also have $|g| \leq 2|\dot{f}|$. The function g depends only on one replica, so we can use (12) with $n = 1$ and $\tau_1 = \tau_2 = 1/2$. We obtain:

$$|\nu(g) - \nu_0(g)| \leq 2K_1(1)\nu^{1/2}(\dot{f}^2)\nu^{1/2}(\dot{R}_{1,2}^2). \tag{32}$$

As we assume that the exponential inequalities (14) are true, we have:

$$\nu^{1/2}(\dot{R}_{1,2}^2) \leq \left(\frac{\hat{L}}{N}\right)^{1/2}.$$

Thus we have:

$$|\nu(g) - \nu_0(g)| \leq 2K_1(1)\left(\frac{\hat{L}}{N}\right)^{1/2}\nu^{1/2}(\dot{f}^2) \overset{\text{def}}{=} \frac{b_0}{N^{1/2}}\nu^{1/2}(\dot{f}^2). \tag{33}$$

Using (7) and $\nu_0(\varepsilon^1 - q_1) = 0$, we have

$$\nu_0(g) = \nu_0((\varepsilon^1 - q_1)(\dot{f} - \dot{f}^-)) = \frac{\nu_0((\varepsilon^1 - q_1)\varepsilon^1)}{N}.$$

Thus

$$|\nu_0(g)| \leq \frac{2}{N}.$$

We then have:

$$\nu(\dot{f}^2) \leq \frac{b_0}{N^{1/2}}\nu^{1/2}(\dot{f}^2) + \frac{2}{N}. \tag{34}$$

Let us set

$$h_2(x) = x^2 - \frac{b_0}{N^{1/2}}x - \frac{2}{N}, \quad \nu^{1/2}(\dot{f}^2) = a.$$

We have $a \geq 0$, $h_2(a) \leq 0$, so that $a \leq x_0$, where x_0 is the positive root of $h_2(x)$. As it is easy to see that $x_0 = \frac{c_0}{N^{1/2}}$, where c_0 is a constant, we necessarily have $\nu(\dot{f}^2) \leq \frac{L}{N}$, if the constant $L \geq c_0^2$. Thus we have $\nu(\dot{f}^2) = O(2)$. □

Corollary 5.1. *We also have:*

$$\nu(|\dot{f}|) = O(1). \tag{35}$$

$$\nu((\dot{f}^-)^2) = O(2). \tag{36}$$

$$\nu(\dot{f}\dot{R}_J) \overset{3}{\sim} \nu_0(\dot{f}^- \dot{R}_J^-). \tag{37}$$

Proof. The first relation is a consequence of Theorem 5.1 and of Hölder's inequalities.

It is also easy to see that $\dot{f}^2 = (\dot{f}^-)^2 + O(2)$. Applying ν to both sides and then using (29), we get (36).

Let us prove the last relation (37). We have:

$$\dot{f}\dot{R}_J - \dot{f}^-\dot{R}_J^- = \dot{f}(\dot{R}_J - \dot{R}_J^-) + (\dot{f} - \dot{f}^-)\dot{R}_J^- = \frac{1}{N}(\dot{f}\varepsilon^J + \varepsilon^1\dot{R}_J^-)\,.$$

Thus we get:

$$|\dot{f}\dot{R}_J - \dot{f}^-\dot{R}_J^-| \leq \frac{1}{N}(|\dot{f}| + |\dot{R}_J^-|)\,.$$

Applying ν to both sides of this inequality and using (35) and (20) with $l = 1$, we get:

$$\nu(\dot{f}\dot{R}_J) \overset{3}{\sim} \nu(\dot{f}^-\dot{R}_J^-)\,. \tag{38}$$

Using now Taylor's formula, we get:

$$\nu(\dot{f}^-\dot{R}_J^-) = \nu_0(\dot{f}^-\dot{R}_J^-) + \nu_t'(\dot{f}^-\dot{R}_J^-)\,.$$

Using (17), we see that

$$|\nu_t'(\dot{f}^-\dot{R}_J^-)| \leq [\nu(|\dot{f}^-\dot{R}_J^-\dot{R}_{K_i}^-|)]^*.$$

Since

$$\nu(|\dot{f}^-\dot{R}_J^-\dot{R}_{K_i}^-|) \leq \nu^{\frac{1}{2}}((\dot{f}^-)^2)\nu^{\frac{1}{2}}((\dot{R}_J^-)^2(\dot{R}_{K_i}^-)^2)\,,$$

we see, using Lemma 3.1 and (36), that

$$|\nu_t'(\dot{f}^-\dot{R}_J^-)| = O(2)\nu^{\frac{1}{2}}((\dot{f}^-)^2) = O(3)\,.$$

Thus we have:

$$\nu(\dot{f}^-\dot{R}_J^-) \overset{3}{\sim} \nu_0(\dot{f}^-\dot{R}_J^-)\,.$$

Using now (38), we get (37). □

5.2. An asymptotic equivalence relation for $\nu(\dot{f}^2)$

Let us describe now the main ways used to prove Theorem 1.1, or a more precise form given by the following

Theorem 5.2. *There exists a constant V such that*

$$\nu(\dot{f}^2) \overset{3}{\sim} \frac{V}{N}\,. \tag{39}$$

We have, by symmetry among sites,

$$\nu(\dot{f}^2) = \nu((\varepsilon^1 - q_1)\dot{f}) = \nu\Big((\varepsilon^1 - q_1)\Big(\dot{f}^- + \frac{\varepsilon^1}{N}\Big)\Big)\,. \tag{40}$$

We will get a sum of two terms, the second one being easy to evaluate, and we will apply Proposition 4.1 to the first one, with

$$g^- = \dot{f}^-\,.$$

We will then have, using Remark 4.1, to study expressions like

$$\nu_0(\dot{f}^-\dot{R}_J^-)$$

or equivalently, using relation (30) above,

$$\nu(\dot{f}\dot{R}_J)\,.$$

But we may write this expression as

$$\nu((\varepsilon^1 - q_1)(\dot{R}_J^- + \varepsilon^J/N))\,.$$

Applying to this quantity the same type of procedure as for equation (40), we finally get expressions like

$$\nu(\dot{R}_J\dot{R}_K) \overset{3}{\sim} C/N\,,$$

for known functions C of the parameters of the model, allowing us to prove Theorem 5.2.

Let us go now into detailed computations.

Proof of Theorem 5.2. Let us continue the computation given in the equation (40). We have

$$\begin{aligned}\nu(\dot{f}^2) &= \nu((\varepsilon^1 - q_1)\dot{f}^-) + \frac{1}{N}\nu(1 - q_1\varepsilon^1)\,,\\ &\overset{\text{def}}{=} T_1 + T_2\,.\end{aligned}$$

As, by (18), we have:

$$\nu(\varepsilon^1) = \nu_0(\varepsilon^1) + O(1) = q_1 + O(1)\,,$$

we find that

$$T_2 = \frac{1 - q_1^2}{N} + O(3)\,.$$

To calculate T_1, we can use Proposition 4.1, with $g^- = \dot{f}^-$ and then $n = 1$. We obtain the following equation:

$$\begin{aligned}T_1 &= \nu((\varepsilon^1 - q_1)\dot{f}^-)\\ &= \beta^2(-\alpha_1\nu_0(\dot{f}^-\dot{R}_{1,2}^-) + \alpha_2\nu_0(\dot{f}^-\dot{R}_{2,3}^-)) + O(2)\nu^{\frac{1}{2}}((\dot{f}^-)^2)\,. \qquad (41)\end{aligned}$$

Using (36) and (37), we have:

$$T_1 \overset{3}{\sim} \beta^2(-\alpha_1\nu(\dot{f}\dot{R}_{1,2}) + \alpha_2\nu(\dot{f}\dot{R}_{2,3}))\,. \qquad (42)$$

We have:

$$\nu(\dot{f}\dot{R}_J) = \nu((\varepsilon^1 - q_1)\dot{R}_J^-) + \nu\Big((\varepsilon^1 - q_1)\frac{\varepsilon^J}{N}\Big)\,.$$

Thus we have, by Proposition 4.1 with $g^- = \dot{R}_J^-$, $\nu^{1/2}((\dot{R}_J^-)^2) = O(1)$:

$$\nu(\dot{f}\dot{R}_J) \overset{3}{\sim} \beta^2\nu_0(\dot{R}_J^-\hat{R}_n^-) + \nu_0\Big((\varepsilon^1 - q_1)\frac{\varepsilon^J}{N})\Big)\,. \qquad (43)$$

We have here the relations:

- If $J=\{1,2\}$, then $\nu_0\Big((\varepsilon^1-q_1)\frac{\varepsilon^J}{N}\Big)=\alpha_1$ and $n=2$.
- If $J=\{2,3\}$, then $\nu_0\Big((\varepsilon^1-q_1)\frac{\varepsilon^J}{N}\Big)=\alpha_2$ and $n=3$.

The quantity $\nu_0(\dot{R}_J^-\hat{R}_n^-)$ is in each case an explicit linear combination of terms of the type

$$\nu_0(\dot{R}_J^-\dot{R}_K^-)\overset{3}{\sim}\nu(\dot{R}_J\dot{R}_K), \tag{44}$$

using (21) for $l=2$. Each term $\nu(\dot{R}_J\dot{R}_K)$ has one of the following types:

$$\nu((\dot{R}_{1,2})^2 \quad \text{if } J=K\,, \tag{45}$$

$$\nu(\dot{R}_{1,2}\dot{R}_{1,3}) \quad \text{if } |J\cap K|=1\,, \tag{46}$$

$$\nu(\dot{R}_{1,2}\dot{R}_{3,4}) \quad \text{if } J\cap K=\emptyset\,. \tag{47}$$

Definition 5.1. Let us set, using the notations of [5, Sections 2.6, 2.7], and more precisely [5, Eqs. (2.268), (2.279), (2.292)]:

- $U=1-th^2(Y)$,
- $A_1^2=\frac{\mathbb{E}(U^2)}{1-\beta^2\,\mathbb{E}(U^2)}$,
- $A^2=\frac{A_1^2}{N}$,
- $B_1^2=\frac{q_2-q_4}{(1-\beta^2\,\mathbb{E}(U^2))(1-\beta^2(1-4q_2+3q_4))}$,
- $B^2=\frac{B_1^2}{N}$,
- $C_1^2=\frac{q_4-q_2^2+\beta^2(q_4-q_2^2)A_1^2+2\beta^2(2q_2+q_2^2-3q_4)B_1^2}{1-\beta^2(1-4q_2+3q_4)}$,
- $C^2=\frac{C_1^2}{N}$.

We also know, using [5, equation (2.265)], that:

$$\nu(\dot{R}_{1,2}\dot{R}_{3,4})\overset{3}{\sim}C^2\,,$$

$$\nu(\dot{R}_{1,2}\dot{R}_{1,3})\overset{3}{\sim}B^2+C^2\,,$$

$$\nu((\dot{R}_{1,2})^2\overset{3}{\sim}A^2+2B^2+C^2.$$

It then follows, thanks to the form of the quantities A^2, B^2, C^2 that we have, V_1 and V_2 being constants, the following equations:

$$\nu_0(\dot{f}^-\dot{R}_{1,2}^-)\overset{3}{\sim}\frac{V_1}{N}\,,$$

$$\nu_0(\dot{f}\dot{R}_{2,3}^-)\overset{3}{\sim}\frac{V_2}{N}\,.$$

Finally, we obtain:

$$\nu(\dot{f}^2)\overset{3}{\sim}\frac{1-q_1^2}{N}+\beta^2\Big(-\alpha_1\frac{V_1}{N}+\alpha_2\frac{V_2}{N}\Big)\,. \tag{48}$$

Thus we proved Theorem 5.2 and then Theorem 1.1. □

5.3. A computation of the constant V

Let us compute more precisely the constant V above, and first the constants V_1 and V_2, as functions of the parameters of the model.

We have, using (24) and (43) when $J = \{1, 2\}$ and $n = 2$:

$$\nu_0(\dot{R}^-_{1,2}\hat{R}^-_2) = \nu_0(\dot{R}^-_{1,2}[\alpha_1(\dot{R}^-_{1,2} - 2\dot{R}_{1,3})^- + \alpha_2(-2\dot{R}^-_{2,3} + 3\dot{R}^-_{3,4})]) .$$

Using now equations (45), (46) and (47), we have:

$$\nu_0(\dot{R}^-_{1,2}\hat{R}^-_2) \overset{3}{\sim} \alpha_1(A^2 + 2B^2 + C^2 - 2(B^2 + C^2)) + \alpha_2(-2(B^2 + C^2) + 3C^2) , \tag{49}$$

$$\overset{3}{\sim} \alpha_1(A^2 - C^2) + \alpha_2(-2B^2 + C^2) . \tag{50}$$

Using now the equation (43), we get:

$$V_1 = \alpha_1 + \beta^2(\alpha_1(A_1^2 - C_1^2) + \alpha_2(-2B_1^2 + C_1^2)) . \tag{51}$$

Let us compute now the constant V_2. In the same way, we have, using the equation (43) when $J = \{2, 3\}$ and $n = 3$:

$$\nu_0(\dot{R}^-_{2,3}\hat{R}^-_3) = \nu_0(\dot{R}^-_{2,3}[\alpha_1(\dot{R}^-_{1,2} + \dot{R}^-_{1,3} - 3\dot{R}^-_{1,4}) + \alpha_2(\dot{R}^-_{2,3} - 3\dot{R}^-_{2,4} - 3\dot{R}^-_{3,4} + 6\dot{R}^-_{4,5}]) .$$

We then have:

$$\nu_0(\dot{R}^-_{2,3}\hat{R}^-_3) \overset{3}{\sim} \alpha_1(2(B^2 + C^2) - 3C^2) + \alpha_2(A^2 + 2B^2 + C^2 - 6(B^2 + C^2) + 6C^2) , \tag{52}$$

$$\overset{3}{\sim} \alpha_1(2B^2 - C^2) + \alpha_2(A^2 + C^2 - 4B^2) . \tag{53}$$

Using equation (43), we obtain:

$$V_2 = \alpha_2 + \beta^2(\alpha_1(2B_1^2 - C_1^2) + \alpha_2(A_1^2 + C_1^2 - 4B_1^2)) . \tag{54}$$

Using now equation (48), we finally get:

$$V = 1 - q_1^2 + \beta^2(-\alpha_1^2 + \alpha_2^2) + \beta^4[\alpha_1^2(A_1^2 - C_1^2) + \alpha_2^2(A_1^2 + C_1^2 - 4B_1^2) + 2\alpha_1\alpha_2(2B_1^2 - C_1^2))] . \tag{55}$$

We will get later an alternative formula giving the value of the constant V.

6. Exponential inequalities for the mean spin

In this section we will prove exponential inequalities for the mean spin, in the form given in the equation (14) in the case of overlaps, then we will prove some consequences of this inequality, given in Corollary 6.1 below. Let us start by giving our main theorem here:

Theorem 6.1. *There exists a constant L such that, for any integer k, we have*

$$\nu(\dot{f}^{2k}) \leq \left(\frac{Lk}{N}\right)^k . \tag{56}$$

Remark 6.1. As $|\dot{f}| \leq 2$, we have $\nu(\dot{f}^{2k}) \leq 4^k$ and the equation (56) is true for $k \geq N$ if we take $L \geq 4$.

Let us describe now the main points of the proof of Theorem 6.1. We will follow the same scheme of proof as for a similar result on overlaps ([5, Theorem 2.5.1]).

We will prove this theorem by induction on k, more precisely we will take $1 \leq k \leq N-1$ and we will assume that there exists a constant $L_0 \geq 4$ such that:

$$\forall l \leq k, \quad \nu(\dot{f}^{2l}) \leq \left(\frac{L_0}{N}\right)^l . \tag{57}$$

We know that this assumption is true when $k = 1$, by Theorem 5.1. We will give first in Lemma 6.1 below some consequences of this assumption, then we will prove our theorem, i.e., that we have, for a convenient choice of the constant L_0, the inequality

$$\nu(\dot{f}^{2k+2}) \leq \left(\frac{L_0(k+1)}{N}\right)^{k+1} . \tag{58}$$

Writing, as in the case of $\nu(\dot{f}^2)$, $\nu(\dot{f}^{2k+2}) = \nu(g_k)$ for a convenient function g_k (see equation (61) below), we will bound the difference $|\nu(g_k) - \nu_0(g_k)|$, using an Hölder's type inequality. This bound will lead to study, as in the case of $\nu(\dot{f}^2)$, a polynomial inequality of degree $2(k+1)$, allowing us, after some calculations, to get our result. This last point is the main difference with the case of overlaps.

Let us give now detailed proofs.

Lemma 6.1. *Under hypothesis* (57), *we have:*

$$\forall j \leq 2k, \qquad \nu(|\dot{f}^j|) \leq \left(\frac{L_0(j+1)/2}{N}\right)^{j/2} = O(j)\,, \tag{59}$$

$$\nu((\dot{f}^-)^{2k} \leq 3\left(\frac{L_0(k+1)}{N}\right)^k = O(2k)\,. \tag{60}$$

Proof. The proof of Lemma 6.1 is almost identical to that of the same results for overlaps ([5, Lemma 2.5.2]), we only have to replace $\dot{R}_{1,2}$ by $\dot{f}$ and to express $\dot{f}^-$ as $\dot{f} - \varepsilon^1/N$. □

Proof of Theorem 6.1. Let us prove now that if the assumption (57) is true and if L_0 is chosen sufficiently large, we have

$$\nu(\dot{f}^{2k+2}) \leq \left(\frac{L_0(k+1)}{N}\right)^{k+1} .$$

We shall proceed in the same way as in the study of $\nu(\dot{f}^2)$. We have

$$\nu(\dot{f}^{2k+2}) = \nu((\varepsilon^1 - q_1)\dot{f}^{2k+1}) \stackrel{\text{def}}{=} \nu(g_k)\,. \tag{61}$$

Using now (12) with $\tau_1 = \frac{2k+2}{2k+1}$, $\tau_2 = 2k+2$ and knowing that $|\varepsilon^1 - q_1| \leq 2$, we have

$$|\nu(g_k) - \nu_0(g_k)| \leq 2K_1(1)\nu^{\frac{2k+1}{2k+2}}(\dot{f}^{2k+2})\nu^{\frac{1}{2k+2}}(\dot{R}_{1,2}^{2k+2}).$$

We know, using (14), that

$$\nu^{\frac{1}{2k+2}}(\dot{R}_{1,2}^{2k+2}) \leq \Big(\frac{\hat{L}(k+1)}{N}\Big)^{1/2},$$

with the constant b_0 defined in equation (33). Then we get:

$$\begin{aligned}|\nu(g_k) - \nu_0(g_k)| &\leq 2K_1(1)\Big(\frac{\hat{L}(k+1)}{N}\Big)^{1/2}\nu^{\frac{2k+1}{2k+2}}(\dot{f}^{2k+2}) \\ &\leq b_0\Big(\frac{k+1}{N}\Big)^{1/2}\nu^{\frac{2k+1}{2k+2}}(\dot{f}^{2k+2}). \qquad (62)\end{aligned}$$

We now evaluate $|\nu_0(g_k)|$. Using (7), we have

$$\nu_0(g_k) = \nu((\varepsilon^1 - q_1)(\dot{f}^{2k+1} - (\dot{f}^-)^{2k+1})).$$

Since we have:

$$\begin{aligned}|\dot{f}^{2k+1} - (\dot{f}^-)^{2k+1})| &\leq (2k+1)|\dot{f} - \dot{f}^-|(\dot{f}^{2k} + (\dot{f}^-)^{2k}), \\ &\leq \frac{2k+1}{N}(\dot{f}^{2k} + (\dot{f}^-)^{2k}),\end{aligned}$$

using (11) and the inequality $|\varepsilon^1 - q_1| \leq 2$, we get:

$$\begin{aligned}|\nu_0(g_k)| &\leq 2\Big(\frac{2k+1}{N}\Big)(\nu_0(\dot{f}^{2k}) + \nu_0(\dot{f}^-)^{2k}), \\ &\leq 2\Big(\frac{2k+1}{N}\Big)K_0(1)(\nu(\dot{f}^{2k}) + \nu(\dot{f}^-)^{2k}). \qquad (63)\end{aligned}$$

Since we know, using (57) and (60), that

$$\nu(\dot{f}^{2k}) \leq \Big(\frac{L_0 k}{N}\Big)^k$$

and that

$$\nu((\dot{f}^-)^{2k}) \leq 3\Big(\frac{L_0(k+1)}{N}\Big)^k,$$

using also the inequality $2k+1 < 2(k+1)$, we get:

$$\begin{aligned}|\nu_0(g_k)| &\leq 4^2 K_0(1)\Big(\frac{k+1}{N}\Big)\Big(\frac{L_0(k+1)}{N}\Big)^k, \\ &\leq 16K_0(1)L_0^k\Big(\frac{(k+1)}{N}\Big)^{k+1}. \qquad (64)\end{aligned}$$

Thus we have, using (62) and (64):

$$\begin{aligned}\nu(\dot{f}^{2k+2}) = \nu(g_k) &\leq |\nu_0(g_k)| + |\nu(g_k) - \nu_0(g_k)|, \\ &\leq 16K_0(1)L_0^k\Big(\frac{k+1}{N}\Big)^{k+1} + b_0\Big(\frac{k+1}{N}\Big)^{1/2}\nu^{\frac{2k+1}{2k+2}}(\dot{f}^{2k+2}). \qquad (65)\end{aligned}$$

Let us set:

$$\hat{b} = b_0(\frac{k+1}{N})^{1/2}, \qquad \hat{c} = 16K_0(1)L_0^k(\frac{k+1}{N})^{k+1},$$
$$a = \nu^{\frac{1}{2k+2}}(\dot{f}^{2k+2}), \qquad \hat{h}(x) = x^{2k+2} - \hat{b}x^{2k+1} - \hat{c}.$$

We have $\hat{h}(0) = -\hat{c} < 0$, and also, by (65), $\hat{h}(a) \leq 0$. It is easy to see that the function $\hat{h}$ decreases when $x < m_0 \stackrel{\text{def}}{=} \frac{2k+1}{2k+2}\hat{b}$, then increases towards $+\infty$ when $x \geq m_0$. Thus $\hat{h}$ has only one positive root called x_0 and we have $0 < a < x_0$.

In fact, we want to prove that we have:

$$a^{2k+2} \leq \Big(\frac{L_0(k+1)}{N}\Big)^{k+1},$$

i.e.,

$$a \leq \Big(\frac{L_0(k+1)}{N}\Big)^{1/2} \stackrel{\text{def}}{=} d. \tag{66}$$

The inequality (66) will be true if it is possible to choose L_0 such that $\hat{h}(d) > 0$. We have:

$$\begin{aligned}\hat{h}(d) &= d^{2k+2} - \hat{b}d^{2k+1} - \hat{c},\\ &= \Big(\frac{L_0(k+1)}{N}\Big)^{k+1} - b_0\Big(\frac{k+1}{N}\Big)^{1/2}\Big(\frac{L_0(k+1)}{N}\Big)^{k+(1/2)}\\ &\quad - 16K_0(1)L_0^k\Big(\frac{(k+1)}{N}\Big)^{k+1},\\ &= L_0^k\Big(\frac{k+1}{N}\Big)^{k+1}[L_0 - b_0L_0^{1/2} - 16K_0(1)]. \end{aligned} \tag{67}$$

Let us set:

$$u(x) = x^2 - b_0x - 16K_0(1). \tag{68}$$

We will have $\hat{h}(d) > 0$ iff $u(L_0^{1/2}) > 0$. This last relation will be true iff $L_0^{1/2} \geq u_0$ (where u_0 is the positive root of $u(x)$). If we take such a constant L_0, then $\hat{h}(d) > 0$, thus Theorem 6.1 is proved. □

Corollary 6.1. *We have the following relations:*

$$\nu(|\dot{f}^-|^j) = O(j). \tag{69}$$

$$\nu(\dot{f}^k) \stackrel{k+1}{\sim} \nu_0((\dot{f}^-)^k). \tag{70}$$

$$\nu(\dot{f}^k\dot{R}_J) \stackrel{k+2}{\sim} \nu_0((\dot{f}^-)^k)\dot{R}_J^-). \tag{71}$$

Proof. By (60), the relation (69) is satisfied when $j = 2k$. When $j = 2k-1$, we have:

$$\nu^{1/j}(|\dot{f}^-|^j) \leq \nu^{1/2k}(|\dot{f}^-|^{2k}).$$

By (60), we have:

$$\nu^{1/2k}(|\dot{f}^-|^{2k}) = O(1).$$

Thus we get (69).

To get (70), we first prove that

$$\nu(\dot{f}^k) \overset{k+1}{\sim} \nu((\dot{f}^-)^k)\,, \tag{72}$$

and then that

$$\nu((\dot{f}^-)^k) \overset{k+1}{\sim} \nu_0((\dot{f}^-)^k)\,. \tag{73}$$

This scheme of proof is also valid to prove (71). We have:

$$\begin{aligned}|\dot{f}^k - (\dot{f}^-)^k| &\le |\dot{f} - \dot{f}^-|(|\dot{f}^{k-1}| + |(\dot{f}^-)^{k-1}|)\,,\\ &\le \frac{k}{N}(|\dot{f}^{k-1}| + |(\dot{f}^-)^{k-1}|)\,.\end{aligned} \tag{74}$$

Then

$$\begin{aligned}(|\nu(\dot{f}^k) - \nu((\dot{f}^-)^k)| &\le \frac{k}{N}\nu(|\dot{f}^{k-1}|) + \nu(|(\dot{f}^-)^{k-1}|)\\ &= O(2)O(k-1) = O(k+1)\,.\end{aligned}$$

Thus we have the relation (72). Using now first (18), then (69), we get:

$$\begin{aligned}\nu((\dot{f}^-)^k) &= \nu_0((\dot{f}^-)^k) + O(1)\nu^{1/2}((\dot{f}^-)^{2k})\\ &\overset{k+1}{\sim} \nu_0((\dot{f}^-)^k)\,.\end{aligned} \tag{75}$$

This is relation (73), thus (70) is proved. In the same way, we have:

$$\begin{aligned}\dot{f}^k \dot{R}_J - (\dot{f}^-)^k \dot{R}_J^- &= (\dot{f}^k - (\dot{f}^-)^k)\dot{R}_J + (\dot{f}^-)^k(\dot{R}_J - \dot{R}_J^-)\\ &= (\dot{f}^k - (\dot{f}^-)^k)\dot{R}_J + (\dot{f}^-)^k\Big(\frac{\varepsilon^J}{N}\Big)\,.\end{aligned}$$

Then we get, using relations (74), (59) , (60), (69) and Schwarz's inequality:

$$\begin{aligned}|\nu(\dot{f}^k \dot{R}_J) - \nu((\dot{f}^-)^k \dot{R}_J^-)| &\le \frac{k}{N}(\nu(|\dot{f}^{k-1}\dot{R}_J|) + \nu(|(\dot{f}^-)^{k-1}\dot{R}_J|)) + \frac{1}{N}\nu(|\dot{f}^-)^k|)\,,\\ &= 2O(2)O(k-1)O(1) + O(2)O(k)\\ &= O(k+2)\,.\end{aligned}$$

Thus we proved that

$$\nu(\dot{f}^k \dot{R}_J) \overset{k+2}{\sim} \nu((\dot{f}^-)^k \dot{R}_J^-)\,. \tag{76}$$

We then have, using Taylor's formula

$$\nu((\dot{f}^-)^k \dot{R}_J^-) = \nu_0((\dot{f}^-)^k \dot{R}_J^-) + \nu_t'((\dot{f}^-)^k \dot{R}_J^-)\,. \tag{77}$$

We know, by (17) that:

$$|\nu_t'((\dot{f}^-)^k \dot{R}_J^-)| \le (\nu(|(\dot{f}^-)^k \dot{R}_J^- \dot{R}_{J_i}^-|))^*\,.$$

Each expression $\nu(|(\dot{f}^-)^k \dot{R}_J^- \dot{R}_{J_i}^-|)$ may be written, using the notations of Section 3, as:

$$\nu(|(\dot{f}^-)^k (h^-)_1^2|)\,.$$

Using now equations (20), (60) and Schwarz's inequality, we get:

$$\nu(|(\dot{f}^-)^k (h^-)_1^2|) = O(k)O(2) = O(k+2)\,.$$

Thus we have:

$$\nu_t'((\dot{f}^-)^k \dot{R}_J^-) = O(k+2)\,.$$

Thus, by (77), we get:

$$\nu((\dot{f}^-)^k \dot{R}_J^-) \overset{k+2}{\sim} \nu_0((\dot{f}^-)^k \dot{R}_J^-)\,. \tag{78}$$

The equations (76) and (78) prove the last inequality (71) of Corollary 6.1. □

7. Study of the k-order moments of the recentered magnetization

Let us set

$$\nu(\dot{f}^k) = m_k\,.$$

Let us define the following quantities:

Definition 7.1. Let us set, using (71):

$$\begin{aligned} X_k^- &= \nu_0((\dot{f}^-)^{k-1}\dot{R}_{1,2}^-) \overset{k+1}{\sim} \nu(\dot{f}^{k-1}\dot{R}_{1,2}) \overset{\text{def}}{=} X_k \\ Y_k^- &= \nu_0((\dot{f}^-)^{k-1}\dot{R}_{2,3}^-) \overset{k+1}{\sim} \nu(\dot{f}^{k-1}\dot{R}_{2,3}) \overset{\text{def}}{=} Y_k\,. \end{aligned}$$

Remark 7.1. We also have, for all J:

$$\nu_0((\dot{f}^-)^{k-1}\dot{R}_J^-) \overset{k+1}{\sim} \nu(\dot{f}^{k-1}\dot{R}_J) = \begin{cases} X_k & \text{if } 1 \in J\,, \\ Y_k & \text{if } 1 \notin J\,. \end{cases}$$

Let us give now an overview of the contents of this section.

In the first part, we prove, in Proposition 7.1 below, that m_k is asymptotically equivalent at order $k+1$ to an explicit linear combination of X_k, Y_k and $\frac{k-1}{N}m_{k-2}$, whose coefficients are independent of k. The proof uses symmetry among sites for the recentered magnetization $\dot{f}$, writing m_k as $\nu((\varepsilon^1 - q_1)(\dot{f}^- + \frac{\varepsilon^1}{N})^{k-1})$, then uses mainly Proposition 4.1 for $g^- = (\dot{f}^-)^{k-1}$ and the asymptotic evaluations such as equation (71) found in the previous section in Corollary 6.1.

To get a "true" induction relationship between m_k and m_{k-2} at order $k+1$, we need to find how X_k and Y_k are related to m_{k-2}. This is done in the second part of the section. To study an expression such as $\nu(\dot{f}^{k-1}\dot{R}_J)$, we use symmetry among sites for the recentered overlap $\dot{R}_J$, writing

$$\nu(\dot{f}^{k-1}\dot{R}_J) = \nu(\dot{f}^{k-1}(\varepsilon^J - q_2))$$

and we make the same kind of asymptotic evaluations as above, using analogs of Proposition 4.1 and Corollary 6.1.

We get two linear asymptotic relations at order $k+1$ between X_k, Y_k and m_{k-2}, given, using a matrix formulation, in Proposition 7.2. Using these linear relations, we get, in Corollary 7.1 below, our "true" induction relationship given by:

$$m_k \overset{k+1}{\sim} \hat{C}\frac{k-1}{N}m_{k-2}\,. \tag{79}$$

Here, the constant $\hat{C}$ is necessarily the constant V found in Theorem 5.2. The next subsection gives explicitely the value of V found by this method, which possibly may appear simpler than equation (55).

At the end of this section, using (79), we will prove Theorem 1.2.

7.1. A first evaluation of m_k

We prove the following

Proposition 7.1. *We have:*

$$m_k \overset{k+1}{\sim} \beta^2(-\alpha_1 X_k + \alpha_2 Y_k) + \frac{k-1}{N}(1-q_1^2)m_{k-2}. \tag{80}$$

Proof. We have:

$$\begin{aligned} m_k &= \nu\Big((\varepsilon^1 - q_1)\Big(\dot{f}^- + \frac{\varepsilon^1}{N}\Big)^{k-1}\Big), \\ &= \nu((\varepsilon^1 - q_1)(\dot{f}^-)^{k-1}) + \frac{k-1}{N}\nu((\varepsilon^1 - q_1)(\dot{f}^-)^{k-2}\varepsilon^1) \\ &\quad + \sum_{r=2}^{k-1} \frac{1}{N^r} C_{k-1}^r \nu((\varepsilon^1 - q_1)(\dot{f}^-)^{k-1-r}(\varepsilon^1)^r). \end{aligned} \tag{81}$$

As we know, by (69), that $\nu(|(\dot{f}^-)^j|) = O(j)$, using also the inequality $|\varepsilon^1 - q_1| \leq 2$, each term of the last sum in (81) is bounded by

$$\begin{aligned} \frac{1}{N^r} C_{k-1}^r 2\nu(|(\dot{f}^-)^{k-1-r}|) &= O(2r)O(k-1-r), \\ &= O(k+1) \text{ when } r \geq 2. \end{aligned}$$

We then have

$$\nu(\dot{f}^k) \overset{k+1}{\sim} \nu((\varepsilon^1 - q_1)(\dot{f}^-)^{k-1}) + \frac{k-1}{N}\nu((\varepsilon^1 - q_1)(\dot{f}^-)^{k-2}\varepsilon^1) \tag{82}$$

$$\overset{\text{def}}{=} T_1 + T_2. \tag{83}$$

Let us first study the expression T_2. We have:

$$\begin{aligned} \nu((\varepsilon^1 - q_1)(\dot{f}^-)^{k-2}\varepsilon^1) &= \nu((1 - q_1\varepsilon^1)(\dot{f}^-)^{k-2}), \\ &= \nu_0((1 - q_1\varepsilon^1)(\dot{f}^-)^{k-2}) + O(1)\nu^{\frac{1}{2}}(\dot{f}^-)^{2(k-2)}, \\ &= \nu_0((1 - q_1\varepsilon^1)(\dot{f}^-)^{k-2}) + O(k-1), \end{aligned}$$

by (69). Thus we have:

$$\begin{aligned} T_2 &\overset{k+1}{\sim} \frac{k-1}{N}\nu_0((1 - q_1\varepsilon^1)(\dot{f}^-)^{k-2}) \\ &= \frac{k-1}{N}(1-q_1^2)\nu_0((\dot{f}^-)^{k-2}). \end{aligned} \tag{84}$$

Using now (70), we have:

$$T_2 \overset{k+1}{\sim} \frac{k-1}{N}(1-q_1^2)m_{k-2}. \tag{85}$$

Let us study now the term T_1. Using first (19), Proposition 4.1 with $g^- = (\dot{f}^-)^{k-1}$, then (69), we have:

$$\begin{aligned}T_1 = \nu((\varepsilon^1 - q_1)(\dot{f}^-)^{k-1}) &= \nu_0'((\varepsilon^1 - q_1)(\dot{f}^-)^{k-1})) + O(2)\nu^{\frac{1}{2}}((\dot{f}^-)^{2(k-1)}),\\ &= \beta^2\nu_0((\dot{f}^-)^{k-1}\hat{R}_1^-) + O(k+1),\\ &\overset{k+1}{\sim} \beta^2\nu_0((\dot{f}^-)^{k-1}(-\alpha_1\dot{R}_{1,2}^- + \alpha_2\dot{R}_{2,3}^-)).\end{aligned}$$

We then have:

$$T_1 \overset{k+1}{\sim} \beta^2(-\alpha_1 X_k + \alpha_2 Y_k)). \tag{86}$$

Using equations (85), (83) and (86), we get Proposition 7.1. □

7.2. Study of X_k and Y_k

Definition 7.2. Let us set:

$$\begin{aligned}a &= 1 - \beta^2(1 - q_2^2),\\ b &= -\beta^2(3q_4 - 2q_2 - q_2^2),\\ c &= -b = \beta^2(3q_4 - 2q_2 - q_2^2),\\ d &= (1 - \beta^2(6q_4 + 1 - 6q_2 - (q_2)^2)).\end{aligned}$$

Let us define the matrix:

$$M = \begin{pmatrix} a & b \\ c & d \end{pmatrix} \tag{87}$$

We will prove the following:

Proposition 7.2. *We have the equation*

$$M\begin{pmatrix} X_k \\ Y_k \end{pmatrix} \overset{k+1}{\sim} \frac{k-1}{N} m_{k-2} \begin{pmatrix} \alpha_1 \\ \alpha_2 \end{pmatrix}.$$

Proof. Let us try to reexpress first X_k. We have:

$$\begin{aligned}\nu(\dot{f}^{k-1}\dot{R}_{1,2}) \overset{\text{def}}{=} X_k &= \nu\Big(\Big(\dot{f}^- + \frac{\varepsilon^1}{N}\Big)^{k-1}(\varepsilon^1\varepsilon^2 - q_2)\Big) \text{ (by symmetry among sites)},\\ &= \nu((\dot{f}^-)^{k-1}(\varepsilon^1\varepsilon^2 - q_2))\\ &\quad + \frac{k-1}{N}\nu((\dot{f}^-)^{k-2}\varepsilon^1(\varepsilon^1\varepsilon^2 - q_2))\\ &\quad + \sum_{r=2}^{k-1}\frac{1}{N^r}C_{k-1}^r\nu((\varepsilon^1)^r(\dot{f}^-)^{k-1-r}(\varepsilon^1\varepsilon^2 - q_2)).\end{aligned}$$

In the last sum, using the inequality $|\varepsilon^1\varepsilon^2 - q_2| \le 2$, the term

$$\frac{1}{N^r}C_{k-1}^r\nu((\varepsilon^1)^r(\dot{f}^-)^{k-1-r}(\varepsilon^1\varepsilon^2 - q_2))$$

is an $O(2r)O(k-1-r) = O(k-1+r) = O(k+1)$ if $r \ge 2$. We then have

$$X_k \overset{k+1}{\sim} T_{1,k}^X + T_{2,k}^X, \tag{88}$$

where

$$T^X_{1,k} = \nu((\dot{f}^-)^{k-1}(\varepsilon^1\varepsilon^2 - q_2))\,,$$
$$T^X_{2,k} = \frac{k-1}{N}\nu((\dot{f}^-)^{k-2}\varepsilon^1(\varepsilon^1\varepsilon^2 - q_2))\,.$$

Let us study first the expression $T^X_{2,k}$. Using (18) and (69), we have:

$$\begin{aligned} T^X_{2,k} &= \frac{k-1}{N}\nu((\dot{f}^-)^{k-2}(\varepsilon^2 - q_2\varepsilon^1)) \\ &= \frac{k-1}{N}(\nu_0((\dot{f}^-)^{k-2}(\varepsilon^2 - q_2\varepsilon^1)) + O(1)O(k-2)) \\ &\overset{k+1}{\sim} \frac{k-1}{N}(\nu_0((\dot{f}^-)^{k-2})\nu_0(\varepsilon^2 - q_2\varepsilon^1)) \\ &= \frac{k-1}{N}(\nu_0((\dot{f}^-)^{k-2})(q_1(1-q_2)))\,. \end{aligned}$$

Thus, using (69) and Definition 4.1, we have:

$$T^X_{2,k} \overset{k+1}{\sim} \frac{(k-1)\alpha_1}{N}m_{k-2}\,. \tag{89}$$

Let us study now the expression $T^X_{1,k}$. As $\nu_0(\varepsilon^1\varepsilon^2 - q_2) = 0$, we have, using (19), (69) and (10) with $n = 2$, the relations:

$$\begin{aligned} T^X_{1,k} &= \nu_0'((\dot{f}^-)^{k-1}(\varepsilon^1\varepsilon^2 - q_2)) + O(2)O(k-1) \\ &\overset{k+1}{\sim} \beta^2\nu_0((\varepsilon^1\varepsilon^2 - q_2)(\dot{f}^-)^{k-1}(\varepsilon^1\varepsilon^2\dot{R}^-_{1,2} - 2\varepsilon^1\varepsilon^3\dot{R}^-_{1,3} \\ &\quad - 2\varepsilon^2\varepsilon^3\dot{R}^-_{2,3} + 3\varepsilon^3\varepsilon^4\dot{R}^-_{3,4}))\,. \end{aligned}$$

Using now (7), we obtain:

$$\begin{aligned} T^X_{1,k} \overset{k+1}{\sim} \beta^2[&(1-q_2^2)\nu_0((\dot{f}^-)^{k-1}\dot{R}^-_{1,2}) - 2(q_2 - q_2^2)\nu_0((\dot{f}^-)^{k-1}\dot{R}^-_{1,3}) \\ &- 2(q_2 - q_2^2)\nu_0((\dot{f}^-)^{k-1}\dot{R}^-_{2,3}) \\ &+ 3(q_4 - q_2^2))\nu_0((\dot{f}^-)^{k-1}\dot{R}^-_{3,4})]\,. \end{aligned}$$

Using Remark 7.1, we obtain:

$$\begin{aligned} T^X_{1,k} &\overset{k+1}{\sim} \beta^2[(1-q_2^2)X_k - 2(q_2 - q_2^2)X_k + (3(q_4 - q_2^2) - 2(q_2 - q_2^2))Y_k] \\ &\overset{k+1}{\sim} \beta^2[(1-q_2)^2X_k + (3q_4 - 2q_2 - q_2^2)Y_k]\,. \end{aligned}$$

Using now equations (88) and (89), we see that we have:

$$X_k(1 - \beta^2(1-q_2^2)) - \beta^2(3q_4 - 2q_2 - q_2^2)Y_k \overset{k+1}{\sim} \frac{(k-1)\alpha_1}{N}m_{k-2}\,. \tag{90}$$

We can proceed in the same way to reexpress Y_k. We have:

$$\nu(\dot{f}^{k-1}\dot{R}_{2,3}) \stackrel{\text{def}}{=} Y_k = \nu\Big(\Big(\dot{f}^- + \frac{\varepsilon^1}{N}\Big)^{k-1}(\varepsilon^2\varepsilon^3 - q_2)\Big) \text{ (by symmetry among sites)},$$
$$= \nu((\dot{f}^-)^{k-1}(\varepsilon^2\varepsilon^3 - q_2)) + \frac{k-1}{N}\nu((\dot{f}^-)^{k-2}\varepsilon^1(\varepsilon^2\varepsilon^3 - q_2))$$
$$+ \sum_{r=2}^{k-1} \frac{1}{N^r} C_{k-1}^r \nu((\varepsilon^1)^r(\dot{f}^-)^{k-1-r}(\varepsilon^2\varepsilon^3 - q_2)).$$

We can easily find, as in the study of X_k, that the last sum is an $O(k+1)$. We then have

$$Y_k \overset{k+1}{\sim} T^Y_{1,k} + T^Y_{2,k}, \tag{91}$$

where

$$T^Y_{1,k} = \nu((\dot{f}^-)^{k-1}(\varepsilon^2\varepsilon^3 - q_2)),$$
$$T^Y_{2,k} = \frac{k-1}{N}\nu((\dot{f}^-)^{k-2}\varepsilon^1(\varepsilon^2\varepsilon^3 - q_2)).$$

Let us study first the expression $T^Y_{2,k}$. Using (19), (69) and (7), we have:

$$T^Y_{2,k} = \frac{k-1}{N}\nu((\dot{f}^-)^{k-2}(\varepsilon^1\varepsilon^2\varepsilon^3 - q_2\varepsilon^1))$$
$$= \frac{k-1}{N}(\nu_0((\dot{f}^-)^{k-2}(\varepsilon^1\varepsilon^2\varepsilon^3 - q_2\varepsilon^1)) + O(1)O(k-2))$$
$$\overset{k+1}{\sim} \frac{k-1}{N}(\nu_0((\dot{f}^-)^{k-2})\nu_0(\varepsilon^1\varepsilon^2\varepsilon^3 - q_2\varepsilon^1))$$
$$= \frac{k-1}{N}(\nu_0((\dot{f}^-)^{k-2})(q_3 - q_1q_2))).$$

Thus, using (69) and Definition 4.1, we have:

$$T^Y_{2,k} \overset{k+1}{\sim} \frac{(k-1)\alpha_2}{N} m_{k-2}. \tag{92}$$

Let us study now the expression $T^Y_{1,k}$. We have, by (19), (69) and (10) (with $n = 3$), the relations:

$$T^Y_{1,k} = \nu_0'((\dot{f}^-)^{k-1}(\varepsilon^2\varepsilon^3 - q_2)) + O(2)O(k-1)$$
$$\overset{k+1}{\sim} \beta^2\nu_0((\varepsilon^2\varepsilon^3 - q_2)(\dot{f}^-)^{k-1}[\varepsilon^1\varepsilon^2\dot{R}^-_{1,2} + \varepsilon^1\varepsilon^3\dot{R}^-_{1,3} + \varepsilon^2\varepsilon^3\dot{R}^-_{2,3}$$
$$- 3(\varepsilon^1\varepsilon^4\dot{R}^-_{1,4} + \varepsilon^2\varepsilon^4\dot{R}^-_{2,4} + \varepsilon^3\varepsilon^4\dot{R}^-_{3,4}) + 6\varepsilon^4\varepsilon^5\dot{R}^-_{4,5}].$$

In the same way as above, using (7) and Remark 7.1, we find:

$$T^Y_{1,k} \overset{k+1}{\sim} \beta^2(\nu_0[(\varepsilon^2\varepsilon^3 - q_2)(\varepsilon^1\varepsilon^2 + \varepsilon^1\varepsilon^3 - 3\varepsilon^1\varepsilon^4)]X_k$$
$$+ \nu_0[(\varepsilon^2\varepsilon^3 - q_2)(\varepsilon^2\varepsilon^3 - 3(\varepsilon^2\varepsilon^4 + \varepsilon^3\varepsilon^4) + 6\varepsilon^4\varepsilon^5)]Y_k).$$

Thus we get

$$T^Y_{1,k} \overset{k+1}{\sim} \beta^2[((q_2)^2 + 2q_2 - 3q_4)X_k + (6q_4 + 1 - 6q_2 - (q_2)^2)Y_k]. \tag{93}$$

We can deduce, using (91) and (92), that:

$$Y_k(1-\beta^2(6q_4+1-6q_2-(q_2)^2)) - X_k\beta^2[((q_2)^2+2q_2-3q_4)] \overset{k+1}{\sim} \frac{(k-1)\alpha_2}{N} m_{k-2}\,. \tag{94}$$

The two equations (90) and (94) give Proposition 7.2. □

Remark 7.2. If $\beta \leq \beta_0$, it is easy to see that the quantities a and d are positive, then the matrix M is invertible in that case. Thus we have:

$$\begin{pmatrix} X_k \\ Y_k \end{pmatrix} \overset{k+1}{\sim} \frac{k-1}{N} m_{k-2} M^{-1} \begin{pmatrix} \alpha_1 \\ \alpha_2 \end{pmatrix}. \tag{95}$$

Corollary 7.1. *We have:*

$$m_k = \hat{C}\frac{k-1}{N} m_{k-2} + R_k\,. \tag{96}$$

In this formula, $\hat{C}$ is the constant

$$\hat{C} = (1-q_1^2) + \beta^2\left(\begin{pmatrix} -\alpha_1 & \alpha_2 \end{pmatrix} M^{-1} \begin{pmatrix} \alpha_1 \\ \alpha_2 \end{pmatrix}\right) \tag{97}$$

and $R_k = O(k+1)$.

Proof. Equation (80) can be written now as:

$$\begin{aligned} m_k &\overset{k+1}{\sim} \beta^2(-\alpha_1 X_k + \alpha_2 Y_k) + \frac{k-1}{N}(1-q_1^2)m_{k-2}\,, \\ &\overset{k+1}{\sim} \beta^2\left(\begin{pmatrix} -\alpha_1 & \alpha_2 \end{pmatrix} M^{-1} \begin{pmatrix} \alpha_1 \\ \alpha_2 \end{pmatrix}\right)\frac{k-1}{N} m_{k-2} + \frac{k-1}{N}(1-q_1^2)m_{k-2}\,. \end{aligned}$$

Corollary 7.1 is then proved. □

7.3. Another expression for the value of V

If we set $k = 2$, we find that $\hat{C}$ must be equal to the constant V which appears in Theorem 1.1. Let us compute $\hat{C}$ using (97). We have:

$$M^{-1} = \frac{1}{ad+b^2}\begin{pmatrix} d & -b \\ b & a \end{pmatrix}. \tag{98}$$

Then

$$\beta^2\left(\begin{pmatrix} -\alpha_1 & \alpha_2 \end{pmatrix} M^{-1} \begin{pmatrix} \alpha_1 \\ \alpha_2 \end{pmatrix}\right) = \beta^2\frac{-d\alpha_1^2 + a\alpha_2^2 + 2b\alpha_1\alpha_2}{ad+b^2}\,. \tag{99}$$

Thus we have:

$$\hat{C} = V = (1-q_1^2) + (1-q_1^2)\beta^2\frac{-d\alpha_1^2 + a\alpha_2^2 + 2b\alpha_1\alpha_2}{ad+b^2}\,. \tag{100}$$

7.4. Proof of Theorem 1.2

When $k = 2p$, by induction over p, we find that:

$$m_{2p} = \left(\frac{V}{N}\right)^p 1.3.5\ldots(2p-1) + S_{2p}\,, \tag{101}$$

where $S_{2p} = O(2p+1)$. In the same way, we can prove that:

$$m_{2p+1} = \left(\frac{V}{N}\right)^p (2.4.6\ldots 2p)m_1 + S_{2p+1} \tag{102}$$

where $S_{2p+1} = O(2p+2)$. As $m_1 = O(2)$, by (27), we then have:

$$m_{2p+1} = O(2p+2)\,.$$

If z is a standard Gaussian random variable we know the identity

$$1.3.5\ldots(2p-1) = \mathbb{E}(z^{2p})\,.$$

Let $[x]$ be the integer part of x. We have, for all k:

$$m_k = \left(\frac{V}{N}\right)^{[k/2]} \mathbb{E}(z^k) + O(k+1)\,. \tag{103}$$

Thus the Central Limit Theorem 1.2 is proved. □

Acknowledgement

I would like to thank Michel Talagrand for his constant interest in this work.

References

[1] Bardina, X., Márquez-Carreras, D., Rovira, C., Tindel, S.: The p-spin interaction model with external field. Potential Anal. **21**(4), 311–362 (2004).

[2] Hanen, A.: Un théorème limite pour les covariances des spins en deux sites dans le modèle de Sherrington-Kirkpatrick avec champ externe. C. R. Math. Acad. Sci. Paris **342**(2), 147–150 (2006).

[3] Hanen, A.: Un théorème limite pour les covariances des spins dans le modèle de Sherrington-Kirkpatrick avec champ externe. Ann. Probab. **35**(1), 141–179 (2007).

[4] Sherrington, D., Kirkpatrick, S.: Solvable model of a spin-glass. Phys. Rev. Lett. **35**(26), 1792–1796 (1975).

[5] Talagrand, M.: Spin glasses: a challenge for mathematicians, *Ergebnisse der Mathematik und ihrer Grenzgebiete. 3. Folge.*, vol. 46. Springer-Verlag, Berlin (2003). Cavity and mean field models.

Albert Hanen
University Paris X Nanterre
200 avenue de la République
F-92000 Nanterre, France
e-mail: `ahanen3752@aol.com`

Part II

Non-mean Field

Progress in Probability, Vol. 62, 205–223

A Percolation-theoretic Approach to Spin Glass Phase Transitions

J. Machta, C.M. Newman and D.L. Stein

Abstract. The magnetically ordered, low temperature phase of Ising ferromagnets is manifested within the associated Fortuin–Kasteleyn (FK) random cluster representation by the occurrence of a single positive density percolating cluster. In this paper, we review our recent work on the percolation signature for Ising spin glass ordering – both in the short-range Edwards–Anderson (EA) and infinite-range Sherrington–Kirkpatrick (SK) models – within a two-replica FK representation and also in the different Chayes–Machta–Redner two-replica graphical representation. Numerical studies of the $\pm J$ EA model in dimension three and rigorous results for the SK model are consistent in supporting the conclusion that the signature of spin-glass order in these models is the existence of a single percolating cluster of maximal density normally coexisting with a second percolating cluster of lower density.

Mathematics Subject Classification (2000). 82B44, 82D30, 82B80, 60K35, 82B43, 05C80.

Keywords. Ising spin glass, percolation, graphical representations, cluster algorithms, Fortuin–Kasteleyn

1. Introduction

The question of whether laboratory spin glasses – or the theoretical models used to represent them – have a thermodynamic phase transition remains unresolved despite decades of work [4, 25, 29]. Although the infinite-range Sherrington–Kirkpatrick (SK) Ising spin glass is easily shown to possess a phase transition [34, 2], the existence of one in the corresponding short-range Edwards–Anderson (EA) [9] Ising model (on the cubic lattice $\mathbf{Z}^d$) has not been established in *any* finite dimension. While some evidence for a transition has been uncovered through high-temperature expansions [11, 37], analytical studies of variable long-range $1d$ models [22], and extensive numerical simulations [31, 32, 4, 21], the issue remains unresolved [26].

Random graph methods, and in particular the Fortuin–Kasteleyn (FK) random cluster (RC) representation [19, 12], provide a set of useful tools for studying phase transitions (more specifically, the presence of multiple Gibbs states arising from broken spin rotational symmetry) in discrete spin models. In these representations spin correlation functions can be expressed through the geometrical properties of associated random graphs. FK and related models are probably best known in the physics literature for providing the basis for powerful Monte Carlo methods for studying phase transitions [35, 36, 40], but they have also proved important in obtaining rigorous results on phase transitions in discrete-spin ferromagnetic (including inhomogeneous and randomly diluted) models (e.g., [1, 15]). Because of complications due to frustration, however, graphical representations have so far played a less important role in the study of spin glasses.

The goal of the studies presented here is to construct a viable approach which uses random graph methods to address the problem of phase transitions and broken symmetry in spin glass models. In earlier papers [23, 24], we studied the "percolation signature" of spin glass ordering within two different graphical representations – the two-replica model of Chayes, Machta and Redner (CMR) [7, 33] and a two-replica version of the FK representation (TRFK), as proposed in Sec. 4.1 of [30]. The purpose of that analysis was to show that FK methods could be utilized to study spin glass phase transitions. The result of this work was the uncovering of strong evidence that the existence of a spin glass transition coincides with the emergence of *doubly* percolating clusters of *unequal* densities. This scenario is more complex than what occurs in ferromagnetic models, where the phase transition coincides with percolation of a single FK cluster.

In what follows, we first review the FK random cluster representation for ferromagnetic models, and then describe our analytical and numerical results for spin glasses.

2. The Fortuin–Kasteleyn random cluster representation

In this section we briefly review the Fortuin–Kasteleyn random cluster representation [19, 12] which relates the statistical mechanics of Ising (or Potts) models to a dependent percolation problem. Our focus will be on Ising models throughout, but the analysis is easily extended to more general Potts models.

2.1. Ferromagnetic models

We start by considering a nearest-neighbor Ising ferromagnet, whose couplings $J_{xy} \geq 0$ are not necessarily identical:

$$\mathcal{H} = -\sum_{\langle xy \rangle} J_{xy}\sigma_x\sigma_y \ , \tag{1}$$

where, as already noted, $J_{xy} \geq 0$ and the sum is over nearest-neighbor pairs of sites in $\mathbf{Z}^d$. Each such coupling is associated with an edge, or bond, $\langle x, y \rangle$, with the set of all such bonds denoted $\mathbf{E}^d$.

The RC approach introduces parameters $\mathcal{P}_{xy} \in [0, 1)$ through the formula:

$$\mathcal{P}_{xy} = 1 - \exp[-2\beta|J_{xy}|], \tag{2}$$

where β is the inverse temperature. One can then define a probability measure $\mu_{\rm RC}$ on $\{0,1\}^{\mathbf{E}^d}$, that is, on 0- or 1-valued bond occupation variables ω_{xy}. It is one of two marginal distributions (the other being the ordinary Gibbs distribution) of a joint distribution on $\Omega = \{-1,+1\}^{\mathbf{Z}^d} \times \{0,1\}^{\mathbf{E}^d}$ of the spins and bonds together (such a joint distribution will be introduced in Sec. 2.3). The marginal distribution $\mu_{\rm RC}$ is given formally by

$$\mu_{\rm RC}(\{\omega_{xy}\}) = Z_{\rm RC}^{-1}\, 2^{\#(\{\omega_{xy}\})}\, \mu_{\rm ind}(\{\omega_{xy}\})\, 1_U(\{\omega_{xy}\}), \tag{3}$$

where $Z_{\rm RC}$ is a normalization constant, $\#(\{\omega_{xy}\})$ is the number of clusters determined by the realization $\{\omega_{xy}\}$, $\mu_{\rm ind}(\{\omega_{xy}\})$ is the Bernoulli product measure corresponding to independent occupation variables with $\mu_{\rm ind}(\{\omega_{xy} = 1\}) = \mathcal{P}_{xy}$, and 1_U is the indicator function on the event U in $\{0,1\}^{\mathbf{E}^d}$ that there exists a choice of the spins $\{\sigma_x\}$ so that $J_{xy}\omega_{xy}\sigma_x\sigma_y \geq 0$ for all $\langle x,y\rangle$ [18, 38, 28].

There are several things to be noted about (3). The factor 2 in the term $2^{\#(\{\omega_{xy}\})}$ arises because we have confined ourselves to Ising models, so that every connected cluster of spins (each such connected cluster consists of all satisfied bonds) can be in one of two states (in the ferromagnetic case, all up or all down); in a q-state Potts model, this term would then be replaced by $q^{\#(\{\omega_{xy}\})}$. More importantly, the indicator function on U, which is the event that there is no frustration in the occupied bond configuration, is always one for the ferromagnet; consequently, this term is superfluous for ferromagnetic models. We include it, however, because it will be needed when we generalize to models with frustration. Finally, we note that finite-volume versions of the above formulas, with specified boundary conditions, can be similarly constructed.

In the case of a ferromagnet, there are general theorems [6] which ensure that when percolation occurs, there is a unique percolating cluster. It then easily follows that RC percolation within the FK representation (or "FK percolation" for short) corresponds to the presence of multiple Gibbs states (in the ferromagnet, magnetization up and magnetization down), and moreover that the onset of percolation occurs at the ferromagnetic critical temperature. To prove this, it is sufficient to show that FK percolation is both necessary and sufficient for the breaking of global spin flip symmetry in the ferromagnet. To see that FK percolation is a necessary condition, note that the contribution to the expectation of σ_0 from any finite RC cluster is zero: if a spin configuration σ is consistent with a given RC bond realization within such a cluster, so is $-\sigma$, and both will be equally likely. As a consequence, $\langle\sigma_0\rangle = 0$ in infinite volume in the absence of RC percolation.

To see that RC percolation is a sufficient condition for the magnetization order parameter to be nonzero, consider a finite volume Λ_L with fixed boundary conditions, i.e., a specification $\overline{\sigma}_x = \pm 1$ for each $\overline{\sigma}_x \in \partial\Lambda_L$. For the ferromagnet, by first choosing all $\overline{\sigma}_x = +1$ and then all $\overline{\sigma}_x = -1$, one can change the sign of the spin σ_0 at the origin even as $L \to \infty$. That is, boundary conditions infinitely far away affect σ_0, which is a signature of the existence of multiple Gibbs states.

When attention is confined to ferromagnetic models, the mapping of the FK formalism to interesting statistical mechanical quantities is straightforward (and intuitive); for example

$$\langle \sigma_x \sigma_y \rangle = \mu_{\mathrm{RC}}(x \leftrightarrow y)\,, \tag{4}$$

where $\langle \sigma_x \sigma_y \rangle$ is the usual Gibbs two-point correlation function and $\mu_{\mathrm{RC}}(x \leftrightarrow y)$ is the RC probability that x and y are in the same cluster. Similarly, with "wired" boundary conditions (i.e., each boundary spin is connected to its neighbors), one has

$$\langle \sigma_x \rangle_+ = \mu_{\mathrm{RC}}(x \leftrightarrow \infty)\,. \tag{5}$$

So for ferromagnets, a phase transition from a unique (paramagnetic) phase at low β to multiple infinite-volume Gibbs states at large β is equivalent to a percolation phase transition for the corresponding RC measure.

2.2. Spin glass models; the TRFK representation

For spin glasses (or other nonferromagnets with frustration) the situation is more complicated. Now for two sites x and y, (4) becomes

$$\langle \sigma_x \sigma_y \rangle = \langle 1_{x \leftrightarrow y} \eta(x,y) \rangle_{\mathrm{RC}}; \;\; \eta(x,y) = \prod_{\langle x'y' \rangle \in \mathcal{C}} \mathrm{sgn}(J_{x'y'})\,, \tag{6}$$

where $\mathcal{C}$ is any path of occupied bonds from x to y. By the definition of U, any two such paths $\mathcal{C}$ and $\mathcal{C}'$ in the *same* cluster will satisfy $\prod_{\langle x'y' \rangle \in \mathcal{C}} \mathrm{sgn}(J_{x'y'}) = \prod_{\langle x'y' \rangle \in \mathcal{C}'} \mathrm{sgn}(J_{x'y'})$.

Just as in the ferromagnet, if percolation of a random cluster occurs in the FK representation, the percolating cluster is unique (in each realization of FK spins and bonds) [13, 30]. In spite of this, RC percolation alone is no longer sufficient to prove broken spin-flip symmetry in the EA spin glass. This is because even in the presence of RC percolation, it is unclear whether there exist any two sets of boundary conditions that can alter the state of the spin at the origin from arbitrarily far away. Although the infinite cluster in any one RC realization is unique, different RC realizations can have different paths from $0 \leftrightarrow \partial \Lambda_L$, and because of frustration this can lead to different signs for σ_0. So percolation might still allow for $\langle \sigma_0 \rangle \to 0$ as $L \to \infty$, independently of boundary condition.

Indeed, it is known that FK bonds percolate well above the spin glass transition temperature. For the three-dimensional Ising spin glass on the cubic lattice, Fortuin–Kasteleyn bonds percolate at $\beta_{\mathrm{FK},p} \approx 0.26$ [3] while the inverse critical temperature is believed to be $\beta_{\mathrm{c}} = 0.89 \pm 0.03$ [20]. Near the spin glass critical temperature, the giant FK cluster includes most of the sites of the system. For this reason, the Swendsen–Wang algorithm, though valid, is inefficient for simulating spin glasses.

However, it is an open question as to whether the presence of "single " (see below) FK percolation would lead to a slower, e.g., power-law, decay of correlation functions even though the Gibbs state is unique. If this were to happen, then the onset of single FK percolation would imply a phase transition in the spin glass, though not multiple Gibbs states and hence no broken spin-flip symmetry. This possibility was suggested in [30]; but to date, no evidence exists to support it.

Nonetheless, single FK percolation remains a *necessary* condition for multiple (symmetry-broken) Gibbs phases in the spin glass, for the same reason as for the ferromagnet. A slightly stronger version of that argument [28] proves that the transition temperature for an EA spin glass, if it exists, is bounded from above by the transition temperature in the corresponding (disordered) ferromagnet.

The essential difference (from the point of view of FK percolation) between the ferromagnet and the spin glass is the factor 1_U in (3). It is somewhat easier to discuss this factor in the context of finite volumes. Let Λ_L denote the L^d cube centered at the origin, and let $\hat{\Lambda}_L$ denote the set of bonds $b = \langle x, y\rangle$ with both x and y in Λ_L. For models containing frustration, U is typically *not* all of $\{0,1\}^{\hat{\Lambda}_L}$, unlike for the ferromagnet; it is the set of all *unfrustrated* bond configurations (so that the product of couplings around any closed loop in such a configuration is always positive).

So is there a way to extend the above considerations to arrive at a sufficient condition for multiple Gibbs states in spin glass models? We begin by noting that FK clusters identify magnetization correlations; but spin glass ordering is manifested by the Edwards–Anderson (EA) order parameter, and not the magnetization, becoming nonzero. The EA order parameter q_{EA} can be defined with respect to two independent replicas of the system, each with the same couplings $\{J_{xy}\}$. Denoting the spins in the two replicas by $\{\sigma_x\}$ and $\{\tau_x\}$, each taking values ± 1, q_{EA} is defined in terms of the overlap,

$$Q = N^{-1} \sum_{\{x\}} \sigma_x \tau_x \ , \tag{7}$$

in the limit as the number of sites $N \to \infty$. In general, Q is a random variable whose maximum possible value is q_{EA}, but in the case where (in the limit $N \to \infty$) $\{\sigma_x\}$ and $\{\tau_x\}$ are drawn from a single pure state, Q takes on only the single value q_{EA}.

Using this as a guide, it appears that one possibility, proposed in [30, 23, 24], for extending FK methods to spin glasses is to use what might be called *double* FK percolation. Here one expands the sample space Ω to include two independent copies of the bond occupation variables (for a given $\mathcal{J}$ configuration), and defines the variable $r_{xy} = \omega_{xy}\omega'_{xy}$, where ω_{xy} and ω'_{xy} are taken from the two copies. One then replaces percolation of $\{\omega_{xy}\}$ in the single RC case with percolation of $\{r_{xy}\}$. It is not hard to see that this would be a sufficient condition for the existence of multiple Gibbs phases (and consequently, for a phase transition).

This is not the only way, however, to use "double" percolation of RC clusters in some form to arrive at a condition for spin glass ordering; the above describes what we denoted in the Introduction as the TRFK approach. In the next subsection we present a closely related approach, the CRM two-replica graphical representation, which requires a more lengthy description. Once this is done, we will discuss how both representations can be used to illuminate the nature of phase transitions in short-range spin glass models.

2.3. The CMR representation

In describing these "double FK" representations, we will find it useful to reformulate our description of the relevant measures slightly using a joint spin-bond distribution introduced by Edwards and Sokal [8]. The statistical weight $\mathcal{W}$ for the Edwards–Sokal distribution on a finite lattice Λ_L is

$$\mathcal{W}(\{\sigma_x\},\{\omega_{xy}\};p) = p^{|\omega|}(1-p)^{N_b-|\omega|}\Delta(\{\sigma_x\},\{\omega_{xy}\}) . \tag{8}$$

Here $|\omega| = \sum_{\langle x,y\rangle}\omega_{xy}$ is the number of occupied bonds and N_b is the total number of bonds on the lattice. The factor $\Delta(\{\sigma_x\},\{\omega_{xy}\})$ is introduced to require that every occupied bond is satisfied; it is defined by

$$\Delta(\{\sigma_x\},\{\omega_{xy}\}) = \begin{cases} 1 & \text{if for every } xy\text{: } \omega_{xy}\sigma_x\sigma_y \geq 0, \\ 0 & \text{otherwise} . \end{cases} \tag{9}$$

With p defined as in (2) and with all $J_{xy} = J > 0$, it is easy to verify that the spin and bond marginals of the Edwards–Sokal distribution are the ferromagnetic Ising model with coupling strength J and the Fortuin–Kasteleyn random cluster model (cf. (3)), respectively.

We can now adapt the above representation to the $\pm J$ Ising spin glass. (With minor modifications, it can also be adapted to Gaussian and other distributions for the couplings.) The corresponding Edwards–Sokal weight is the same as that given in Eq. (8). The Δ factor must still enforce the rule that all occupied bonds are satisfied,

$$\Delta(\{\sigma_x\},\{\omega_{xy}\};\{J_{xy}\}) = \begin{cases} 1 & \text{if for every } xy\text{: } J_{xy}\omega_{xy}\sigma_x\sigma_y \geq 0, \\ 0 & \text{otherwise} . \end{cases} \tag{10}$$

so (cf. (3)) $1_U = \max_{\{\sigma_x\}}\Delta(\{\sigma_x\},\{\omega_{xy}\};\{J_{xy}\})$. The spin marginal of the corresponding Edwards–Sokal distribution is now the Ising spin glass with couplings $\{J_{xy}\}$. But the relationship between spin-spin correlations and bond connectivity is complicated by the presence of antiferromagnetic bonds. Now one has

$$\begin{aligned} &\langle\sigma_x\sigma_y\rangle \\ &\quad = \text{Prob}\{x \text{ and } y \text{ connected by even number of antiferromagnetic bonds}\} \\ &\qquad - \text{Prob}\{x \text{ and } y \text{ connected by odd number of antiferromagnetic bonds}\}. \end{aligned} \tag{11}$$

The two-replica CMR graphical representation, introduced in [7, 33], incorporates, in addition to the spin variables $\{\sigma_x\}$ and $\{\tau_x\}$, two sets of bond variables $\{\omega_{xy}\}$ and $\{\eta_{xy}\}$, each taking values in $\{0,1\}$. Now the Edwards–Sokal weight is

$$\begin{aligned} &\mathcal{W}(\{\sigma_x\},\{\tau_x\},\{\omega_{xy}\},\{\eta_{xy}\};\{J_{xy}\}) \\ &\quad = B_{\text{blue}}(\{\omega_{xy}\})B_{\text{red}}(\{\eta_{xy}\})\Delta(\{\sigma_x\},\{\tau_x\},\{\omega_{xy}\};\{J_{xy}\})\Gamma(\{\sigma_x\},\{\tau_x\},\{\eta_{xy}\}) \end{aligned} \tag{12}$$

where the B's are Bernoulli factors for the two types of bonds,

$$B_{\text{blue}}(\{\omega_{xy}\}) = \mathcal{P}_{\text{blue}}^{|\omega|}(1 - \mathcal{P}_{\text{blue}})^{N_b - |\omega|}, \tag{13}$$

$$B_{\text{red}}(\{\eta_{xy}\}) = \mathcal{P}_{\text{red}}^{|\eta|}(1 - \mathcal{P}_{\text{red}})^{N_b - |\eta|}, \tag{14}$$

and the bond occupation probabilities are

$$\mathcal{P}_{\text{blue}} = 1 - \exp(-4\beta|J|), \tag{15}$$

$$\mathcal{P}_{\text{red}} = 1 - \exp(-2\beta|J|). \tag{16}$$

So $B_{\text{blue}}(\{\omega_{xy}\}) = \mu_{\text{ind}}(\{\omega_{xy}\})$. The Δ and Γ factors constrain where the two types of occupied bonds are allowed,

$$\Delta(\{\sigma_x\}, \{\tau_x\}, \{\omega_{xy}\}; \{J_{xy}\}) = \begin{cases} 1 & \text{if for every } xy\text{: } J_{xy}\omega_{xy}\sigma_x\sigma_y \geq 0 \text{ and } J_{xy}\omega_{xy}\tau_x\tau_y \geq 0, \\ 0 & \text{otherwise}, \end{cases} \tag{17}$$

$$\Gamma(\{\sigma_x\}, \{\tau_x\}, \{\eta_{xy}\}) = \begin{cases} 1 & \text{if for every } xy\text{: } \eta_{xy}\sigma_x\sigma_y\tau_x\tau_y \leq 0, \\ 0 & \text{otherwise}. \end{cases} \tag{18}$$

We refer to the ω-occupied bonds as "blue" and the η-occupied bonds as "red". The Δ constraint says that blue bonds are allowed only if the bond is satisfied in both replicas. The Γ constraint says that red bonds are allowed only if the bond is satisfied in exactly one replica.

It is straightforward to verify that the spin marginal of the two-replica Edwards–Sokal weight is that for two independent Ising spin glasses with the same couplings,

$$\sum_{\{\omega\}\{\eta\}} \mathcal{W}(\{\sigma_x\}, \{\tau_x\}, \{\omega_{xy}\}, \{\eta_{xy}\}; \{J_{xy}\}) = \text{const} \times \exp\left[\beta \sum_{\langle xy \rangle} J_{xy}(\sigma_x\sigma_y + \tau_x\tau_y)\right]. \tag{19}$$

Connectivity by occupied bonds in the two-replica representation is related to correlations of the local spin glass order parameter,

$$Q_x = \sigma_x\tau_x. \tag{20}$$

It is straightforward to verify that

$$\langle Q_x Q_y \rangle = \text{Prob}\{x \text{ and } y \text{ connected by even number of red bonds}\} - \text{Prob}\{x \text{ and } y \text{ connected by odd number of red bonds}\}. \tag{21}$$

As in the case of the FK representation, a minus sign complicates the relationship between correlations and connectivity but in a conceptually different way. The second term in Eq. (21) is independent of the underlying couplings in the model and is present for both spin glasses and ferromagnetic models.

As noted earlier, for ferromagnets (in the absence of non-translation-invariant boundary conditions), the signature of ordering is a *single* percolating FK cluster. For spin glasses, the situation is more complicated as there can be more than one percolating cluster. However, if the CMR graphical representation displays a single percolating blue cluster of largest density, one can similarly show broken symmetry, for either EA or SK models. This is because one can impose "agree" or "disagree" boundary conditions between those σ_x and τ_x boundary spins belonging to the maximum density blue network. In the infinite volume limit, these two boundary conditions give different Gibbs states for the σ-spin system (for fixed τ) related to each other by a global spin flip (of σ).

It is a separate matter, however, to relate in a simple way the EA order parameter to the density difference in blue (or more generally, doubly percolating) clusters. Intuitively, it seems that there should be a simple correspondence, and in fact such a relation is easy to show for the SK model: here the overlap Q is exactly the difference in density between two percolating blue clusters [24]. But it is not immediately obvious that a similar density difference can be simply related to the EA order parameter (although the arguments above imply that here also a density difference implies a nonzero Q.) This is because in a two-replica situation, it is not immediately obvious that there is no contribution from finite clusters in short-range models.

For the TRFK representation, similar reasoning shows that the occurrence of exactly two doubly-occupied percolating FK clusters with different densities implies broken symmetry for the spin system [30] and that Q should equal (and once again, does equal in the SK model) the density difference. In [24] we presented preliminary numerical evidence that there is such a nonzero density difference below the spin glass transition temperature for the $d = 3$ EA $\pm J$ spin glass (for both the TRFK and CMR representations). We will discuss these results further in Sec. 4. But now we turn to a review of some rigorous results, particularly for the SK model, that appeared in [23] and [24].

3. Rigorous results

The SK model permits a fairly extensive rigorous analysis of both the CMR and TRFK representations which, when combined with other known results, permits a sharp picture to emerge of the connection between double FK percolation (we use this more general term to refer to any representation that relies on two FK replicas, such as CMR or TRFK), and a phase transition to a low-temperature spin glass phase with broken spin-flip symmetry. The surprising feature to emerge from this analysis is that in the CMR representation, there already exist well *above* the transition temperature two percolating networks of blue bonds, of equal density and in all respects macroscopically indistinguishable. Below the critical temperature T_c, the indistinguishability is lifted: the two infinite clusters assume different densities.

On the other hand, there is no double FK percolation at all above T_c in the TRFK representation, but below T_c there are again two percolating double clusters of unequal density.

We do not yet know whether this difference in the two pictures above T_c persists in the EA model, but numerical evidence (to be discussed in the next section) so far appears to indicate that it does not: in the EA spin glass, both representations seem to behave similarly to CMR in the SK model.

The SK Hamiltonian for an N-spin system is

$$\mathcal{H}_N = -\frac{1}{\sqrt{N}} \sum_{1 \leq i < j \leq N} J_{ij} \sigma_i \sigma_j \ , \tag{22}$$

where $i, j = 1, \ldots, N$ are vertices on a complete graph. The couplings J_{ij} are i.i.d. random variables chosen from a probability measure ρ satisfying the following properties:

1) The distribution is symmetric: $\rho(u) = \rho(-u)$.
2) ρ has no δ-function at $u = 0$.
3) The moment generating function is finite: $\int_{-\infty}^{\infty} \mathrm{d}\rho(u)\ \mathrm{e}^{-t|u|} < \infty$ for all real t.
4) The second moment of ρ is finite; we normalize it to one: $\int_{-\infty}^{\infty} \mathrm{d}\rho(u)\ u^2 = 1$.

So the results of this section hold for Gaussian and many other distributions (although not diluted ones, because of requirement (2)), but the analysis is simplest for the $\pm J$ model, to which we confine our attention for the remainder of this section. We therefore assume that $J_{ij} = \pm 1$ with equal probability; for this distribution (or any other satisfying property (3)) $\beta_c = 1$ [34, 4].

It is relatively easy to see, using a heuristic argument presented in [24], at what temperature a single FK cluster will form. The SK energy per spin, u, is given above the critical temperature by $u = -\beta/2$. Therefore, for large N the fraction f_s of satisfied edges is

$$f_s \sim \frac{1}{2} - uN^{-1/2} \ . \tag{23}$$

From (2) it follows that a fraction $\mathcal{P}_{\mathrm{FK}} = 1 - \exp(-2\beta N^{-1/2}) \approx 2\beta N^{-1/2}$ of satisfied edges are occupied. According to the theory of random graphs (see [5]), a giant cluster forms in a random graph of N vertices when a fraction x/N of edges is occupied with $x > 1$, and there is then a single giant cluster [10]. So if edges are satisfied independently (which of course they're not – this is why this argument is heuristic only), then single replica FK giant clusters should form when $\beta = xN^{-1/2}$ when $x > 1$. The single replica FK percolation threshold is therefore at

$$\beta_{\mathrm{FK},p} = N^{-1/2} \ , \tag{24}$$

and above this threshold, there should be a single giant FK cluster.

This simple argument can be made rigorous [30], but the basic ideas are already displayed above. The rigorous argument obtains upper and lower bounds for the conditional probability that an edge $\{x_0 y_0\}$ is satisfied, given the satisfaction status of all the other edges. If these bounds are close to each other (for large N)

then treating the satisfied edges as though chosen independently can be justified. We now describe that argument.

To avoid the problem of non-independence of satisfied edges, we fix all couplings J_{xy} but one, which we denote $J_{x_0y_0}$, and ask for the conditional probability of its sign given the configuration of all other couplings and all spins σ_x. The ratio Z_+/Z_- of the partition functions with $J_{x_0y_0} = +N^{-1/2}$ and $J_{x_0y_0} = -N^{-1/2}$ satisfies

$$\exp(-2\beta N^{-1/2}) \leq |Z_+/Z_-| \leq \exp(2\beta N^{-1/2}) . \tag{25}$$

The conditional probabilities $P_\pm$ that $J_{x_0y_0} = \pm N^{-1/2}$ therefore satisfy

$$\mathrm{e}^{-4\beta/\sqrt{N}} \leq \mathrm{e}^{-2\beta/\sqrt{N}}|Z_-/Z_+| \leq P_+/P_- \leq \mathrm{e}^{2\beta/\sqrt{N}}|Z_-/Z_+| \leq \mathrm{e}^{4\beta/\sqrt{N}} . \tag{26}$$

Let P_s (P_u) be the conditional probability for any edge $\{x_0y_0\}$ to be satisfied (unsatisfied) given the satisfaction status of all other edges. These must then be bounded as follows:

$$\mathrm{e}^{-4\beta/\sqrt{N}} \leq P_s/P_u \leq \mathrm{e}^{4\beta/\sqrt{N}} , \tag{27}$$

and therefore

$$\frac{1}{2} - \mathrm{O}(\beta/\sqrt{N}) = (\mathrm{e}^{4\beta/\sqrt{N}}+1)^{-1} \leq P_s \leq (\mathrm{e}^{-4\beta/\sqrt{N}}+1)^{-1} = \frac{1}{2} + \mathrm{O}(\beta/\sqrt{N}) . \tag{28}$$

One now obtains rigorously the same conclusions as before – i.e., (24) is valid with a single giant FK cluster for $\beta = \beta_N \geq xN^{-1/2}$ with any $x > 1$.

Essentially the same argument is used for the more interesting case of double percolation. Here there are two spin replicas denoted by σ and τ. We are now interested in percolation of doubly satisfied edges; the spins at the vertices of each such edge must satisfy $\sigma_x\tau_x = \sigma_y\tau_y$ (and then will be satisfied for exactly one of the two signs of J_{xy}). We note an immediate difference between the CMR and TRFK representations: for the former, doubly satisfied edges are occupied with probability $\mathcal{P}_{\mathrm{CMR}} = 1 - \exp(-4\beta N^{-1/2}) \sim 4\beta N^{-1/2}$, while for the latter, $\mathcal{P}_{\mathrm{TRFK}} = [1 - \exp(-2\beta N^{-1/2})]^2 \sim 4\beta^2/N$. It seems likely that this difference occurs only for the SK model; the various factors of N are absent in the EA model.

We can proceed much as in the single-replica case by dividing all (σ, τ) configurations into two sectors – the *agree* (where $\sigma_x = \tau_x$) and the *disagree* sectors (where $\sigma_x = -\tau_x$). We also denote by N_a and N_d the numbers of sites in the sectors and denote by $D_a = N_a/N$ and $D_d = N_d/N$ the sector densities (so that $D_a + D_d = 1$). The spin overlap Q is then just

$$Q = \frac{1}{N}\sum_x \sigma_x\tau_x = \frac{N_a - N_d}{N} = D_a - D_d . \tag{29}$$

For $\beta \leq \beta_{\mathrm{c}} = 1$, $Q \to 0$ as $N \to \infty$ (because the EA order parameter is zero in the paramagnetic phase) while for $\beta > \beta_{\mathrm{c}} = 1$, $\overline{\langle Q^2 \rangle} > 0$ as $N \to \infty$, where $\overline{(\cdot)}$ denotes an average over couplings. So it must be that $D_a = D_d$ for $\beta \leq 1$ while $D_a \neq D_d$ for $\beta > 1$.

The arguments for the single replica case can be repeated separately within the agree and disagree sectors. Letting $\bar{P}_\pm$ denote the conditional probabilities that $J_{x_0y_0} = \pm N^{-1/2}$ given the other J_{xy}'s and all σ_x's and τ_x's, we have within either of the two sectors that

$$e^{-8\beta/\sqrt{N}} \leq e^{-4\beta/\sqrt{N}} |Z_-/Z_+|^2 \leq \bar{P}_+/\bar{P}_- \leq e^{4\beta/\sqrt{N}} |Z_-/Z_+|^2 \leq e^{8\beta/\sqrt{N}} \tag{30}$$

so that the conditional probability *within a single sector* P_{ds} for x_0y_0 to be doubly satisfied is $(1/2) + O(\beta N^{-1/2})$. For $\beta \leq \beta_c$, we have $D_a = 1/2$, $D_d = 1/2$ (in the limit $N \to \infty$) and so in either sector, double FK percolation is approximately a random graph model with $N/2$ sites and bond occupation probability $(1/2)4\beta^2 N^{-1} = \beta^2 (N/2)^{-1}$; thus double FK giant clusters do not occur for $\beta^2 \leq 1$ in the TRFK representation.

In the CMR representation, blue percolation corresponds to bond occupation probability $(1/2)4\beta N^{-1/2} = \beta N^{1/2}(N/2)^{-1}$ and so the threshold for blue percolation is given by $\beta_{\text{CMR},p} = N^{-1/2}$. But now there are *two* giant clusters, one in each of the two sectors, and as noted above they satisfy $D_a = D_d$ when $\beta \leq \beta_c = 1$. For $\beta > \beta_c$, $D_a \neq D_d$. Since $\beta N^{1/2} \to \infty$ for $\beta > \beta_c$ (indeed for any fixed $\beta > 0$), it follows from random graph theory that each giant cluster occupies the entire sector so that D_a and D_d are also the cluster percolation densities of the two giant clusters.

In the case of two-replica FK percolation for $\beta > \beta_c$, let us denote by $D_{\max}$ and $D_{\min}$ the larger and smaller of D_a and D_d, so that $D_{\max} + D_{\min} = 1$ and $D_{\max} - D_{\min} = Q$. Then for $\beta > \beta_c$, the bond occupation probability in the larger sector is $\beta^2(N/2)^{-1} = 2\beta^2 D_{\max}(D_{\max}N)^{-1}$ with $2\beta^2 D_{\max} > 1$ and there is a (single) giant cluster in that larger sector. There will be another giant cluster (of lower density) in the smaller sector providing $2\beta^2 D_{\min} = \beta^2(1-Q) > 1$. Since $Q \leq q_{\text{EA}}$, for this to be the case it suffices if for $\beta > \beta_c$,

$$q_{\text{EA}} < 1 - \frac{1}{\beta^2} \,. \tag{31}$$

The estimated behavior of q_{EA} both as $\beta \to 1+$ and as $\beta \to \infty$ [4] suggests that this is always valid. In any case, we have rigorously proved that there is a unique maximal density double FK cluster for $\beta > \beta_c$.

An important feature of spin glass order is ultrametricity, which is believed to be true at least for the SK model [27], although it has not yet been proved rigorously. For the spin overlaps coming from three replicas put into rank order, $Q_{(1)} \geq Q_{(2)} \geq Q_{(3)}$, ultrametricity is the property that $Q_{(2)} = Q_{(3)}$. The issue of percolation signatures for ultrametricity in the SK model is treated at length in [23]. We only mention here one of those signatures, which concerns the *four* percolating clusters that arise in a CMR representation of three replicas when considering bonds that are simultaneously blue *both* for replicas one and two as well as for replicas one and three. Denoting the four cluster densities in rank order as $x_{(1)} \geq x_{(2)} \geq x_{(3)} \geq x_{(4)}$, the percolation version of the ultrametric property is that $x_{(1)} > x_{(2)}$ *and* $x_{(3)} = x_{(4)}$. The case $x_{(2)} = x_{(3)}$ (resp., $x_{(2)} > x_{(3)}$) corresponds to $Q_{(1)} > Q_{(2)}$ (resp, $Q_{(1)} = Q_{(2)}$). See [23] for more details.

4. Numerical methods and results

We have carried out numerical simulations to explore the properties of two-replica FK representations described in the previous sections for the case of the three-dimensional $\pm J$ EA spin glass. The Monte Carlo algorithm that we use also makes use of the CMR representation in addition to parallel tempering and Metropolis sweeps. Similar methods have been previously applied by Swendsen and Wang [36, 38, 39] and others [14, 16, 17]. The algorithm is described in more detail in [24], here we provide some additional details about the CMR component of the algorithm. The CMR cluster algorithm alternates between "bond moves" and "spin moves." The bond move takes a pair of EA spin configurations $\{\sigma_x\}$ and $\{\tau_x\}$, both in the same realization of disorder, $\{J_{xy}\}$, and populates the bonds of the lattice with red and blue CMR bonds. Each bond that is satisfied in both replicas is occupied with a blue bond with probability $\mathcal{P}_{\rm blue}$ and each bond that is satisfied in only one replica is occupied with a red bond with probability $\mathcal{P}_{\rm red}$, defined in (15) and (16), respectively. The bond move is described by the conditional probability, $\mathcal{T}(\{\omega_{xy}\},\{\eta_{xy}\}|\{\sigma_x\},\{\tau_x\};\{J_{xy}\})$ for bond configuration $\{\omega_{xy}\},\{\eta_{xy}\}$ given spin configuration $\{\sigma_x\},\{\tau_x\}$. To simplify the notation, we omit the lattice indices and the dependence on J_{xy} in the following equations except where explicitly needed. For the bond move we have,

$$\mathcal{T}(\{\omega\},\{\eta\}|\{\sigma\},\{\tau\}) = \frac{B_{\rm blue}(\{\omega\})B_{\rm red}(\{\eta\})\Delta(\{\sigma\},\{\tau\},\{\omega\})\Gamma(\{\sigma\},\{\tau\},\{\eta\})}{\sum_{\{\omega'\}\{\eta'\}} B_{\rm blue}(\{\omega'\})B_{\rm red}(\{\eta'\})\Delta(\{\sigma\},\{\tau\},\{\omega'\})\Gamma(\{\sigma\},\{\tau\},\{\eta'\})}. \tag{32}$$

The spin move takes a blue and red CMR bond configuration and produces a pair of spin configurations $\{\sigma_x\}$ and $\{\tau_x\}$. All spin configurations obeying the constraints that blue bonds are doubly satisfied and red bonds are singly satisfied have the same probability. The spin move is described by the conditional probability, $\mathcal{T}(\{\sigma_x\},\{\tau_x\}|\{\omega_{xy}\},\{\eta_{xy}\})$:

$$\mathcal{T}(\{\sigma\},\{\tau\}|\{\omega\},\{\eta\}) = \frac{\Delta(\{\sigma\},\{\tau\},\{\omega\})\Gamma(\{\sigma\},\{\tau\},\{\eta\})}{\sum_{\{\sigma'\},\{\tau'\}} \Delta(\{\sigma'\},\{\tau'\},\{\omega\})\Gamma(\{\sigma'\},\{\tau'\},\{\eta\})}. \tag{33}$$

A bond move followed by a spin move constitutes a single sweep of the CMR cluster algorithm. We now demonstrate the validity of the CMR algorithm by showing that detailed balance and ergodicity are satisfied. Ergodicity is clearly satisfied since there is a non-vanishing probability that no red or blue bonds will be created during a bond move so that the spin move can then produce any spin configuration. Detailed balance with respect to the equilibrium spin distribution can be stated as

$$\sum_{\{\omega''\},\{\eta''\}} \mathcal{T}(\{\sigma\},\{\tau\}|\{\omega''\},\{\eta''\})\mathcal{T}(\{\omega''\},\{\eta''\}|\{\sigma'\},\{\tau'\}) \times \exp\Big[\beta \sum_{\{xy\}} J_{xy}(\sigma'_x\sigma'_y + \tau'_x\tau'_y)\Big] \tag{34}$$

$$= \sum_{\{\omega''\},\{\eta''\}} \mathcal{T}(\{\sigma'\},\{\tau'\}|\{\omega''\},\{\eta''\})\mathcal{T}(\{\omega''\},\{\eta''\}|\{\sigma\},\{\tau\})$$
$$\times \exp\left[\beta \sum_{\{xy\}} J_{xy}(\sigma_x\sigma_y + \tau_x\tau_y)\right].$$

This equation must hold for all choices of $\{\sigma_x\}$, $\{\tau_x\}$, $\{\sigma'_x\}$ and $\{\tau'_x\}$. Note that by (12) and (19) the denominator in (32) is simply the Boltzmann weight for two independent EA spin glasses, cancelling the same factors in the numerator of the detailed balance equation. Thus, the LHS of (34) can be written as

$$\sum_{\{\omega''\},\{\eta''\}} \frac{\Delta(\{\sigma\},\{\tau\},\{\omega''\})\Gamma(\{\sigma\},\{\tau\},\{\eta''\})}{\sum_{\{\sigma''\},\{\tau''\}} \Delta(\{\sigma''\},\{\tau''\},\{\omega''\})\Gamma(\{\sigma''\},\{\tau''\},\{\eta''\})} \times B_{\mathrm{blue}}(\{\omega\})B_{\mathrm{red}}(\{\eta\})\Delta(\{\sigma'\},\{\tau'\},\{\omega''\})\Gamma(\{\sigma'\},\{\tau'\},\{\eta''\}). \tag{35}$$

Note that this expression is symmetric under the exchange of $\{\sigma_x\}$, $\{\tau_x\}$ and $\{\sigma'_x\}$, $\{\tau'_x\}$ and thus equal to the RHS of (34), demonstrating that detailed balance holds. The bond configurations observed after the bond move are, in the same fashion, easily shown to be equilibrium CMR bond configurations as described by the bond marginal of the Edwards–Sokal weight (12).

Although the CMR algorithm correctly samples equilibrium configurations of EA spin glasses, in 3D it is not efficient. To obtain a more efficient algorithm, we also make use of parallel tempering and Metropolis sweeps. The CMR algorithm employs two replicas at a single temperature while parallel tempering exchanges replicas at different temperatures. Here we use 20 inverse temperatures equally spaced between $\beta = 0.16$ to $\beta = 0.92$. The phase transition temperature of the system was recently measured as $\beta_c = 0.89 \pm 0.03$ [20]. A single sweep of the full algorithm consists of a CMR cluster sweep for each pair of replicas at each temperature, a parallel tempering exchange between replicas at each pair of neighboring temperatures and a Metropolis sweep for every replica.

We simulated the three-dimensional $\pm J$ Edwards–Anderson model on skew periodic cubic lattices for system sizes 6^3, 8^3, 10^3 and 12^3. For each size we simulated 100 realizations of disorder for 50,000 Monte Carlo sweeps of which the first 1/4 of the sweeps were for equilibration and the remaining 3/4 for data collection. The quantities that we measure are the fraction of sites in the largest blue cluster, $\mathcal{C}_1$ and second largest blue cluster, $\mathcal{C}_2$ and the number of blue CMR wrapping clusters, w_{CMR}, and the number of TRFK "wrapping" clusters, w_{TRFK}. A cluster is said to wrap if it is connected around the system in any of the three directions.

Figure 1 shows the average number $\overline{w_{\mathrm{CMR}}}$ of CMR blue wrapping clusters as a function of inverse temperature β. The curves are ordered by system size with largest size on the bottom for the small β and on top for large β. The data suggests that there is a percolation transition at some $\beta_{\mathrm{CMR},p}$. For $\beta > \beta_{\mathrm{CMR},p}$ there are *two* wrapping clusters while for $\beta < \beta_{\mathrm{CMR},p}$ there are none. Near and above the spin glass transition at $\beta_c \approx 0.89$ the expected number of wrapping clusters falls

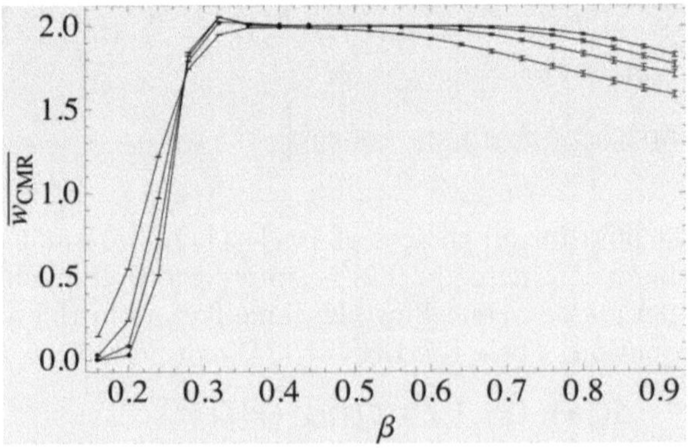

FIGURE 1. Average number of wrapping CMR clusters, $\overline{w_{\text{CMR}}}$ vs. β for the 3D EA model.

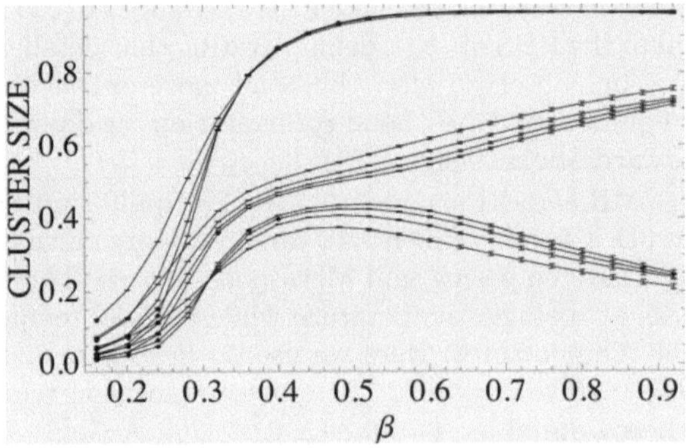

FIGURE 2. $\mathcal{C}_1$ (middle set), $\mathcal{C}_2$ (bottom set) and $\mathcal{C}_1 + \mathcal{C}_2$ (top set) vs. β for the CMR graphical representation for the 3D EA model.

off but the fall-off diminishes as system size increases. This figure suggests that in the large size limit there are exactly two spanning clusters near the spin glass transition both above and below the transition temperature.

Figure 2 shows the fraction of sites in the largest CMR blue cluster, $\mathcal{C}_1$, second largest CMR blue cluster, $\mathcal{C}_2$ and the sum of the two, $\mathcal{C}_1 + \mathcal{C}_2$. The middle set of four curves is $\mathcal{C}_1$ for sizes 6^3, 8^3, 10^3 and 12^3, ordered from top to the bottom at $\beta = 0.5$. The bottom set of curves is $\mathcal{C}_2$ with system sizes ordered from smallest on bottom to largest on top at $\beta = 0.5$. The difference between the fraction of sites in the two largest clusters, $\mathcal{C}_1 - \mathcal{C}_2$ is approximately the spin glass order parameter. As the system size increases, this difference decreases below the transition suggesting that $\mathcal{C}_1 = \mathcal{C}_2$ for $\beta < \beta_c$ in the thermodynamic limit. On the other hand, the sum of the two largest clusters is quite constant independent of system size. Near the transition, approximately 96% of the sites are in the two largest clusters.

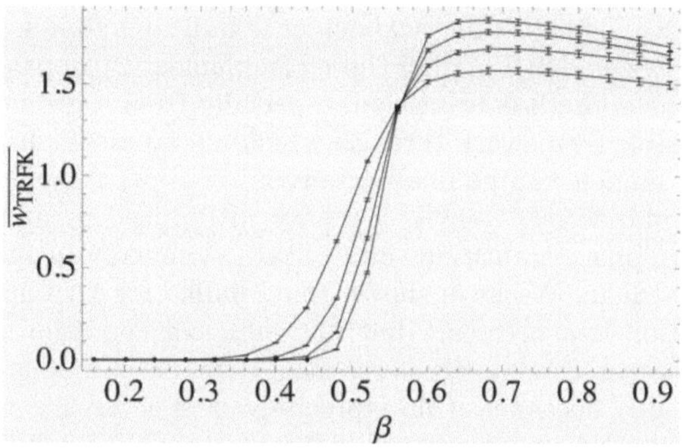

FIGURE 3. Average number of doubly occupied wrapping Fortuin–Kasteleyn clusters, $\overline{w_{\mathrm{TRFK}}}$ vs. β for the 3D EA model.

The large fraction of sites in the two largest clusters makes the CMR cluster moves inefficient. If all sites were in the two largest clusters then the cluster moves would serve only to flip all spins in one or both clusters or exchange the identity of the two replicas. Equilibration depends on the small fraction of spins that are not part of the two largest clusters.

Figure 3 shows the average number of wrapping TRFK clusters $\overline{w_{\mathrm{TRFK}}}$ as a function of inverse temperature. The largest system size is on the bottom for the small β and on top for the large β. As for the case of CMR clusters, the data suggests a transition at some $\beta_{\mathrm{TRFK},p}$ from zero to two wrapping TRFK clusters. Although the number of TRFK wrapping clusters is significantly less than two for all β and all system sizes, the trend in system size suggests that it might approach two for large systems and $\beta > \beta_{\mathrm{TRFK},p}$.

The percolation signature for both CMR and TRFK clusters is qualitatively similar in three dimensions. In both cases two giant clusters with opposite values of the local order parameter appear at a temperature substantially above the phase transition temperature. In the high temperature phase, the two giant clusters have the same density and the phase transition is marked by the onset of different densities of the two clusters. The numerical evidence, however, suggests that the transition from equal to unequal giant cluster densities is quite broad for the small system sizes explored here.

5. Discussion

We begin by noting that our results, even for the SK infinite-range spin glass, are not used to prove a phase transition (which has already been proved using other techniques [2]). Rather, we use as an input the fact that such a transition exists and that the EA order parameter is zero above T_c and nonzero below, and then show that the transition coincides exactly with the onset of a density difference in doubly occupied FK clusters. Numerical evidence indicates that something similar is occurring in the EA model. Together these complementary approaches shed light on the nature of the spin glass transition, especially from a geometric viewpoint, and suggest a possible framework through which a spin glass phase transition in realistic spin glass models can be finally proved.

We now summarize our results. We have introduced a random cluster approach for studying phase transitions and broken symmetry in spin glasses, both short- and infinite-range. We have shown that, unlike for ferromagnetic models, single FK percolation is a necessary but not sufficient condition for broken spin flip symmetry. However, double FK percolation (with a unique largest cluster) is sufficient and probably necessary. (More precisely, it *is* necessary in the SK model, because otherwise the EA order parameter is zero; and it is probably necessary in the EA model.)

In the SK model, there is a difference above T_c between the CMR and TRFK approaches, but not below. In the former, there already is double percolation below β_c, above an onset inverse temperature of $\beta_{\mathrm{CMR},p} = N^{-1/2}$, below which there are no giant clusters. Between this inverse temperature and β_c (which equals one in the class of models we study) there are exactly two giant clusters of equal density, which become $1/2$ in the $N \to \infty$ limit.

More importantly, there is a second transition occurring at exactly the SK spin glass critical value $\beta_c = 1$. Above this threshold, the two giant clusters take on unequal densities, whose sum is one (i.e., every bond and spin belongs to one of the two giant clusters). It could be the case that at some even higher β there is only a single giant cluster, but our methods so far are unable to determine whether this is the case.

For TRFK percolation there are no giant clusters above $\beta_{\mathrm{TRFK},p} = 1$. For $\beta > 1$ there are exactly two giant clusters with unequal densities, and the picture then becomes similar to that of the CMR representation.

The numerical simulations of the 3D EA model suggest a scenario similar to what we find in the SK model. We observe a sharp percolation transition for both the CMR and TRFK representations at a temperature well above the spin glass transition temperature. At this percolation transition, two giant clusters form with opposite values of the EA order parameter. These clusters are nearly equal in density and together occupy most of the system. As the temperature is decreased toward the presumed location of the spin glass transition, the densities of the two clusters becomes increasingly unequal. Although this transition in the densities of the two largest clusters appears quite rounded for the small systems investigated numerically, we believe it is sharp in the thermodynamic limit. The results for both the SK model and the connection between the EA order parameter and the density difference between giant clusters strongly suggests that the spin glass transition in finite dimensions is marked by the onset of this density difference in both the CMR and TRFK representations. These results provide an interesting geometric avenue for investigating the phase transition in the EA model and related models with frustration such as the random bond Ising model or the Potts spin glass.

Acknowledgement

JM was supported in part by NSF DMR-0242402. CMN was supported in part by NSF DMS-0102587 and DMS-0604869. DLS was supported in part by NSF DMS-0102541 and DMS-0604869. Simulations were performed on the Courant Institute of Mathematical Sciences computer cluster. CMN and DLS thank the organizers of the 2007 Paris Summer School "Spin Glasses", from which these lectures were adapted. JM and DLS thank the Aspen Center for Physics, where some of this work was done.

References

[1] Aizenman, M., Chayes, J.T., Chayes, L., Newman, C.M.: The phase boundary in dilute and random Ising and Potts ferromagnets. J. Phys. A Lett. **20**, L313–L318 (1987).

[2] Aizenman, M., Lebowitz, J.L., Ruelle, D.: Some rigorous results on the Sherrington-Kirkpatrick spin glass model. Commun. Math. Phys. **112**, 3–20 (1987).

[3] de Arcangelis, L., Coniglio, A., Peruggi, F.: Percolation transition in spin glasses. Europhysics Letters **14**, 515–519 (1991).

[4] Binder, K., Young, A.P.: Spin glasses: experimental facts, theoretical concepts, and open questions. Rev. Mod. Phys. **58**, 801–976 (1986).

[5] Bollobás, B.: Random Graphs, 2nd ed. Cambridge Univ. Press, Cambridge (2001).

[6] Burton, R.M., Keane, M.: Density and uniqueness in percolation. Commun. Math. Phys. **121**, 501–505 (1989).

[7] Chayes, L., Machta, J., Redner, O.: Graphical representations for Ising systems in external fields. J. Stat. Phys. **93**, 17–32 (1998).

[8] Edwards, R.G., Sokal, A.D.: Generalizations of the Fortuin-Kasteleyn-Swendsen-Wang representation and Monte Carlo algorithm. Phys. Rev. D **38**, 2009–2012 (1988).

[9] Edwards, S., Anderson, P.W.: Theory of spin glasses. J. Phys. F **5**, 965–974 (1975).

[10] Erdös, P., Rényi, E.: On the evolution of random graphs. Magyar Tud. Akad. Mat. Kutató Int. Közl **5**, 17–61 (1960).

[11] Fisher, M.E., Singh, R.R.P.: Critical points, large-dimensionality expansions and Ising spin glasses. In: G. Grimmett, D.J.A. Welsh (eds.) Disorder in Physical Systems, pp. 87–111. Clarendon Press, Oxford (1990).

[12] Fortuin, C.M., Kasteleyn, P.W.: On the random-cluster model. I. Introduction and relation to other models. Physica **57**, 536–564 (1972).

[13] Gandolfi, A., Keane, M., Newman, C.M.: Uniqueness of the infinite component in a random graph with applications to percolation and spin glasses. Probab. Th. Rel. Fields **92**, 511–527 (1992).

[14] Houdayer, J.: A cluster Monte Carlo algorithm for 2-dimensional spin glasses. European Physical Journal B **22**, 479–484 (2001).

[15] Imbrie, J., Newman, C.M.: An intermediate phase with slow decay of correlations in one dimensional $1/|x-y|^2$ percolation, Ising and Potts models. Commun. Math. Phys. **118**, 303–336 (1988).

[16] Jörg, T.: Cluster Monte Carlo algorithms for diluted spin glasses. Prog. Theor. Phys. Suppl. **157**, 349–352 (2005).

[17] Jörg, T.: Critical behavior of the three-dimensional bond-diluted Ising spin glass: Finite-size scaling functions and universality. Phys. Rev. B **73**, 224431, 1–9 (2006).

[18] Kasai, Y., Okiji, A.: Percolation problem describing $\pm J$ ising spin glass system. Prog. Theor. Phys. **79**, 1080–1094 (1988).

[19] Kasteleyn, P.W., Fortuin, C.M.: Phase transitions in lattice systems with random local properties. J. Phys. Soc. Jpn. [Suppl.] **26**, 11–14 (1969).

[20] Katzgraber, H.G., Körner, M., Young, A.P.: Universality in three-dimensional Ising spin glasses: A Monte Carlo study. Phys. Rev. B **73**, 224432, 1–11 (2006).

[21] Kawashima, N., Young, A.P.: Phase transition in the three-dimensional $\pm J$ Ising spin glass. Phys. Rev. B **53**, R484–R487 (1996).

[22] Kotliar, G., Anderson, P.W., Stein, D.L.: A one-dimensional spin glass model with long-range random interactions. Phys. Rev. B **27**, 602–605 (1983).

[23] Machta, J., Newman, C.M., Stein, D.L.: Percolation in the Sherrington-Kirkpatrick spin glass. In: V. Sidoravicious, M.E. Vares (eds.) Progress in Probability Vol. 60: In and Out of Equilibrium 2, pp. 527–542. Birkhäuser, Boston (2008).

[24] Machta, J., Newman, C.M., Stein, D.L.: The percolation signature of the spin glass transition. J. Stat. Phys **130**, 113–128 (2008).

[25] Marinari, E., Parisi, G., Ricci-Tersenghi, F., Ruiz-Lorenzo, J.J., Zuliani, F.: Replica symmetry breaking in spin glasses: Theoretical foundations and numerical evidences. J. Stat. Phys. **98**, 973–1047 (2000).

[26] Marinari, E., Parisi, G., Ritort, F.: On the $3D$ Ising spin glass. J. Phys. A **27**, 2687–2708 (1994).

[27] Mézard, M., Parisi, G., Sourlas, N., Toulouse, G., Virasoro, M.: Nature of the spin glass phase. Phys. Rev. Lett. **52**, 1156–1159 (1984).

[28] Newman, C.M.: Disordered Ising systems and random cluster representations. J. Phys.: Cond. Mat. **15**, R1319 – R1364 (1992).

[29] Newman, C.M., Stein, D.L.: Topical Review: Ordering and broken symmetry in short-ranged spin glasses. J. Phys.: Cond. Mat. **15**, R1319–R1364 (2003).

[30] Newman, C.M., Stein, D.L.: Short-range spin glasses: Results and speculations. In: E. Bolthausen, A. Bovier (eds.) Spin Glasses (Lecture Notes In Mathematics, v. 100), pp. 159–175. Springer, Berlin (2007).

[31] Ogielski, A.T.: Dynamics of three-dimensional spin glasses in thermal equilibrium. Phys. Rev. B **32**, 7384–7398 (1985).

[32] Ogielski, A.T., Morgenstern, I.: Critical behavior of the three-dimensional Ising spin-glass model. Phys. Rev. Lett. **54**, 928–931 (1985).

[33] Redner, O., Machta, J., Chayes, L.F.: Graphical representations and cluster algorithms for critical points with fields. Phys. Rev. E **58**, 2749–2752 (1998).

[34] Sherrington, D., Kirkpatrick, S.: Solvable model of a spin glass. Phys. Rev. Lett. **35**, 1792–1796 (1975).

[35] Sweeny, M.: Monte Carlo study of weighted percolation clusters relevant to the Potts model. Phys. Rev. B **27**, 4445–4455 (1983).

[36] Swendsen, R.H., Wang, J.S.: Replica Monte Carlo simulations of spin glasses. Phys. Rev. Lett. **57**, 2607–2609 (1986).

[37] Thill, M.J., Hilhorst, H.J.: Theory of the critical state of low-dimensional spin glass. J. Phys. I **6**, 67–95 (1996).

[38] Wang, J.S., Swendsen, R.H.: Low temperature properties of the $\pm$ J Ising spin glass in two dimensions. Phys. Rev. B **38**, 4840–4844 (1988).

[39] Wang, J.S., Swendsen, R.H.: Replica Monte Carlo simulation (revisited). Prog. Theor. Phys. Suppl. **157**, 317–323 (2005).

[40] Wolff, U.: Collective Monte Carlo updating for spin systems. Phys. Rev. Lett. **62**, 361–364 (1989).

J. Machta
Physics Department
University of Massachusetts
Amherst, MA 01003, USA
e-mail: `machta@physics.umass.edu`

C.M. Newman
Courant Institute of Mathematical Sciences
New York University
New York, NY 10012, USA
e-mail: `newman@courant.nyu.edu`

D.L. Stein
Physics Department and Courant Institute
of Mathematical Sciences
New York University
New York, NY 10012, USA
e-mail: `daniel.stein@nyu.edu`

Progress in Probability, Vol. 62, 225–231

Fluctuations in Finite-dimensional Spin-glass Dynamics

Claudio Chamon and Leticia F. Cugliandolo

Abstract. We present a very short summary of a theory of dynamic fluctuations in glassy systems that is based on the assumption that a symmetry, time-reparametrization invariance, develops asymptotically in these systems and that it is responsible for spatio-temporal fluctuations. Here we focus primarily on the application of these ideas to disordered spin models with an energy function.

Mathematics Subject Classification (2000). 82C05, 82C20, 34F05.

Keywords. Non-equilibrium systems, dynamics, glasses, spin glasses

1. Introduction

The out-of-equilibrium relaxation of glasses is well understood at the mean-field level. Fully-connected disordered spin systems, finite-dimensional manifolds evolving in infinite-dimensional transverse spaces with quenched random potentials, and mode-coupling-like approximations to models of interacting particles with short-ranged potentials have been successfully analyzed. The results obtained are consistent with the alternative analysis of metastable states that uses an effective free-energy density (Thouless–Anderson–Palmer approach) and the replica trick extended to access metastability and not only equilibrium properties. All these formalisms allow one to analyse global or macroscopic observables and correlation functions measured over the full system [1].

In super-cooled liquids and glasses glassy systems dynamic fluctuations are expected to be very important [2]. It has been only recently that theoretical attention has turned to their analytic description, notably in the super-cooled liquid regime [3]. In this note we shall present a very short summary of a theory of dynamic fluctuations in the *glassy* regime that is based on the assumption that time-reparametrization invariance develops asymptotically in these systems and that it is responsible for spatio-temporal fluctuations (for a recent review see [4]).

We provide a rather complete list of references on this approach that should help the reader find all the details that are omitted here. In this note we focus on the application of these ideas to disordered spin models with an energy function. The proposal has also been discussed in the context of kinetically facilitated spin systems and models of particles in interaction.

2. Mean-field disordered models

The p-spin spherical disordered model mimics glassiness in the so-called fragile glasses [1]. It is defined by its energy function:

$$E_J(\vec{s}) = \sum_{i_1 \neq \cdots \neq i_p} J_{i_1 \ldots i_p} s_{i_1} \ldots s_{i_p} + z\Big(\sum_i s_i^2 - N\Big) ,$$

where z is a Lagrange multiplier enforcing the spherical constraint on the 'vector' spin $\vec{s} = (s_1, \ldots, s_N)$. Each component can take any real value. The $J_{i_1 \ldots i_p}$ are quenched i.i.d. Gaussian random variables with $[\, J_{i_1 \ldots i_p}\,] = 0$ and $[\, J^2_{i_1 \ldots i_p}\,] \propto N^{1-p}$ with the square brackets denoting an average of the disorder strength distribution function. The integer p defines the model and it characterizes different 'universality classes' depending on $p = 2$ or $p > 2$.

The system is coupled to its environment that generates stochastic dynamics for s_i. Since the spin-components are continuous variables one proposes a Langevin dynamics in the overdamped limit

$$\gamma \dot{s}_i(t) = -\frac{\delta E_J(\vec{s})}{s_i(t)} + \xi_i(t) ,$$

with ξ a Gaussian white noise:

$$\langle\, \xi_i(t)\,\rangle = 0 , \qquad \langle\, \xi_i(t)\xi_j(t')\,\rangle = 2\gamma k_B T \delta_{ij} \delta(t - t') .$$

γ is the friction coefficient, T is the temperature of the bath and k_B is the Boltzmann constant ($k_B = 1$ henceforth). A rapid quench from high temperature is mimicked by a random initial conditions, $s_i(0)$, taken, e.g., from a Gaussian pdf. Such an initial condition is uncorrelated with the quenched randomness $J_{i_1 \ldots i_p}$.

In the $N \to \infty$ limit the causal dynamics can be described with the global correlation function

$$C(t, t_w) = N^{-1} \sum_{i=1}^{N} [\, \langle s_i(t) s_i(t_w) \rangle\,] ,$$

the associated linear response function

$$R(t, t_w) = N^{-1} \sum_{i=1}^{N} \left. \frac{\langle\, \delta s_i(t)\,\rangle}{\delta h_i(t_w)} \right|_{h=0} ,$$

or its integral over time $\chi(t,t_w) = \int_{t_w}^{t} \mathrm{d}t' R(t,t')$. In the $N \to \infty$ limit exact causal Schwinger-Dyson equations

$$(\partial_t - z_t)C(t,t_w) = 2TR(t',t) + \int \mathrm{d}t' \, [\Sigma(t,t')C(t',t_w) + D(t,t')R(t_w,t')] \, ,$$

$$(\partial_t - z_t)R(t,t_w) = \delta(t-t_w) + \int \mathrm{d}t' \, \Sigma(t,t')R(t',t_w) \, ,$$

where the self-energy and vertex are functions of C and R:

$$D(t,t_w) = \frac{p}{2}C^{p-1}(t,t_w) \, , \quad \Sigma(t,t_w) = \frac{p(p-1)}{2}C^{p-2}(t,t_w)\, R(t,t_w) \, ,$$

and the time-dependent Lagrange multiplier z_t is fixed by imposing $C(t,t) = 1$ [5, 6]. These equations can be solved numerically but also analytically in the long t_w limit [5, 7] if one uses a few assumptions.

Below a critical temperature $T_d(p)$ the system cannot equilibrate with its environment and relaxes out of equilibrium. The correlation and linear response age (stationary is lost). A separation of time scales controlled by t_w develops and is illustrated in Figure 1 (a). The relaxation below the plateau q scales as

$$C^s(t,t_w) \approx q f_{\rm C}\left(\frac{L(t)}{L(t_w)}\right) \, , \qquad \partial_t C^s(t,t_w) \ll C^s(t,t_w) \, , \tag{1}$$

with $f_{\rm C}(1) = 1$ and $f_{\rm C}(\infty) \to 0$, and it is very slow. One can then approximate the dynamic equations by dropping the time-derivatives and approximating the integrals. The equations for the slow correlation and linear response then become invariant under time-reparametrization. For example, taking $t - t_w \gg t_w$, using $z_t \to z_\infty$, dropping $\partial_t R$ and separating the fast contributions to the integrals the equation for the linear response becomes

$$\tilde{z}_\infty R^s(t,t_w) \sim \int_{t_w}^{t} \mathrm{d}t' \; D'[C^s(t,t')]\, R^s(t,t')\, R^s(t',t_w) \, . \tag{2}$$

$\tilde{z}_\infty$ differs from z_∞ in that it got contributions from the integrals. Equation (2) is invariant under the transformation

$$t \to h_t \equiv h(t) \, , \qquad \begin{cases} C^s(t,t_w) \to C^s(h_t, h_{t_w}) \, , \\ R^s(t,t_w) \to \frac{\mathrm{d}h_{t_w}}{\mathrm{d}t_w} R^s(h_t, h_{t_w}) \, , \end{cases}$$

with h_t any positive-definite and monotonic function of time.

The methods described in [1, 5] allow one to compute analytically $f_{\rm C}$ and $\chi(C)$

$$\chi(t,t_w) \equiv \int_{t_w}^{t} \mathrm{d}t' \, R(t,t') \sim \frac{1-q}{T} + \frac{1}{T_{\rm eff}} C^s(t,t_w) = \chi(C)$$

at times t and t_w such that $1 < L(t)/L(t_w)$, but not the 'clock' $L(t)$. Note that the slow part of χ, $1/T_{\rm eff} C^s(t,t_w)$ is finite since $T_{\rm eff} < +\infty$ [8].

In finite-dimensional models numerical simulations show that the global correlation function also shows a separation of time-scales stationary-aging, although the plateau is less clearly established [9]. A renormalization-group like argument

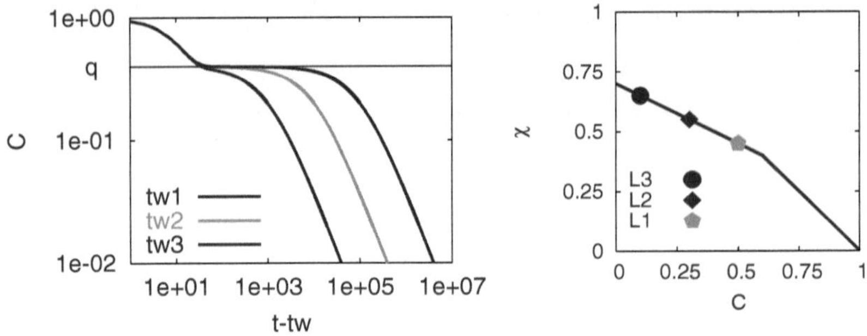

FIGURE 1. (a) Sketch of the decay of the two-time correlation below $T_d(p)$ for three waiting-times $t_{w1} < t_{w2} < t_{w3}$. (b) With a solid line the parameteric construction $\chi(C)$ for t_w fixed and t running from t_w to ∞. The three points are the values obtained at a pair (t, t_w) using different functions $L_1(t) < L_2(t) < L_3(t)$.

based on the assumption that a separation of time-scales exists allows one to show the approximate asymptotic invariance of the slow part of the action S_{slow} in the Martin–Siggia–Rose generating functional of the Langevin equation under *global* time-reparametrizations [10, 11, 12, 13]:

$$t \to h_t \equiv h(t)\,, \qquad \begin{cases} C_r^s(t,t_w) \to C_r^s(h_t, h_{t_w})\,, \\ R_r^s(t,t_w) \to \frac{\mathrm{d}h_{t_w}}{\mathrm{d}t_w}\, R_r^s(h_t, h_{t_w})\,. \end{cases}$$

Symmetry breaking terms become less important as $t_w \to \infty$ and $t - t_w \to \infty$.

The idea is to use this invariance to characterize the fluctuations measured on different boxes with volume V_r centered at sites $\vec{r}$ in the sample:

$$\begin{aligned} C_r(t,t_w) &\equiv \frac{1}{V_r} \sum_{i \in V_r} s_i(t) s_i(t_w)\,, \\ \chi_r(t,t_w) &\equiv \frac{1}{V_r} \sum_{i \in V_r} \int_{t_w}^{t} \mathrm{d}t' \left. \frac{\delta s_i(t)}{\delta h_i(t')} \right|_{h=0}. \end{aligned} \tag{3}$$

These are local coarse-grained two-time functions and the proposal is that they scale as

$$C_r^s(t,t_w) \approx q_r f_{\mathrm{C}} \left(\frac{L_r(t)}{L_r(t_w)} \right) \tag{4}$$

with f_{C} the same scaling function as the one in the global correlation. For instance, different regions can have different $L_1 = \ln(\frac{t}{t_0})$, $L_2 = \frac{t}{t_0}$, $L_3 = \mathrm{e}^{\ln^a(\frac{t}{t_0})}$ with $a > 1$, as sketched in Figure 2 (a). The reason for this is that the time-reparametrization invariance makes it easy to modify the 'clock' from region-to-region (massless fluctuations). Instead, the scaling function f_{C} is hard to change (massive fluctuations).

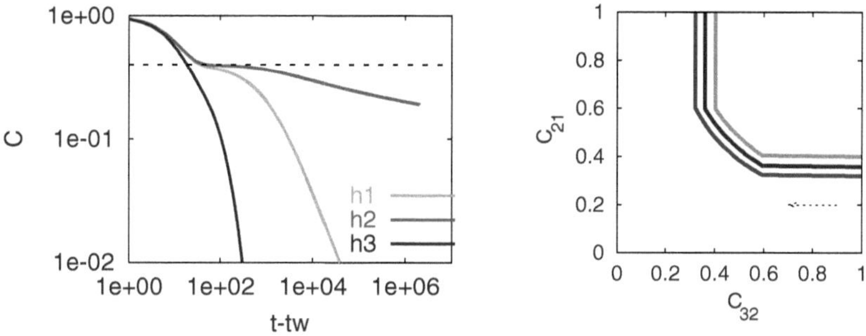

FIGURE 2. (a) Decay of the local two-time correlation, at fixed t_w as a function $t - t_w$, on different slower and faster regions. (b) Sketch of the triangular relations between correlation functions measured at three times on local regions in space.

A number of consequences of this proposal is relatively easy to put to the test numerically or even experimentally and has been summarized in [4]. One of the most striking one is the extension of the triangular relations [14] between global correlation functions measured at times $t_1 \leq t_2 \leq t_3$, taken by pairs $C(t_i, t_j)$ with $i, j = 1, 2, 3$, to the local ones [15]. Indeed, the scaling (3) implies that the local correlations should be related by the *same* triangular relation as the global ones. A sketch of the behaviour expected is shown in Figure 2 (b). Each curve is traced for a region using the intermediate time as a parameter. Each region has its own different value of C_{13}^r. This fact was checked in [15] for the $3d$ Edwards–Anderson model.

In order to go further one should obtain an effective action for the local ages $L_r(t)$ – a sigma model. Of course this is a very difficult task. A possible family of models in which this action could be computed are spin-glass models with Kac-type interactions [17]. In practice, in the past we have just proposed an action $S[L_r]$ requiring it to be

(i) global time-reversal invariant;
(ii) local in space;
(iii) positive definite [10, 11, 12, 16].

We have derived from it some predictions that we checked numerically in disordered finite-dimensional spin models [10, 11, 12, 15] and kinetically constrained models [16].

The analysis of simple coarsening models (the $O(N)$ model in the large N limit) [18] suggests that dynamic fluctuations in simple coarsening systems might be different from the ones in glassy problems. At least for this mean-field model there is no invariance under generic time-reparametrization as the one discussed above and this result seems to be strongly related to the fact that the effective temperature [8], as defined from the deviations from the equilibrium fluctuation dissipation theorem in the out-of-equilibrium relaxation, is infinite in this case. This suggestion should be confirmed by further calculations and numerical studies in other domain growth problems and critical dynamics. One should be very careful, though, and analyse consequences of the time-reparametrization invariance scenario that are not simply due to the existence of a growing length scale (see the analysis and discussion of the dynamics of the $3d$ random field Ising model in [19]). Two candidates are the local triangular relations and the local fluctuation-dissipation relation. It would be interesting to extend the numerical study of Lennard–Jones mixtures [20] to the analysis of these local properties.

Acknowledgement

The results summarized in this paper have been obtained in a series of papers in collaboration with C. Aron, J.J. Arenzon, A.J. Bray, S. Bustingorry, H.E. Castillo, P. Charbonneau, D. Domínguez, S. Franz, L.D.C. Jaubert, M.P. Kennett, J.L. Iguain, M. Picco, D.R. Reichman, M. Sellitto, A. Sicilia, and H. Yoshino. We wish to express our gratitude to all of them.

References

[1] L.F. Cugliandolo, Lecture notes in *Slow Relaxation and non equilibrium dynamics in condensed matter*, Les Houches Session 77 July 2002, J.-L. Barrat, J. Dalibard, J. Kurchan, M.V. Feigel'man eds. (Springer-Verlag, 2003); cond-mat/0210312.

[2] M.D. Ediger, Annu. Rev. Phys. Chem. **51**, 99 (2000).

[3] Different points of view are exposed in the following references. S. Franz and G. Parisi, J. Phys. I **5**, 1401 (1995), Phys. Rev. Lett. **79**, 2486 (1997). C. Donati, S. Franz, G. Parisi, and S.C. Glotzer, Phil. Mag. B **79**, 1827 (1999). G. Biroli and J.-P. Bouchaud, Europhys. Lett. **67**, 21 (2004). J.P. Garrahan and D. Chandler, Proc. Natl. Acad. Sci. USA **100**, 9710 (2003). G. Tarjus, S.A. Kivelson, Z. Nussinov, and P. Viot, *The frustration-based approach of supercooled liquids and the glass transition: a review and critical assessment*, cond-mat/0509127 and references therein. V. Lubchenko and P.G. Wolynes, Ann. Rev. Phys. Chem. **58**, 235–266 (2007).

[4] C. Chamon and L.F. Cugliandolo, J. Stat. Mech. (2007) P07022.

[5] L.F. Cugliandolo and J. Kurchan, Phys. Rev. Lett. **71**, 173 (1993).

[6] G. Ben Arous, A. Dembo, and A. Guionnet, Probability theory and related fields **120**, 1–67 (2001).

[7] See A. Dembo, A. Guionnet, and C. Mazza, J. Stat. Phys. **126**, 781–815 (2007) for a rigorous analysis of the high temperature dynamics of these models.

[8] L.F. Cugliandolo, J. Kurchan, and L. Peliti, Phys. Rev. E **55**, 3898 (1997).

[9] H. Rieger, J. Phys. A **26** (1993) L615. F. Belletti *et al.* (the Janus collaboration), *Simulating an Ising spin-glass for* 0.1 *seconds with Janus*, arXiv:0804.1471.

[10] C. Chamon, M.P. Kennett, H.E. Castillo, and L.F. Cugliandolo, Phys. Rev. Lett. **89**, 217201 (2002).

[11] H.E. Castillo, C. Chamon, L.F. Cugliandolo, and M.P. Kennett, Phys. Rev. Lett. **88**, 237201 (2002).

[12] H.E. Castillo, C. Chamon, L.F. Cugliandolo, J.L. Iguain, and M.P. Kennett, Phys. Rev. B **68**, 134442 (2003).

[13] H.E. Castillo, *Time reparametrization symmetry in spin glass models*, arXiv: 0801.0014.

[14] L.F. Cugliandolo and J. Kurchan, J. Phys. A **27**, 5749 (1994).

[15] L.D.C. Jaubert, C. Chamon, L.F. Cugliandolo, and M. Picco, cond-mat/0701116, J. Stat. Mech. (2007) P05001.

[16] C. Chamon, P. Charbonneau, L.F. Cugliandolo, D.R. Reichman, and M. Sellitto, J. of Chem. Phys. **121**, 10120 (2004).

[17] S. Franz, J. Stat. Phys. **126**, 765–780 (2007), and references therein.

[18] C. Chamon, L.F. Cugliandolo, and H. Yoshino, J. Stat. Mech (2006) P01006.

[19] C. Aron, C. Chamon, L.F. Cugliandolo, and M. Picco, J. Stat. Mech. (2008) P05016.

[20] A. Parsaeian, H.E. Castillo, *Growth of spatial correlations in the aging of a simple structural glass* arXiv:cond-mat/0610789; *Equilibrium and non-equilibrium fluctuations in a glass-forming liquid*, arXiv:0802.2560. H.E. Castillo and A. Parsaeian, Nature Phyics **3** 26–28 (2007).

Claudio Chamon
Physics Department
Boston University
590 Commonwealth Avenue
Boston, MA 02215, USA
e-mail: `chamon@bu.edu`

Leticia F. Cugliandolo
Université Pierre et Marie Curie – Paris VI
Laboratoire de Physique Théorique et Hautes Énergies
5ème étage, Tour 25
4 Place Jussieu
F-75252 Paris Cedex 05, France
e-mail: `leticia@lpthe.jussieu.fr`

Part III

Disordered Pinning Models

Progress in Probability, Vol. 62, 235–270

Renewal Sequences, Disordered Potentials, and Pinning Phenomena

Giambattista Giacomin

Abstract. We give an overview of the state of the art of the analysis of disordered models of pinning on a defect line. This class of models includes a number of well-known and much studied systems (like polymer pinning on a defect line, wetting of interfaces on a disordered substrate and the Poland–Scheraga model of DNA denaturation). A remarkable aspect is that, in absence of disorder, all the models in this class are exactly solvable and they display a localization-delocalization transition that one understands in full detail. Moreover the behavior of such systems near criticality is controlled by a parameter and one observes, by tuning the parameter, the *full spectrum* of critical behaviors, ranging from first-order to infinite-order transitions. This is therefore an ideal set-up in which to address the question of the effect of disorder on the phase transition, notably on critical properties. We will review recent results that show that the physical prediction that goes under the name of *Harris criterion* is indeed fully correct for pinning models. Beyond summarizing the results, we will sketch most of the arguments of proof.

Mathematics Subject Classification (2000). 82B44, 60K37, 60K35.

Keywords. Directed polymers, renewal processes, pinning models, disorder, Harris criterion, finite size estimates, rare-stretch strategies, fractional moment estimates.

1. Pinning and disorder: models and motivations

1.1. The basic example: pinning of simple random walks

It is somewhat customary to introduce pinning models by talking of pinning of simple random walks (SRW). This is due to a number of reasons, like the widespread grasp on SRW, or the fact that modeling several pinning phenomena naturally leads to random walk pinning, as we will see. However, we will see also that, in a sense, the SRW case is the hardest to deal with: nonetheless, we are going to follow the tradition and start from SRW pinning.

Let $S := \{S_n\}_{n=0,1,\ldots}$ be a sequence of random variables such that $S_0 = 0$ and such that $\{S_n - S_{n-1}\}_{n=1,2,\ldots}$ are i.i.d. (*i.e.*, independent and identically distributed) symmetric random variables taking only the values $+1$ and -1. The *disorder* is given by a sequence $\omega := \{\omega_n\}_{n=1,2,\ldots}$ of real numbers and we will play with two real parameters β and h. We actually assume that ω is a realization of an i.i.d. sequence of standard Gaussian variables (see Remark 1.4 for some comments on generalizations). We call $\mathbb{P}$ the law of ω and we denote by θ the left-shift operator on $\mathbb{R}^{\mathbb{N}}$: $(\theta\omega)_n = \omega_{n+1}$, where $\mathbb{N} = \{1, 2, \ldots\}$. Our aim is to study the probability measure $P_{N,\omega}$ (N is a positive integer: we will be interested in the limit $N \to \infty$) defined as

$$P_{N,\omega}(s_0, s_1, \ldots, s_N) := \frac{1}{Z_{N,\omega}} \exp\left(\sum_{n=1}^{N} \Big(\beta\omega_n + h\Big) \mathbf{1}_{s_n=0}\right) P(s_0, s_1, \ldots, s_N), \tag{1}$$

where

1. $P(s_1, s_2, \ldots, s_N) = (1/2)^N$ if and only if $s_0 = 0$ and $|s_n - s_{n-1}| = 1$ for $n = 1, 2, \ldots, N$;
2. $Z_{N,\omega}$ is the normalization constant (*partition function*), that is

$$Z_{N,\omega} = \sum_{s_0, s_1, \ldots, s_N} \exp\left(\sum_{n=1}^{N} (\beta\omega_n + h)\, \mathbf{1}_{s_n=0}\right) P(s_0, s_1, \ldots, s_N). \tag{2}$$

We will actually prefer a slightly different definition of the model, namely given the sequence $s = \{s_0, s_1, \ldots\}$ we set

$$\frac{\mathrm{d}\mathbf{P}_{N,\omega}}{\mathrm{d}\mathbf{P}}(s) := \frac{1}{Z_{N,\omega}} \exp\left(\sum_{n=1}^{N} (\beta\omega_n + h)\, \mathbf{1}_{s_n=0}\right). \tag{3}$$

Notice that this time $\mathbf{P}_{N,\omega}$ is a measure on (infinite) sequences, namely the trajectory of the walk all the way to infinity, while in (1) we had defined a measure only up to *step* (or *time*) N. As a matter of fact, if we consider *cylindrical* events of the type $E = \{s = \{s_n\}_{n=0,1,\ldots} : s_0 = t_0, s_1 = t_1, \ldots, s_N = t_N\}$, then the measure of E under $\mathbf{P}_{N,\omega}$ defined in (3) coincides with $P_{N,\omega}(t_0, t_1, \ldots, t_N)$.

Remark 1.1. It is worth stressing that, unless $\beta = 0$, in this model there are two sources of randomness: the polymer chain is modeled by a random walk with law $\mathbf{P}$ and the disorder is a typical realization of the random sequence ω with law $\mathbb{P}$. These two sources of randomness are treated in very different ways: ω is *quenched*, that is chosen once and for all, while the polymer location fluctuates and in fact we study the distribution of S under $\mathbf{P}_{N,\omega}$.

As is well known, the Markov process S is null-recurrent, namely every site of the state space $\mathbb{Z}$ is visited (infinitely often) $\mathbf{P}$-almost surely, but the expectation of the time between successive visits to a given site is infinite. Let us be more explicit about this last concept and let us introduce, for $m \in \mathbb{Z}$,

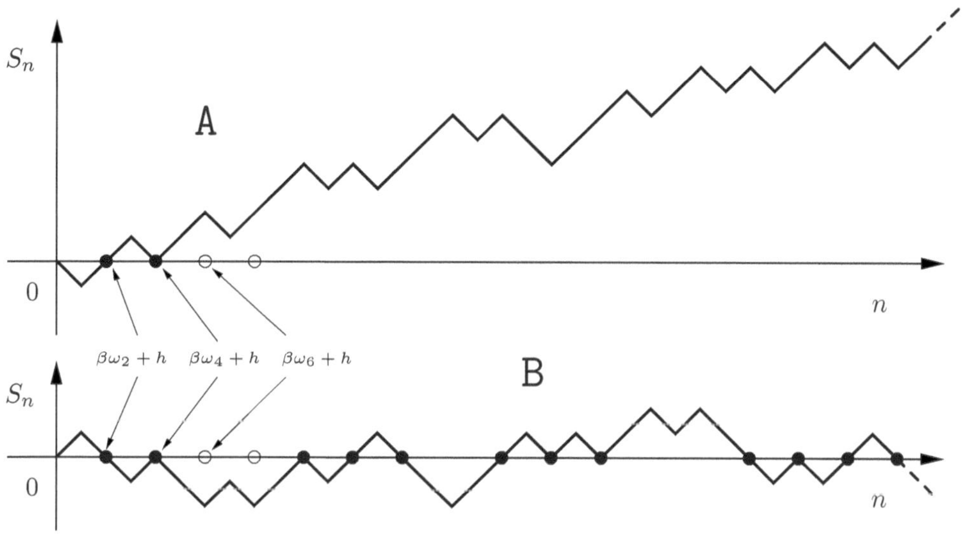

FIGURE 1. Two trajectories sampled from $\mathbf{P}_{N,\omega}$ for different values of β and h, with N *very large*, and represented as directed polymers, in the sense that we plot the function $n \mapsto S_n$ or, equivalently, we look at the walk $\{(n, S_n)\}_{n=0,1,\dots}$ with one deterministic component. The case A sketches a delocalized trajectory that is what one observes when h is, for example, negative and large (one should first think of the homogeneous case $\beta = 0$). Since the SRW is a periodic Markov chain, the origin is visited only at even times, so only $\omega_2, \omega_4, \omega_6, \dots$ play a role and they are the only *charges* (a charge at n is the quantity $\beta\omega_n + h$) marked in the drawing, with a filled circle if they are visited and with an empty circle if they are not visited: not all the unvisited charges are marked). The distinctive aspect of case A is that there are just a few *contacts*, *i.e.*, visits to the origin, and then the walk resembles a walk conditioned not to hit zero. The case B is instead what one observes when h is positive and large: the number of contacts is large, as a matter of fact the drawing wants to suggest that there is a positive density of contacts. It is natural to call such a regime *localized*, in contrast to the previous one that we call *delocalized*. Note that both cases A and B are atypical for the free walk ($\beta = h = 0$) in which there is a zero density of contacts but they are spread through the system and the walk certainly does not stay on one side of the axis as in case A. It is important to remark that we have a full up–down symmetry and this implies that in the delocalized regime A the walk is either delocalized above or below the axis with probability 1/2. Of course at this stage it is highly unclear that one observes either localized or delocalized trajectories (for typical ω) and to a certain extent this is not correct because one has to exclude the so-called *critical regime*, which however appears only at exceptional values of β and h (the *phase transition* point, or *critical* point). We have of course avoided the delicate issue of the role of disorder, at the hearth of this presentation. Here we will simply content ourselves with pointing out that, for example, even when h is very large and negative (pushing thus toward delocalization) any amount of disorder, *i.e.*, $\beta > 0$, yields a positive density of sites in which $\beta\omega_n + h$ is positive and therefore attractive. There could therefore be a *smart targeting strategy* of the polymer in placing the contacts at these sites, leading thus to localization.

the random variable $\tau_1(m) := \inf\{n > 0 : S_n = m\}$ and, for $j > 1$, also $\tau_j(m) := \inf\{n > \tau_{j-1}(m) : S_n = m\}$. It is then a direct consequence of the (strong) Markov property that $\{\tau_{j+1}(m) - \tau_j(m)\}_{j=1,2,\ldots}$ is a sequence of i.i.d. random variables. It should be also clear that the law of $\{\tau_{j+1}(m) - \tau_j(m)\}_{j=1,2,\ldots}$ does not depend on the value of m: we are going to denote $\tau_j(0)$ simply by τ_j and, since $S_0 = 0$, we are setting $\tau_0 = 0$. The recurrent character of S boils simply down to the fact that $\sum_n \mathbf{P}(\tau_1 = n) = 1$ and the recurrence is of null type because $\mathbf{E}[\tau_1] = +\infty$: the distribution of τ_1 is known in detail [23, Ch. III] and in particular

$$\mathbf{P}(\tau_1 = 2n) \overset{n\to\infty}{\sim} \frac{c}{n^{3/2}}, \tag{4}$$

where $c = 1/\sqrt{4\pi}$ and we have introduced the notation

$$a_n \overset{n\to\infty}{\sim} b_n \quad \text{for} \quad \lim_{n\to\infty} \frac{a_n}{b_n} = 1. \tag{5}$$

Since (clearly) $\mathbf{E}[\tau_1] = \infty$, the classical Renewal Theorem (see (12) below) tells us that the expected number of visits to 0 of S up to time N is $o(N)$ (a more precise analysis shows that it is of the order of $\sqrt{N}$, see [23, Ch. III] or Theorem 1.1 below).

As we shall see, the trajectories of the process S are very strongly affected if β or h are non zero and, except for *critical cases*, what happens is roughly that, in the limit as $N \to \infty$, under $\mathbf{P}_{N,\omega}$ the expected number of the visits paid by S to 0 is of the order of N, or it is much smaller than $\sqrt{N}$ (in some cases one can show that they are $O(1)$). A first glimpse at these different scenarios can be found in Figure 1.

1.2. The general model: renewal pinning

We have introduced the τ sequence in the previous subsection in order to give some intuition about the model, but its interest goes well beyond. A look at (3) suffices to realize that $Z_{N,\omega}$ can be expressed simply in terms of τ:

$$Z_{N,\omega} = \mathbf{E} \exp\left(\sum_{n=1}^{N} (\beta\omega_n + h) \mathbf{1}_{n\in\tau} \right), \tag{6}$$

where we have introduced a notation that comes from looking at $\tau = \{\tau_j\}_{j=0,1,\ldots}$ as a random subset of $\mathbb{N} \cup \{0\}$, so that $n \in \tau$ means that there exists j such that $\tau_j = n$. Therefore the model in (3) is just a particular case of when τ is a general discrete renewal:

$$\frac{\mathrm{d}\mathbf{P}^{\mathtt{f}}_{N,\omega}}{\mathrm{d}\mathbf{P}}(\tau) := \frac{1}{Z^{\mathtt{f}}_{N,\omega}} \exp\left(\sum_{n=1}^{N} (\beta\omega_n + h)\, \mathbf{1}_{n\in\tau} \right), \tag{7}$$

where the superscript $\mathtt{f}$, that stands for *free*, has been introduced because a slightly different version of the model is going to be relevant too:

$$\frac{\mathrm{d}\mathbf{P}^{\mathtt{c}}_{N,\omega}}{\mathrm{d}\mathbf{P}}(\tau) := \frac{1}{Z^{\mathtt{c}}_{N,\omega}} \exp\left(\sum_{n=1}^{N} (\beta\omega_n + h)\, \mathbf{1}_{n\in\tau} \right) \mathbf{1}_{N\in\tau}, \tag{8}$$

and c stands for *constrained.* Let us stress that by (discrete) renewal process τ we simply mean a sequence of random variables with (positive and integer-valued) i.i.d. increments: we call these increments *inter-arrival* variables.

We will not be interested in the most general discrete renewal, but we will rather focus on the case in which

$$K(n) := \mathbf{P}(\tau_1 = n) = \frac{L(n)}{n^{1+\alpha}}, \quad n \in \mathbb{N} = \{1, 2, \ldots\} \tag{9}$$

where α is a positive number and

$$\lim_{n\to\infty} L(n) = c_K > 0. \tag{10}$$

We call $K(\cdot)$ *inter-arrival distribution.* Note that we always assume $K(0) = 0$. Moreover we will assume that that $\sum_{n\in\mathbb{N}} K(n) \le 1$: the case $\sum_{n\in\mathbb{N}} K(n) < 1$ has to be interpreted as the case of a *terminating* renewal, in the sense that $K(\infty) := 1 - \sum_{n\in\mathbb{N}} K(n)$ is the probability that $\tau_1 = +\infty$. Therefore, if $K(\infty) > 0$, the cardinality $|\tau|$ of the random set τ is almost surely finite. The case of $K(\infty) = 0$ is instead the case of a *persistent* renewal, and $|\tau| = \infty$ almost surely. But persistent renewals are of two different kinds: they are *positive* persistent if $\sum_n nK(n)(= \mathbf{E}\tau_1) < \infty$, or *null* persistent if the same quantity diverges. This terminology reflects the fact that for any *aperiodic* renewal (aperiodicity refers to the fact that τ_1 does not concentrate on a sublattice of $\mathbb{N}$) the law of large numbers ensures that almost surely

$$\lim_{n\to\infty} \frac{1}{n}|\tau \cap [0, n]| = \frac{1}{\mathbf{E}\tau_1} \in [0, 1], \tag{11}$$

so that if $\mathbf{E}\tau_1 = \infty$ we are facing a *zero density* renewal. We stress that the last statement holds also for terminating renewals for which $\mathbf{E}[\tau_1] = \sum_{n\in\mathbb{N}} nK(n) + \infty K(\infty) = \infty$ (note on the way that, for us, $\sum_n \cdots$ never includes $n = \infty$).

Very relevant for the analysis of renewal processes is the renewal function $n \mapsto \mathbf{P}(n \in \tau)$, that is the probability that the site n is visited by the renewal. We call such a function $K(\cdot)$*-renewal function* when we want to be more precise. The asymptotic behavior of the renewal function is captured by the so-called Renewal Theorem (for a proof see, *e.g.*, [8]). This theorem says that if τ is an aperiodic renewal (the generalization to the periodic case is immediate) we have

$$\lim_{n\to\infty} \mathbf{P}(n \in \tau) = \frac{1}{\mathbf{E}\tau_1} \in [0, 1]. \tag{12}$$

Note the link with (11), but note also that this is little informative if $\mathbf{E}\tau_1 = \infty$. The leading asymptotic behavior in such a case is summed up in the following statement that calls for the definition of the Gamma function: $\Gamma(x) = \int_0^\infty e^{-t} t^{x-1} dt$, for $x > 0$. Recall that in our set-up $\mathbf{E}[\tau_1] = \infty$ either because $K(\infty) > 0$ (terminating renewal), regardless of the value of α, or because $\alpha \le 1$.

Proposition 1.1. *Assuming* (9) *and* (10) *we have:*

1. *If* $K(\infty) > 0$ *then*

$$\mathbf{P}(n \in \tau) \stackrel{n\to\infty}{\sim} \frac{K(n)}{K(\infty)^2}. \tag{13}$$

2. *If* $K(\infty) = 0$ *and* $\alpha \in (0,1)$ *then*

$$\mathbf{P}(n \in \tau) \stackrel{n\to\infty}{\sim} \left(\frac{\alpha \sin(\pi\alpha)}{\pi C_K}\right) n^{\alpha-1}. \tag{14}$$

Proposition 1.1(1) is a classical result detailed for example in [9] or [29, § A.6]. Proposition 1.1(2) is instead more delicate and while the case $\alpha \in (1/2, 1)$ is under control since [27], the full case is instead a rather recent result [21]. Note however that if full proofs are non-trivial, one can rather easily find intuitive arguments suggesting the validity of Proposition 1.1 [29, § A.6].

The asymptotic behavior for the case $\alpha = 1$ is known too, but this case is a bit anomalous and, in this review, we will often skip the results for $\alpha = 1$ that would make the exposition heavier.

Remark 1.2. In full generality, given a renewal τ one can find a Markov process S on a state space containing a point 0 such that $\tau_1 = \inf\{n \in \mathbb{N} : S_n = 0\}$. Still in full generality, the state space can be chosen equal to $\mathbb{N} \cup \{0\}$, see [29, App. A.5] for details. Therefore, with this remark in mind, one could go back to the original Definition (3) without loss of generality.

Remark 1.3. Everything we are going to present works assuming simply that $K(\,\cdot\,)$ is *regularly varying* or, equivalently, that $L(\,\cdot\,)$ is *slowly varying*. Examples of slowly varying functions include $\log(n)^c$, c a real number, or any product of powers of iterated logarithms (see [9] for full definitions and properties or [29, § A.4] for a quick sum-up). Regularly varying functions are a natural set-up for pinning models also because some natural cases do involve slowly varying functions: for example the law of τ_1 for the two-dimensional simple random walk one has $K(n) \stackrel{n\to\infty}{\sim} c/(n(\log n)^2)$, for an explicit value of $c > 0$ [40].

In this subsection we have focused on the behavior of the free system: $\beta = h = 0$. The observations made in the caption of Figure 1 do apply to the general case too (that is to $(\beta, h) \neq (0,0)$), with, nevertheless, two distinctions:

1. If $\alpha > 1$ and if the renewal is persistent, the free renewal is already localized, since by the Renewal Theorem the contact points have a positive density. We will see that this affects the discussion in the caption of Figure 1 only for what concerns the critical case, but the general picture still holds.
2. If τ is terminating it is of course harder to localize the process, but in reality it is rather easy to show that the model can be mapped to a persistent τ case, precisely if one sets $\widetilde{K}(n) = K(n)/(1 - K(\infty))$ and if $\widetilde{\tau}$ is the (persistent!)

$\widetilde{K}(\cdot)$-renewal

$$Z^{\mathtt{c}}_{N,\omega} = \mathbf{E}\left[\exp\left(\sum_{n=1}^{N}(\beta\omega_n + h + \log(1-K(\infty)))\mathbf{1}_{n\in\widetilde{\tau}}\right); N\in\widetilde{\tau}\right], \tag{15}$$

and the same is true also at the level of the measure $\mathbf{P}_{N,\omega}$ itself. The proof of such a statement is absolutely elementary and it is detailed for example in [29, Ch. 1]: note that, from a mathematical standpoint, this allows us to restrict ourselves to persistent renewals τ.

1.3. A gallery of applications

The localization mechanism captured by class of models we have just introduced comes up in modeling a variety of phenomena. Here we just extract some examples and cite some references.

1.3.1. Polymers and defect lines. The interaction between polymers, chains of elementary units called monomers, and the surrounding medium or other polymers is omnipresent in physics, chemistry and biology. We cite for example [28], but it is of course impossible to account for the literature in such a direction. The case we are interested in is the one in which a polymer is fluctuating in a neutral medium except for a line, or a tube, with which the polymer interacts. Actually, also cases in which the line is for example a surface or even simply a point may be modeled by the type of pinning models we are considering. The key point is that polymers are often modeled by self-avoiding random walks and a simplified way to impose the self-avoiding condition is considering directed walks (see Figure 1). So the polymer pinning model becomes precisely the renewal pinning we are considering (we refer to [17, 24, 26, 42, 57] for examples of the phenomena that are modeled via directed walk pinning). Here we just stress that the dimensionality of the problem enters the renewal pinning only via the exponent α: for example a polymer in three dimensions pinned to a line can be modeled by $\{(n, S_n)\}_{n=0,1,\ldots}$, where S is a random walk in two dimensions, for which $\alpha = 0$ (see Remark 1.3). The general case of a physical space of $d+1$ dimensions leads to $\alpha = (d/2) - 1$, for $d \geq 2$, and of course $\alpha = 1/2$ if $d = 1$.

1.3.2. Wetting phenomena. Modeling interfaces in two-dimensional media by random walks has a long history [1] that is somewhat summed up also in [29, App. C] in which one can find the explanation of why anisotropic Ising models do reduce in a suitable limit to the renewal pinning model with $\alpha = 1/2$. A particular choice of the boundary conditions leads to the so-called wetting problem [13, 46], which is just the case in which the random walk trajectories that one considers are only the ones above (and touching) the axis: with reference to Figure 1, to obtain allowed trajectories one has to flip over the negative excursions. As it is explained in detail in [29, Ch. 1], this problem just corresponds to renewal pinning with $\alpha = 1/2$ and $K(\infty) = 1/2$ and, at the level of contact points, the process can be mapped to the case in Figure 1 with h replaced by $h - \log 2$ (see (15)).

1.3.3. DNA denaturation: the Poland–Scheraga model. Two-stranded DNA has been often modeled by two directed walks with pinning potentials, see, *e.g.*, [47] and references therein. Since the difference of two independent random walks is still a random walk, we are back to renewal pinning. However directed walk models lead to values of α that are in contrast with observation, so that a considerable amount of work has been put into understanding whether (in our language) renewal pinning is a reasonable model and which α should be chosen (see in particular [41], but once again we refer to [29] for a more complete bibliography). We note that inhomogeneous or disordered modeling is really required in this context, because the pinning strength does depend on the type of base pair, see Figure 2. The renewal pinning model with inhomogeneous charges has been and is extensively used for the study of DNA denaturation [11, 17]: appropriate values of α are close to 1.15.

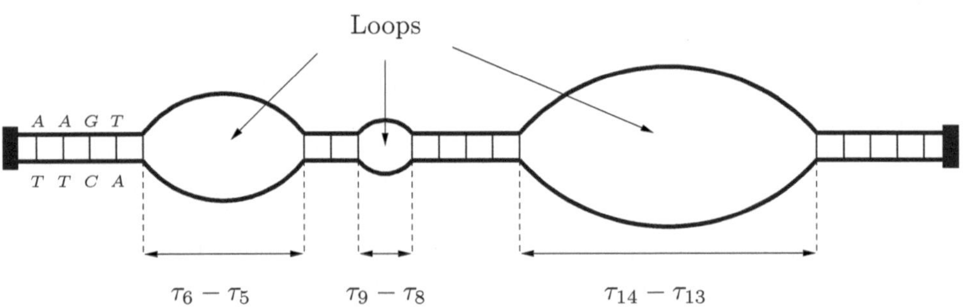

FIGURE 2. The two thick lines are the DNA strands. They may be paired, gaining thus energetic contributions that depend on whether the base pair is A-T or G-C (the model is therefore inhomogeneous). There are then sections of unpaired bases (the *loops*) to which an entropy is associated. The DNA portion in the drawing corresponds to the renewal model trajectory with $\tau_j - \tau_{j-1} = 1$ except three τ-interarrivals (so the loops correspond to inter-arrival of length 2 or more).

Remark 1.4. For DNA denaturation taking ω_1 Gaussian is not appropriate. In this case ω_1 should rather be a binary variable. To be more precise one should also take into account *stacking* energies, that is energies depending on blocks of two pairs, and probably one should also study correlated sequences of bases. Sticking to the issue of binary variables versus Gaussian ones, we take this occasion to stress that much of the mathematical literature is written for rather general charge distribution (say, with finite exponential moments of all orders).

2. The homogeneous case

The full solution of the non disordered ($\beta = 0$), or *homogeneous*, case is crucial and, at the same time, it is rather elementary once it is phrased in the renewal theory language. Such a solution has been repeatedly presented in the physical literature in several particular instances, by using what a probabilist would call *generating function techniques*: in particular one can find a very nice and complete presentation in [24]. However the presentation we are going to outline in detail here is different and much more direct.

In this section, but also later, with abuse of notation we will denote by $Z^{\mathtt{a}}_{N,h}$ ($\mathtt{a} = \mathtt{c}$, $\mathtt{f}$) the partition function $Z^{\mathtt{a}}_{N,\omega}$ when $\beta = 0$. Let us start by observing that we can write

$$Z^{\mathtt{c}}_{N,h} = \sum_{n=1}^{N} \sum_{\substack{\ell \in \mathbb{N}^n: \\ \sum_{j=1}^n \ell_j = N}} \prod_{j=1}^{n} \exp(h) K(\ell_j). \tag{16}$$

Note that if $h = 0$ then $Z^{\mathtt{c}}_{N,h} = \mathbf{P}(N \in \tau)$, *i.e.*, the partition function is just the $K(\cdot)$-renewal function. The right-hand side of (16) is still a renewal function if $\mathrm{e}^h K(\cdot)$ is an inter-arrival law. And indeed it is if $\sum_{n\in\mathbb{N}} \mathrm{e}^h K(n) \leq 1$ and in this case $Z^{\mathtt{c}}_{N,h}$ is the $\mathrm{e}^h K(\cdot)$-renewal function: its asymptotic behavior is hence given in Theorem 1.1, but we prefer to delay such a result since a unified approach holds for every h. In fact if $\sum_{n\in\mathbb{N}} \mathrm{e}^h K(n) \geq 1$ we can renormalize the expression by introducing an exponential correction, going back to a renewal function (times an exponentially growing factor). Precisely we call b (≥ 0) the (unique) solution of

$$\sum_{n\in\mathbb{N}} \exp(-bn + h) K(n) = 1, \tag{17}$$

and we set $K_b(n) := \exp(-bn + h)K(n)$. We have therefore defined a function $h \mapsto b(h)$ for h such that $\sum_{n\in\mathbb{N}} \mathrm{e}^h K(n) \geq 1$, that is for $h \geq h_c(0)$, with

$$h_c(0) := -\log \sum_{n\in\mathbb{N}} K(n). \tag{18}$$

For $h < h_c(0)$ we set instead $b(h) = 0$ and $K_0(n) := \exp(h)K(n)$ (of course the latter notation is poor since h is not explicit). With these notations we can write

$$Z^{\mathtt{c}}_{N,h} = \exp(bN) \sum_{n=1}^{N} \sum_{\substack{\ell \in \mathbb{N}^n: \\ \sum_{j=1}^n \ell_j = N}} \prod_{j=1}^{n} K_b(\ell_j) = \exp(bN)\mathbf{P}_h(N \in \tau), \tag{19}$$

where, under $\mathbf{P}_h$, τ is a $K_{b(h)}(\cdot)$-renewal. By the Renewal Theorem

$$\lim_{N\to\infty} \mathbf{P}_h(N \in \tau) = \frac{1}{\mathbf{E}_h[\tau_1]}, \tag{20}$$

which is a positive constant if $h > h_c(0)$, but it is zero if $h < h_c(0)$ because the $K_{b(h)}(\cdot)$-renewal is terminating. For $h = h_c(0)$ this limit may or may not be zero, but let us postpone this issue to Remark 2.1. Let us focus for now on the fact that for $h < h_c(0)$ the Renewal Theorem does not yield the leading behavior, but thanks to Proposition 1.1(1) we see that

$$\mathbf{P}_h(N \in \tau) \overset{N\to\infty}{\sim} \frac{K(N)}{\left(1 - \exp(h)(1 - K(\infty))\right)^2}. \tag{21}$$

With these results in our hands we see

1. that since $N^{-1}\log \mathbf{P}_h(N\in\tau)$ vanishes as $N\to\infty$, we have therefore proven that
$$\mathsf{F}(0,h) := \lim_{N\to\infty}\frac{1}{N}\log Z^{\mathtt{c}}_{N,h} = b(h), \tag{22}$$
for every h. The quantity $\mathsf{F}(0,h)$ is usually called *free energy* (*per unit volume*) and of course we use such a notation because later there will be $\mathsf{F}(\beta,h)$;
2. that (19) goes well beyond Laplace asymptotics: this is very relevant and allows us for example to compute the limit of
$$\begin{aligned}&\mathbf{P}^{\mathtt{c}}_{N,h}\left(\tau_1=n_1,\tau_2=n_1+n_2,\ldots,\tau_j=n_1+\cdots+n_j\right)\\ &\quad=\prod_{i=1}^{j}(\mathrm{e}^h K(n_i))\frac{Z^{\mathtt{c}}_{N-n_1-\ldots-n_j,h}}{Z^{\mathtt{c}}_{N,h}},\end{aligned} \tag{23}$$
as $N\to\infty$. For example when $h>h_{\mathrm{c}}(0)$ the ratio of partition functions in the right-hand side converges to $\exp(-(n_1+\ldots+n_j)\,\mathsf{F}(0,h))$ and therefore all the expression converges to $\prod_{i=1}^{j}K_{\mathsf{F}(0,h)}(n_i)$. It is rather easy to see that the same holds also for $h<h_{\mathrm{c}}(0)$.

Remark 2.1. In the above list we have been a bit clumsy about the critical case $h=h_{\mathrm{c}}(0)$, but in reality what happens at $h=h_{\mathrm{c}}(0)$ is clear too. First of all, $\sum_n K_{\mathsf{F}(0,h_{\mathrm{c}}(0))}(n)=1$, so that the associated renewal is persistent. More precisely $K_{\mathsf{F}(0,h_{\mathrm{c}}(0))}(\cdot)=K(\cdot)$ if $\sum_n K(n)=1$, and $Z^{\mathtt{c}}_{N,\omega}=\mathbf{P}(N\in\tau)$, and otherwise $K_{\mathsf{F}(0,h_{\mathrm{c}}(0))}(\cdot)$ is just a multiple of $K(\cdot)$ and $Z^{\mathtt{c}}_{N,\omega}$ coincides with the $K_{\mathsf{F}(0,h_{\mathrm{c}}(0))}(\cdot)$-renewal function computed in N. Recall now that the $K_{\mathsf{F}(0,h_{\mathrm{c}}(0))}(\cdot)$-renewal function converges to a positive constant if $\alpha>1$ and to zero otherwise. But when it converges to zero there is Proposition 1.1(2) that comes to our help so that once again we know the sharp asymptotic behavior of $Z^{\mathtt{c}}_{N,\omega}$. In particular $\mathsf{F}(0,h_{\mathrm{c}}(0))=0$.

All these remarks are telling us in particular that (17) is a formula for the free energy, in the sense that $\mathsf{F}(0,h)=b$ if there exists a positive solution b to (17) and otherwise $\mathsf{F}(0,h)=0$. From such a formula one can extract a number of consequences that are summed up in the caption of Figure 3. In particular the behavior of the free energy near criticality is trivial for $h<h_{\mathrm{c}}(0)$, but it is not for $h>h_{\mathrm{c}}(0)$: let us make explicit the behavior of $\mathsf{F}(0,h)$ as $h\searrow h_{\mathrm{c}}(0)$. If $\sum_n nK(n)<\infty$ and if $h_{\mathrm{c}}(0)=0$ (which we may assume without loss of generality: recall (15)!)
$$\sum_n(1-\exp(-b(h)n))K(n) = 1-\exp(-h) \overset{h\searrow 0}{\sim} h. \tag{24}$$

The asymptotic behavior of the left-hand side is easily obtained since for every fixed n the limit of $(1-\exp(-b(h)n)/b$ as $b\searrow 0$ is n. On the other hand $1-\exp(-x)\le x$ for every $x\ge 0$, so that the Dominated Convergence Theorem yields that the left-most side in (24) is asymptotically equivalent to $b\sum_n nK(n)$, and therefore $b(h)\sim$

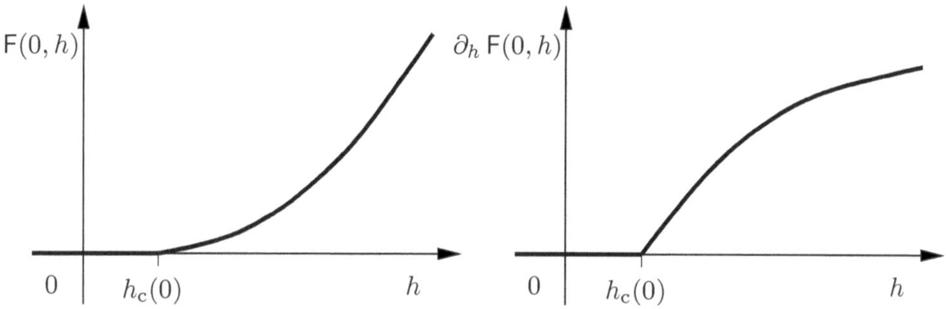

FIGURE 3. The function $h \mapsto \mathsf{F}(0,h)$ is non decreasing, convex and non negative (convexity follows either from (17) or (22)). It is therefore equal to 0 up to $h_c(0) = \sup\{h : \mathsf{F}(0,h) = 0\}$ and after this point it is positive and strictly increasing. Of course $h = h_c(0)$ is a point of non analyticity of the map $h \mapsto \mathsf{F}(0,h)$: this map is of course analytic on $(-\infty, h_c(0))$. It is analytic also on $(h_c(0), \infty)$, by the Implicit Function Theorem for analytic functions. The graph of $\partial_h \mathsf{F}(0,h)$ indicates that we are considering the case $\alpha = 1/2$: $\mathsf{F}(0,\cdot)$ is C^1 but not C^2. For $\alpha \in (1/2, 1]$ the slope of $h \mapsto \partial_h \mathsf{F}(0,h)$ at $h_c(0)$ is infinite and for $\alpha > 1$ a jump discontinuity appears. We stress that $\partial_h \mathsf{F}(0,h)$ is the contact fraction of the system, see Remark 2.4, and therefore such an observable has a jump at the transition for $\alpha > 1$.

$h/\sum_n nK(n)$. If instead $\alpha \in (0,1)$ formula (24) still holds, but the asymptotic behavior of the left-hand side is gotten by Riemann sum approximation:

$$\sum_n (1 - \exp(-b(h)n))\, K(n) \overset{h \searrow 0}{\sim} b^\alpha c_K b \sum_n \frac{1 - \exp(-b(h)n)}{(bn)^{1+\alpha}}$$
$$\sim b^\alpha c_K \int_0^\infty \frac{1-\exp(-x)}{x^{1+\alpha}} \mathrm{d}x = b^\alpha c_K \frac{\Gamma(1-\alpha)}{\alpha}, \quad (25)$$

so that it suffices to invert the asymptotic relation $(b(h))^\alpha c_K (\Gamma(1-\alpha)/\alpha) \overset{h \searrow 0}{\sim} h$ and this of course gives that $b(h)$ is asymptotically proportional to $h^{1/\alpha}$.

The arguments that we have developed directly lead to the following statement (see [29, Ch. 2] for a more complete statement and for more details on the proof):

Theorem 2.1. *The critical behavior of the free energy is given by*

$$\mathsf{F}(0, h_c(0) + \delta) \overset{\delta \searrow 0}{\sim} \begin{cases} c_1 \delta & \text{if } \alpha > 1, \\ c_2 \delta^{1/\alpha} & \text{if } \alpha < 1, \end{cases} \quad (26)$$

with

$$c_1 := \frac{1 - K(\infty)}{\sum_{n \in \mathbb{N}} nK(n)} \quad \text{and} \quad c_2 := \left(\frac{\alpha(1 - K(\infty))}{c_K \Gamma(1-\alpha)} \right)^{1/\alpha}. \quad (27)$$

Moreover in full generality, as $N \to \infty$, $\mathbf{P}^{\mathtt{c}}_{N,h}$ (that denotes the measure $\mathbf{P}^{\mathtt{c}}_{N,\omega}$ when $\beta = 0$) converges weakly in the product topology of $\mathbb{R}^{\mathbb{N}}$ to a limit measure $\mathbf{P}_h$. The *limit process is a* $K_{\mathsf{F}(0,h)}(\cdot)$*-renewal, namely:*

$$\mathbf{P}_h\left(\tau_1 = \ell_1, \tau_2 = \ell_1 + \ell_2, \ldots, \tau_k = \ell_1 + \ldots + \ell_k\right) = \prod_{j=1}^{k} K_{\mathsf{F}(0,h)}(\ell_j), \tag{28}$$

where

$$K_{\mathsf{F}(0,h)}(n) = \begin{cases} \exp(h - n\,\mathsf{F}(0,h))K(n), & \text{if } \mathsf{F}(0,h) > 0, \\ \exp(h)K(n), & \text{if } \mathsf{F}(0,h) = 0. \end{cases} \tag{29}$$

Therefore $\mathsf{F}(0,h) > 0$ *implies that the limit process is positive persistent (in fact, the inter-arrival distribution decays exponentially), while instead if* $\mathsf{F}(0,h) = 0$ *the limit inter-arrival distribution has power law decay and, if* $h < h_c(0)$*, the* $K_{\mathsf{F}(0,h)}(\cdot)$*-renewal is terminating.*

Remark 2.2. Once sharp results for the constrained case are obtained, one can deduce sharp results on the free case. This is just based on the elementary formula

$$Z^{\mathtt{f}}_{N,h} = \sum_{n=0}^{N} Z^{\mathtt{c}}_{n,h} \overline{K}(N-n), \tag{30}$$

that we have written in the case $\sum_{n\in\mathbb{N}} K(n) = 1$ and we have introduced

$$\overline{K}(N) := \sum_{n\in\mathbb{N}:\, n>N} K(n). \tag{31}$$

For example if $h > h_c(0)$ from (19) we have

$$Z^{\mathtt{f}}_{N,h} = \exp(\mathsf{F}(0,h)N) \sum_{n=0}^{N} \mathbf{P}_h(n \in \tau) \exp(-\,\mathsf{F}(0,h)(N-n))\overline{K}(N-n). \tag{32}$$

Since $\exp(-\,\mathsf{F}(0,h)(N-n))\overline{K}(N-n)$ is bounded from below by

$$\sum_{j>N-m} \exp(-\,\mathsf{F}(0,h)j)K(j), \tag{33}$$

and since $\sum_{n=0}^{N} \mathbf{P}(n \in \tau)\overline{K}(N-n) = 1$ for any persistent $K(\cdot)$-renewal we obtain that for every N

$$Z^{\mathtt{f}}_{N,h} \geq \exp(-h)\exp(\mathsf{F}(0,h)N). \tag{34}$$

A (rough) bound in the other direction is obtained by neglecting $\mathbf{P}_h(n \in \tau)\overline{K}(N-n)$ in the right-hand side of (32), so that

$$Z^{\mathtt{f}}_{N,h} \leq \frac{1}{1-\mathsf{F}(0,h)} \exp(\mathsf{F}(0,h)N), \tag{35}$$

which holds once again for every N. The sharp asymptotic result is

$$\begin{aligned} Z^{\mathtt{f}}_{N,h} &\overset{N\to\infty}{\sim} \frac{\exp(\mathsf{F}(0,h)N)}{\sum_n nK_{\mathsf{F}(0,h)}(n)} \sum_{n=0}^{N} \exp(-\,\mathsf{F}(0,h)n)\overline{K}(n) \\ &\sim \frac{(1-\exp(-h))\partial_h \mathsf{F}(0,h)}{1-\exp(-\,\mathsf{F}(0,h))} \exp(\mathsf{F}(0,h)N). \end{aligned} \tag{36}$$

This type of estimates directly leads to computing the limit behavior of $\mathbf{P}^{\mathtt{f}}_{N,h}$, see [29, Ch. 2] for details.

Remark 2.3. A key concept in statistical mechanics (and a key concept here) is the notion of *correlation length.* For example a natural correlation length of the system for $h > h_c(0)$ is given by the rate of exponential decay, as $n \to \infty$, of $\mathbf{P}_h(n \in \tau)$ to its limit value $1/\mathbf{E}_h[\tau_1]$. One can show [30] in particular that if h is sufficiently close to $h_c(0)$ then $\mathbf{P}_h(n \in \tau) - 1/\mathbf{E}_h[\tau_1] > 0$ and

$$\lim_{n\to\infty} \frac{1}{n} \log \left(\mathbf{P}_h(n \in \tau) - \frac{1}{\mathbf{E}_h[\tau_1]} \right) = -\mathsf{F}(0,h), \tag{37}$$

which says that the correlation length coincides with $1/\mathsf{F}(0,h)$. Even if one takes a *finite volume viewpoint*, $1/\mathsf{F}(0,h)$ appears as a natural correlation length, for example because in (19) one sees that is only when N is of the order of $1/\mathsf{F}(0,h)$ that one starts observing the exponential growth of the partition function. One could push these arguments a bit further and see that if N is *much smaller* than $1/\mathsf{F}(0,h)$ (of course this has a precise sense only when $h \searrow h_c(0)$) then $\mathbf{P}^{\mathtt{c}}_{N,h}$ resembles $\mathbf{P}$, while for N *much larger* than $1/\mathsf{F}(0,h)$ the measure $\mathbf{P}^{\mathtt{c}}_{N,h}$ starts exhibiting localization. The fact that the inverse of the free energy is the correlation length still holds also in presence of disorder [35, 52, 53], even if a full understanding of this important issue is still elusive.

Remark 2.4. The density of contacts, or contact fraction, that is the limit as $N \to \infty$ of $N^{-1}\mathbf{E}_h[\sum_{n=1}^N \mathbf{1}_{n\in\tau}]$ coincides by Theorem 2.1 with $\lim_{n\to\infty} \mathbf{P}^{\mathtt{c}}_h(n \in \tau) = 1/\mathbf{E}_h\tau_1$. Note moreover that $N^{-1}\mathbf{E}_h[\sum_{n=1}^N \mathbf{1}_{n\in\tau}] = N^{-1}\partial_h \log Z^{\mathtt{c}}_{N,h}$, so that the contact fraction is equal to $\partial_h \mathsf{F}(0,h)$ (except, possibly, at $h = h_c(0)$). Moreover, by simple conditioning arguments one easily sees for example that for $h \neq h_c(0)$

$$\lim_{M\to\infty} \lim_{N\to\infty} \max_{n:M\le n\le N-M} \left| \mathbf{P}^{\mathtt{c}}_{N,h}(n \in \tau) - \frac{1}{\mathbf{E}_h\tau_1} \right| = 0, \tag{38}$$

and, by arguing like in Remark 2.2, one directly sees that the same statement holds for the free case.

3. The disordered case

3.1. The quenched free energy

An elementary observation that turns out to be really crucial for us at several instances is that for every $M = 0, 1, \ldots, N$

$$\log Z^{\mathtt{c}}_{N,\omega} \ge \log Z^{\mathtt{c}}_{M,\omega} + \log Z^{\mathtt{c}}_{N-M,\theta^M\omega}. \tag{39}$$

It is simply proven by restricting the renewal trajectories, in the expression for $Z^{\mathtt{c}}_{N,\omega}$, to the ones that contain the contact site M and by using the renewal property. By averaging over the disorder one sees that $\{\mathbb{E}\log Z^{\mathtt{c}}_{N,\omega}\}_{N=0,1,\ldots}$ is super-additive and this entails [43] the existence of the limit

$$\lim_{N\to\infty}\frac{1}{N}\mathbb{E}\log Z^{\mathtt{c}}_{N,\omega} =: \mathsf{F}(\beta,h), \tag{40}$$

and the limit of this sequence coincides with its supremum:

$$\mathsf{F}(\beta,h) = \sup_{N\in\mathbb{N}}\frac{1}{N}\mathbb{E}\log Z^{\mathtt{c}}_{N,\omega}. \tag{41}$$

The super-additive property (39) can however be exploited further in order to obtain results about the limit of the non averaged sequence $\{\log Z^{\mathtt{c}}_{N,\omega}\}_{N=0,1,\ldots}$, by applying the tools available for super-additive ergodic sequences, and notably the celebrated Kingman's Theorem [43]. Alternatively one can stick to the super-additive character of $\{\mathbb{E}\log Z^{\mathtt{c}}_{N,\omega}\}_{N=0,1,\ldots}$ and establish a concentration property of the non averaged sequence, either by using concentration inequalities, *e.g.*, [45, 51], or even by more elementary tools [29, Ch. 4]. In all cases the result that one obtains is the existence and the self-averaging character of the free energy of pinning systems:

Theorem 3.1. *The sequence* $\left\{N^{-1}\log Z^{\mathtt{c}}_{N,\omega}\right\}_{N\geq 1}$ *converges to* $\mathsf{F}(\beta,h)$ *both* $\mathbb{P}(\mathrm{d}\omega)$-*almost surely and in the* L^1 *sense.*

Remark 3.1. It is not difficult [29, Ch. 4] to show that for every $K(\cdot)$, β and h there exists $c>0$ such that

$$Z^{\mathtt{c}}_{N,\omega} \leq Z^{\mathtt{f}}_{N,\omega} \leq c\,N Z^{\mathtt{c}}_{N,\omega}, \tag{42}$$

uniformly in ω. We can therefore restate Theorem 3.1 replacing the superscript $\mathtt{c}$ with $\mathtt{f}$.

Another elementary central fact is that

$$\begin{aligned} Z^{\mathtt{c}}_{N,\omega} &\geq \mathbf{E}\left[\exp\left(\sum_{n=1}^{N}(\beta\omega_n+h)\mathbf{1}_{n\in\tau}\right); \tau\cap[1,N]=\{N\}\right] \\ &= \exp(\beta\omega_N+h)K(N), \end{aligned} \tag{43}$$

and therefore

$$\mathsf{F}(\beta,h) \geq 0. \tag{44}$$

We now partition the parameter space of the system into:

$$\mathcal{L} := \{(\beta,h): \mathsf{F}(\beta,h)>0\} \quad \text{and} \quad \mathcal{D} := \{(\beta,h): \mathsf{F}(\beta,h)=0\}. \tag{45}$$

$\mathcal{L}$ and $\mathcal{D}$ stand respectively for a "localized" and a "delocalized" regime, a nomenclature that calls for further explanations (see § 3.2 just below), but for the moment we just point out that one of our main aim is to characterize these regions as precisely as possible. And a substantial help is given by the fact that the function $(\beta,h)\mapsto\mathsf{F}(\beta,h)$ is convex, as limit of convex functions, and it is monotonic non decreasing in both variables (monotonicity in h is immediate, in β it is instead a consequence of convexity and of the fact that $\partial_h \mathbb{E} \log Z^{\mathtt{c}}_{N,\omega} = 0$ for $\beta = 0$). Since by (44) we see that $\mathcal{D}$ coincides with $\{(\beta, h) : \mathsf{F}(\beta, h) = 0\}$ so that $\mathcal{D}$ is a convex set.

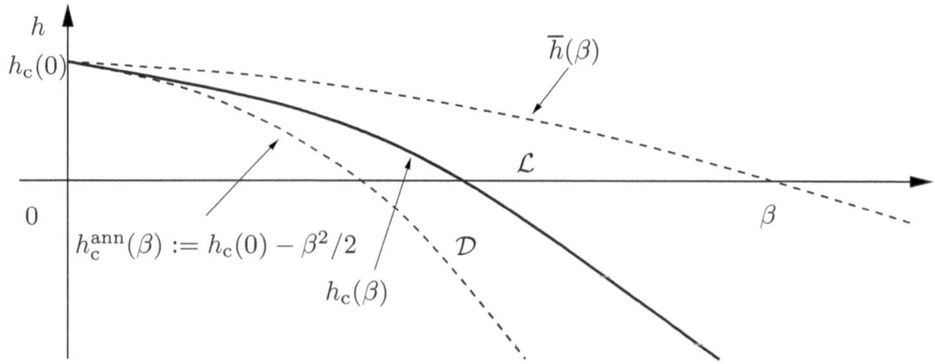

FIGURE 4. The critical curve $\beta \mapsto h_c(\beta)$ that separates $\mathcal{D}$ and $\mathcal{L}$ is concave decreasing. This follows from the fact that $\mathcal{D}$ is a convex set and from the explicit bounds we have on the critical curve. The upper bound $\overline{h}(\beta)$ is less explicit than the lower bound, but we stress that $\overline{h}(\beta) < h_c(0)$ for every $\beta > 0$ and this shows that disorder may induce localization and it never suppresses it. The lower bound comes from the standard annealing procedure. Note that the annealed partition function $\mathbb{E} Z^{\mathtt{a}}_{N,\omega}$, $\mathtt{a} = \mathtt{c}, \mathtt{f}$, is just the homogeneous partition function with pinning potential $h + \beta^2/2$.

One can go beyond: Jensen inequality (*annealing*) yields

$$\mathbb{E} \log Z^{\mathtt{c}}_{N,\omega} \le \log \mathbb{E} Z^{\mathtt{c}}_{N,\omega} = \log \mathbf{E}\left[\exp\left((h+\beta^2/2)\sum_{n=1}^{N} \mathbf{1}_{n\in\tau}\right); N \in \tau\right], \qquad (46)$$

so that

$$\mathsf{F}(\beta, h) \le \mathsf{F}(0, h + \beta^2/2), \qquad (47)$$

and if we recall that $\mathsf{F}(\beta, h) \ge \mathsf{F}(0, h)$ we directly get

$$h_c^{\rm ann}(\beta) := h_c(0) - \frac{\beta^2}{2} \le h_c(\beta) \le h_c(0). \qquad (48)$$

As a matter of fact the upper bound can be made strict, that is $h_c(\beta) < h_c(0)$ as soon as $\beta > 0$, in full generality (the generality here refers to the choice of $K(\cdot)$, [6]) and in the framework that we consider here one can show also that, given $K(\cdot)$, for every $\beta_0 > 0$ one can find an explicit constant $c \in (0, 1/2)$ such that $h_c(\beta) \le h_c(0) - c\beta^2$ for $\beta \in (0, \beta_0]$ [29, Ch. 5]. Instead, showing that $h_c(\beta) > h_c^{\rm ann}(\beta)$ is a more delicate issue (and it is not true in general!). These bounds are summed up in Figure 4 and they imply that, since $\mathcal{D}$ is a convex set, then $h_c(\cdot)$ is concave and, since it is bounded, it is continuous.

Let us sum up the outcome of the arguments we have just outlined:

Proposition 3.1. *If we set* $h_c(\beta) = \inf\{h : \mathsf{F}(\beta, h) > 0\}$ *then*

$$h_c(\beta) = \sup\{h : \mathsf{F}(\beta, h) = 0\}$$

and

$$\mathcal{L} = \{(\beta, h) : h > h_c(\beta)\} \quad \text{and} \quad \mathcal{D} = \{(\beta, h) : h \leq h_c(\beta)\} \tag{49}$$

Moreover the function $\beta \mapsto h_c(\beta)$ *is concave, decreasing and* (48) *holds for every* β.

Remark 3.2. From (41) we actually extract the important observation that localization can be observed in finite volume, in the sense that $(\beta, h) \in \mathcal{L}$ if and only if there exists N such that $\mathbb{E} \log Z^{\mathtt{c}}_{N,\omega} > 0$.

3.2. On path behavior

Characterizing localization and delocalization simply by looking at whether the free energy is positive or zero may look, from a mathematical standpoint, rather cheap. This is not the case as one can first see by observing that

$$\partial_h \frac{1}{N} \log Z^{\mathtt{c}}_{N,\omega} = \mathbf{E}^{\mathtt{c}}_{N,\omega} \left[\sum_{n=1}^{N} \mathbf{1}_{n \in \tau} \right], \tag{50}$$

which, by exploiting the convexity of the free energy and Theorem 3.1, tells us that $\mathbb{P}(\mathrm{d}\omega)$-a.s.

$$\partial_h \mathsf{F}(\beta, h) = \lim_{N \to \infty} \mathbf{E}^{\mathtt{c}}_{N,\omega} \left[\frac{1}{N} \sum_{n=1}^{N} \mathbf{1}_{n \in \tau} \right], \tag{51}$$

when $\partial_h \mathsf{F}(\beta, h)$ exists. By convexity, such a derivative exists except possibly at a countable set of points and in any case (51) can be extended to a (standard) suitable statement also if the derivative does not exist, in terms of right and left derivatives [29] (as a matter of fact, in [35] it is shown that $\mathsf{F}(\beta, \cdot)$ is C^∞ except possibly at $h_c(\beta)$). In the end (51) is telling us that the *contact fraction, i.e.*, the right-hand side in (51), of our system is zero if $h < h_c(\beta)$ (that is, in the interior of $\mathcal{D}$) and it is positive if $h > h_c(\beta)$ (that is, in the whole of $\mathcal{L}$). By itself this fully justifies our definition of (de)localization.

However (51) is still a poor result and plenty of questions could be asked about the limit of the sequence $\{\mathbf{P}^a_{N,\omega}\}_{N=0,1,\ldots}$, $a = \mathtt{c}$ or $a = \mathtt{f}$, starting with the existence of such a limit. And of course the question is: *how close can one get to the very sharp description of the limit measure available for homogeneous systems?*

Work has been done in this direction, but we will not concentrate on this aspect. We just point out that

1. The localized phase is, to a certain extent, rather well understood. In the sense that if $(\beta, h) \in \mathcal{L}$ then one can show that the weak limit as N tends to infinity of the sequence of probability measures $\{\mathbf{P}_{N,\omega}\}_N$ exists $\mathbb{P}(\mathrm{d}\omega)$-a.s. and the limit process is a point process with a positive density of points [35]. One can show also other estimates going toward the completely clear picture that

emerges from the homogeneous case. Intriguing differences however do arise, naturally connected to the existence of exceptional deviations in the sequence of charges. Moreover a number of open questions still stand (see, *e.g.*, [36]).
2. Progress has been made only recently on the delocalized phase, at least away of criticality [32] (see [52] for some estimates at criticality). Essentially one now knows that in the delocalized non critical regime the number of contacts for a system of size N is $O(\log N)$ and such a result has been achieved by a subtle argument combining concentration bounds and super-additivity properties of $\log Z^{c}_{N,\omega}$. Such a result still leaves open intriguing questions in the direction, for example, of the precise results proven in [16, 39] in the homogeneous or weakly inhomogeneous context, see for example in the bibliographic complements at the end of [29, Ch. 8].

3.3. The role of disorder

The main questions we want to address are:

1. How does the disorder affect the phase diagram? Namely can we determine $h_c(\beta)$, for $\beta > 0$, beyond the bounds in Figure 4?
2. What can one say about the critical behavior of the free energy? This amounts to estimating how $\mathsf{F}(\beta, h)$ vanishes as $h \searrow h_c(\beta)$.

It is particularly interesting to raise such questions because we know $h_c(0)$ and we know the sharp asymptotic behavior of $\mathsf{F}(0, h)$ for h close to $h_c(\beta)$ (see Theorem 2.1), so that in our framework inquiring about the role of the disorder makes perfect sense. At this point it is important to underline that such questions do find partial (non rigorous) answers in the physical literature: the rest of this subsection is devoted to explaining what one expects on the basis of formal expansions, following some renormalization group ideas. We must say that the arguments that follow are adaptation to the pinning model context of an argument developed by A.B. Harris [37] in the context of the Ising model with random bond defects. Harris' argument is based on the idea that the behavior of a system near criticality should become rather independent of fine details, so in particular one can replace the system by a coarse grained one without changing substantially the properties. What one actually tries to do is defining a renormalization transformation, like decimation or block summation, that, once applied repeatedly at criticality, transforms the system into a limit model. Harris' work aims at determining whether introducing the disorder modifies the fixed point of the renormalization transformation: if the renormalization transformation suppresses the disorder and the limit point is like in the homogeneous case, then one says that disorder is irrelevant. If instead disorder is enhanced one says that disorder is relevant and most probably the renormalization transformation flow leads to a fixed point which is different from the one obtained in the homogeneous case. It should be noted on the one hand that at the border between relevance and irrelevance the renormalization transformation, to first order, neither decreases nor increases the disorder: this is the so-called marginal case. On the other hand, Harris argument is just a small disorder expansion and as such it does not apply to the whole range of the parameters and, above all, it does not characterize the limit fixed point when disorder is relevant.

Harris' ideas have been first applied in the pinning model context by G. Forgacs, J.M. Luck, Th.M. Nieuwenhuizen and H. Orland [25] and then by B. Derrida, V. Hakim and J. Vannimenus [20] with predictions that differ somewhat in a sense that we are going to explain just below.

Let us start with an expansion that is freely inspired by [25]. Without loss of generality we assume $h_c(0) = 0$ (recall (15)). Moreover the argument does not feel the boundary condition: we work it out in the free case. In what follows $\delta := h + \beta^2/2 \geq 0$: this change of variable is a natural one because in particular

$$\mathbb{E}\left[Z^{\mathtt{f}}_{N,\omega}\right] = \mathbf{E}\left[\exp\left(\delta \sum_{n=1}^{N} \mathbf{1}_{n\in\tau}\right)\right] = Z^{\mathtt{f}}_{N,\delta}. \tag{52}$$

We set $\zeta_n = \exp(\beta\omega_n - \beta^2/2) - 1$ and let us note that

$$\begin{aligned}\mathbb{E}\log\frac{Z^{\mathtt{f}}_{N,\omega}}{\mathbb{E}Z^{\mathtt{f}}_{N,\omega}} &= \mathbb{E}\log\mathbf{E}^{\mathtt{f}}_{N,\delta}\left[\exp\left(\sum_{n=1}^{N}(\beta\omega_n - \beta^2/2)\mathbf{1}_{n\in\tau}\right)\right] \\ &= \mathbb{E}\log\mathbf{E}^{\mathtt{f}}_{N,\delta}\left[\prod_{n=1}^{N}(1+\zeta_n\mathbf{1}_{n\in\tau})\right] \\ &= \mathbb{E}\log\left(1+\sum_{n}\zeta_n\mathbf{P}^{\mathtt{f}}_{N,\delta}(n\in\tau) + \sum_{n_1<n_2}\zeta_{n_1}\zeta_{n_2}\mathbf{P}^{\mathtt{f}}_{N,\delta}(\{n_1,n_2\}\subset\tau)+\cdots\right).\end{aligned} \tag{53}$$

Let us now expand the logarithm and let us use the fact that the ζ random variables are centered and i.i.d. with variance equal to $\exp(\beta^2) - 1$ to see that

$$\mathbb{E}\log\frac{Z^{\mathtt{f}}_{N,\omega}}{\mathbb{E}Z^{\mathtt{f}}_{N,\omega}} = -\frac{1}{2}\left(\exp(\beta^2)-1\right)\sum_{n=1}^{N}\mathbf{P}^{\mathtt{f}}_{N,\delta}(n\in\tau)^2 + \cdots \tag{54}$$

By Remark 2.4, for $\delta > 0$ and as long as n and $N - n$ are large, $\mathbf{P}^{\mathtt{f}}_{N,\delta}(n \in \tau)$ is close to $\partial_\delta \mathsf{F}(0, \delta)$ so that from (53) we extract

$$\begin{aligned}\mathsf{F}(\beta, h_c^{\mathrm{ann}}(\beta)+\delta) &= \mathsf{F}(\beta, \delta - \beta^2/2) \\ &= \mathsf{F}(0,\delta) - \frac{1}{2}\left(\exp(\beta^2)-1\right)\left(\partial_\delta\mathsf{F}(0,\delta)\right)^2 + \cdots\end{aligned} \tag{55}$$

Of course this expansion is only formal and in order to make it rigorous one has to control the rest. Let us note on the way that one can in principle try to compute all the terms in this expansion, but the issue of controlling the rest is still there and convergence issues may very well require β to be small (note that we are expanding using as small parameter the variance of ζ, but aiming at capturing the critical behavior, hence h is small too). All the same, (55) is compatible with $h_c(\beta) = h_c(0)$ if $\mathsf{F}(0, \delta)$ vanishes much slower than $(\partial_\delta \mathsf{F}(0, \delta))^2$ as $\delta \searrow 0$ (β possibly small, but fixed). But by Remark (51) (or directly by taking the h derivative in (17)) we see that $\partial_\delta \mathsf{F}(0,\delta) = 1/\mathbf{E}_\delta \tau_1$ and by direct computation (similar to (25)) one sees that $\partial_\delta \mathsf{F}(0,\delta) \stackrel{\delta \searrow 0}{\sim} (c_2/\alpha)\delta^{-1+1/\alpha}$ for $\alpha \in (0,1)$ (c_2 is given in Theorem 2.1, but the precise value does not play a role here), while the contact fraction is bounded away from zero when $\alpha > 1$ even approaching criticality. So (55) is compatible with $h_c(\beta) = h_c(0)$ if

$$\delta^{1/\alpha} \stackrel{\delta \searrow 0}{\gg} \delta^{2(-1+1/\alpha)} \iff \alpha < \frac{1}{2}. \tag{56}$$

This argument therefore suggests that disorder is irrelevant for $\alpha < 1/2$.

If $\alpha > 1/2$ the expansion we have performed looks hopeless, but we may argue that this is just due to the fact that $h_c(\beta) > h_c^{\mathrm{ann}}(\beta)$ and we are expanding around the *wrong point*. Of course we do know that $\mathsf{F}(\beta, h_c(\beta)) = 0$ and therefore (55) suggests that for β small the shift of the quenched critical point $\delta_c(\beta) := h_c(\beta) - h_c^{\mathrm{ann}}(\beta)$ is found by equating the two terms in the rightmost side of (55) and this procedure suggests $\delta_c(\beta) \approx \beta^{2\alpha/(2\alpha-1)}$.

A second approach is instead inspired by [20]. If we aim at analyzing whether the annealed system is close to the quenched system one could sit at the annealed critical point ($h = h_c(0) - \beta^2/2 = -\beta^2/2$, *i.e.*, $\delta = 0$) and study the variance of $Z_{N,\omega}^{\mathtt{f}}$ (once again, the argument would go through also with constrained boundary condition). Divergence of the variance, as $N \to \infty$, would be a sign that quenched and annealed systems are not close. Since at $\delta = 0$ we have $\mathbf{E} Z_{N,\omega}^{\mathtt{f}} = 1$ and

$$\begin{aligned}\mathsf{var}_{\mathbb{P}}(Z_{N,\omega}^{\mathtt{f}}) &= \mathbb{E}\big[(Z_{N,\omega}^{\mathtt{f}})^2 - 1\big] \\ &= \mathbb{E}\mathbf{E}^{\otimes 2}\Big[\exp\Big(\sum_n (\beta\omega_n - \beta^2/2)(\mathbf{1}_{n\in\tau} + \mathbf{1}_{n\in\tau'})\Big) - 1\Big],\end{aligned} \tag{57}$$

with τ and τ' independent copies of the same renewal process. Integrating out the ω variables we obtain

$$\mathsf{var}_{\mathbb{P}}(Z_{N,\omega}^{\mathtt{f}}) = \mathbf{E}^{\otimes 2}\left[\exp\left(\beta^2 \sum_{n=1}^N \mathbf{1}_{n\in\tau\cap\tau'}\right) - 1\right]. \tag{58}$$

This expression can be evaluated in a sharp way because the random set $\tau \cap \tau'$ is still a renewal process and therefore the variance that we are evaluating is the partition function of a homogeneous pinning model (minus one). And the first relevant question is whether $\tau \cap \tau'$ is a terminating or a persistent renewal. The inter-arrival law of $\tau \cap \tau'$ can be expressed in terms of the inter-arrival law of τ only in an implicit way, but the renewal function of $\tau \cap \tau'$ is explicit in terms of the renewal function of τ:

$$\mathbf{P}^{\otimes 2}(n \in \tau \cap \tau') = \mathbf{P}(n \in \tau)^2, \tag{59}$$

and $\tau \cap \tau'$ is terminating (respectively, persistent) if $\sum_n \mathbf{P}(n \in \tau)^2 < \infty$ (respectively, $\sum_n \mathbf{P}(n \in \tau)^2 = \infty$) and, by Proposition 1.1, we see that

$$\gamma_2 := \sum_{n=1}^\infty \mathbf{P}(n \in \tau)^2 < \infty \iff \sum_n \frac{1}{n^{2(1-\alpha)}} < \infty \iff \alpha < \frac{1}{2}. \tag{60}$$

By the general solution of the homogeneous model, cf. Section 2, we see that if $\tau \cap \tau'$ is persistent, then for every $\beta > 0$ the variance of $Z^{\mathtt{f}}_{N,\omega}$ grows exponentially, while if $\tau \cap \tau'$ is terminating then $X := |\tau \cap \tau'| - 1$ is a geometric random variable (this is just a consequence of the renewal property) of expectation γ_2, that is $\mathbf{P}(X = n) = (\gamma_2/(1+\gamma_2))^n (1/(1+\gamma_2))$, $n = 0, 1, \ldots$. Therefore, as long as $\beta < \beta_0 := \sqrt{\log((1+\gamma_2)/\gamma_2)}$, with $p_2 := 1/(1+\gamma_2)$ we have

$$\lim_{N\to\infty} \mathsf{var}_{\mathbb{P}}\left(Z^{\mathtt{f}}_{N,\omega}\right) = \frac{p_2}{1-(1-p_2)\exp(\beta^2)} - 1 = \gamma_2\beta^2 + \cdots \tag{61}$$

where the expansion is for β small. Therefore if $\tau \cap \tau'$ is terminating (note that τ and τ' are persistent since we are assuming $h_c(0) = 0$) the variance of $Z^{\mathtt{f}}_{N,\omega}$, at the critical annealed point, stays bounded and it is small if β is small. To complement (61) note that

$$\sup_N \mathsf{var}_{\mathbb{P}}\left(Z^{\mathtt{f}}_{N,\omega}\right) \le \frac{p_2}{1-(1-p_2)\exp(\beta^2)} - 1 \overset{\beta\le\beta_0/2}{\le} \widetilde{c}\,\beta^2, \tag{62}$$

for some $\widetilde{c} > 0$.

Let us sum up the outcome of the arguments we have just outlined:

1. Both approaches suggest that disorder is irrelevant if $\alpha < 1/2$ (and, as a consequence, relevant if $\alpha > 1/2$, with the case $\alpha = 1/2$ as marginal one), as one can read from (56) and (60). Moreover the arguments do suggest that the annealed system is very close to the quenched one, in particular $h_c(\beta) = h_c(0)$, at least for β not too large. This observation may be considered as the Harris criterion prediction for pinning models.
2. There is a difference between (56) and (60) in the case $\alpha = 1/2$ that we cannot appreciate since we are assuming (10). In the more general framework of Remark 1.3 one sees that the fact that $L(\cdot)$ diverges at infinity does not imply that $\tau \cap \tau'$ is terminating, while it is sufficient to conclude that $\mathsf{F}(0,\delta)$ is much larger that $(\partial_\delta \mathsf{F}(0,\delta))^2$ for δ small. As a matter of fact, we are dealing with the marginal case in the renormalization group sense. This is a very subtle issue, still unresolved even on a purely heuristic level. We should stress that the steps that we have just presented here are just a part of the arguments in [25] and [20], in particular [25] aims at an expansion to all orders and [20] contains a subtle attempt to study the renormalization group flow for δ close to 0. Both [25] and [20] consider only the case of $L(\cdot)$ asymptotically constant and, for this case, their predictions differ.

3.4. Relevance and irrelevance of the disorder: the results

The heuristic picture outlined in the previous subsection has now been made rigorous. A summary of these rigorous results is given in the next three theorems. We recall that $h_c^{\mathrm{ann}}(\beta) = h_c(0) - \beta^2/2$ and that the annealed free energy is $\mathsf{F}(0, h + \beta^2/2)$, so that the annealed critical behavior is obtained by looking at $\mathsf{F}(0, h_c^{\mathrm{ann}}(\beta) - (\beta^2/2) + \delta) = \mathsf{F}(0, h_c(0) + \delta)$ as $\delta \searrow 0$.

Theorem 3.2. *Choose $\alpha \in (0, 1/2)$ and $K(\cdot)$ satisfying (9) and (10). Then there exists $\beta_0 > 0$ such that $h_c(\beta) = h_c^{\mathrm{ann}}(\beta)$ for $\beta \le \beta_0$. Moreover, for the same values of β the critical behavior of the quenched free energy coincides with the critical behavior of the annealed free energy:*

$$\log \mathsf{F}(\beta, h_c(\beta) + \delta) \overset{\delta \searrow 0}{\sim} \log \mathsf{F}(0, h_c(0) + \delta). \tag{63}$$

Theorem 3.2 has been first proven in [4] by using a modified second moment method that we are going to outline in Section 4. It has then been proven also in [54], by interpolation techniques. Both works contain more detailed results than just Theorem 3.2, in particular (63) has been established by showing that the stronger statement (70) holds. As a matter of fact in [36, Th. 2.3] it has been proven that

$$\lim_{\beta \searrow 0} \limsup_{\delta \searrow 0} \left| \frac{\mathsf{F}(0, \delta) - \mathsf{F}(\beta, h_c^{\mathrm{ann}}(\beta) + \delta)}{(\beta^2/2)(\partial_\delta \mathsf{F}(0, \delta))^2} - 1 \right| = 0, \tag{64}$$

written for $h_c(0) = 0$ for sake of compactness. Note that (64) is in agreement with (55) and in fact the first step in justifying that expansion.

Theorem 3.3. *Choose $K(\cdot)$ satisfying (9) and (10). If $\alpha > 1/2$ we have $h_c(\beta) > h_c^{\mathrm{ann}}(\beta)$. Moreover:*

1. *If $\alpha \in (1/2, 1)$ we have that for every $\varepsilon > 0$ there exists $c_\varepsilon > 0$ such that*

$$h_c(\beta) - h_c^{\mathrm{ann}}(\beta) \ge c_\varepsilon \beta^{\frac{2\alpha}{2\alpha-1}+\varepsilon}, \tag{65}$$

   for every $\beta \le 1$.

2. *For every $K(\cdot)$ such that if $\alpha > 1$ there exists $c > 0$ such that*

$$h_c(\beta) - h_c^{\mathrm{ann}}(\beta) \ge c\beta^2, \tag{66}$$

   for every $\beta \le 1$.

The results in Theorem 3.3 are *almost* sharp, because in [4, 54] it is proven that for every $K(\cdot)$ such that $\alpha \in (1/2, 1)$ there exists $C > 0$ such that

$$h_c(\beta) - h_c^{\mathrm{ann}}(\beta) \le C\beta^{\frac{2\alpha}{2\alpha-1}}, \tag{67}$$

for every $\beta \le 1$. On the other hand the bound in Theorem 3.3(2) is already optimal (in the same sense) in view of the bounds summed up in the caption of Figure 4. The result (65) has now been improved to match precisely (67), *i.e.*, it has been shown in [7] that in (65) one can take $\varepsilon = 0$ and c_0 is still positive.

Theorem 3.3 has been proven in [19] and we give an outline of the proof in Section 5. The method is based on estimating fractional moments of the free energy, while (67) is derived by adapting the techniques yielding Theorem 3.2 (and a sketch of the proof is in Section 4). In [19] the case $\alpha = 1$ is not considered, but in [10] it is shown that the fractional moment method can be generalized to establish in particular that $h_c(\beta) > h_c^{\mathrm{ann}}(\beta)$ also for $\alpha = 1$.

For what concerns the critical behavior we have the following:

Theorem 3.4. *For every $K(\cdot)$ we have*

$$\left(0 \le \mathsf{F}(\beta,h) =\right) \mathsf{F}(\beta,h) - \mathsf{F}(\beta,h_c(\beta)) \le \frac{1+\alpha}{\beta^2}\left(h-h_c(\beta)\right)^2, \tag{68}$$

for every h.

Of course this result is non-trivial only for $h > h_c(\beta)$.

The result in Theorem 3.4 has been established in [34] for rather general charge distribution. The proof that we give in Section 6 uses rather heavily the Gaussian character of the charges and it is close to the argument sketched in [33].

Let us point out that Theorem 3.4, coupled with Theorem 2.1, shows that the critical behavior of quenched and annealed systems differ as soon as $\alpha > 1/2$, in full agreement with the Harris criterion:

$$\begin{aligned}\liminf_{\delta \searrow 0} & \frac{\log\left(\mathsf{F}(\beta,h_c(\beta)+\delta) - \mathsf{F}(\beta,h_c(\beta))\right)}{\log \delta} \\ & \stackrel{\alpha>1/2}{>} \frac{\log\left(\mathsf{F}(0,h_c(0)+\delta) - \mathsf{F}(0,h_c(0))\right)}{\log \delta},\end{aligned} \tag{69}$$

since the left-hand side is bounded below by 2, by Theorem 3.4, and the right-hand side is equal to $\max(1, 1/\alpha)$, by Theorem 2.1.

Remark 3.3. Theorem 3.4 therefore shows that the disorder, for $\alpha > 1/2$, has a smoothing effect on the transition. The Harris criterion in principle is just suggesting that there is no reason to believe that the critical behavior is the same. There is a general belief that disorder smoothes the transitions: this is definitely the case for a number of statistical mechanics models to which a celebrated result of M. Aizenman and J. Wehr applies [3] (see also [38]). It should however be remarked that the Aizenman–Wehr smoothing mechanism does not yield smoothing for the pinning model and that the argument leading to Theorem 3.4 is very different from the argument in [3] (for more on this issue see the caption of Figure 6).

Remark 3.4. The amount of smoothing proven by Theorem 3.4 was not fully expected. In fact in [48] it is claimed that for $\alpha > 1$ the transition is still of first order, in disagreement for example with [17, 18]. Needless to say that it would be very interesting to understand what the value of the exponent for $\alpha > 1/2$ is really and how it depends on α. In [5] it is shown that pinning models based on exponentially decaying inter-arrival laws may not exhibit smoothing.

Remark 3.5. The results we have presented do not consider the case $\alpha = 1/2$. The results in this case are incomplete and not conclusive under hypothesis (10). In [4, 54] it is shown that under (10) one has $h_c(\beta) - h_c^{\mathrm{ann}}(\beta) \le c\exp(-1/(c\beta^2))$ for some $c > 0$ and $\beta \le 1$. This bound matches the prediction in [20]. It is not known whether $h_c(\beta) - h_c^{\mathrm{ann}}(\beta) > 0$, leaving open the possibility for the prediction in [25], *i.e.*, $h_c(\beta) = h_c^{\mathrm{ann}}(\beta)$ for β small, to be the right one. One has to point out, however, that the approach in [25] is an expansion in powers of β

that cannot capture contributions beyond all orders. This issue remains open and debated also at a heuristic level. About the critical behavior, the smoothing result in Theorem 3.4 is once again not conclusive (but it does imply for example that disorder smoothes the transition as soon as $\lim_{n\to\infty} L(n) = 0$, if $L(\cdot)$ is chosen as in Remark 1.3). So, under assumption (10) (which is the one that arises in the basic example, *cf.* (4)), the issue of whether $\alpha = 1/2$ is *marginally relevant* or *marginally irrelevant* is open.

4. Free energy lower bounds and irrelevant disorder estimates

This section is mostly devoted to giving the main ideas of the proof of Theorem 3.2, but in § 4.2 we will also explain why such arguments yield also (67).

4.1. The case of $\alpha < 1/2$: the irrelevant disorder regime

As we have already pointed out, the annealed bound already yields, in full generality, that $\mathsf{F}(\beta, h) \leq \mathsf{F}(0, h+\beta^2/2)$, and hence $h_c(\beta) \geq h_c(0) - \beta^2/2 = h_c^{\mathrm{ann}}(\beta)$. In order to pin, for $\alpha < 1/2$, that $h_c(\beta) \leq h_c^{\mathrm{ann}}(\beta)$ we need to prove a lower bound on the free energy showing that $\mathsf{F}(\beta, h) > 0$ whenever $\mathsf{F}(0, h+\beta^2/2) > 0$. We are actually aiming at capturing also the critical behavior of the free energy; in fact we are aiming at showing that for every $\varepsilon > 0$ there exists $\beta_\varepsilon > 0$ such that for every $\beta \in (0, \beta_\varepsilon)$ we have

$$\liminf_{h \searrow 0} \frac{\mathsf{F}(\beta, h)}{\mathsf{F}(0, h+\beta^2/2)} \geq 1 - \varepsilon, \tag{70}$$

which yields $h_c(\beta) \leq h_c^{\mathrm{ann}}(\beta)$ and is a result (sensibly) stronger than (63). Note that by what we have seen on the expansion of the free energy (55), or by the rigorous result (64), we cannot aim at proving that the quenched free energy coincides with the annealed one. This actually casts some doubts about the applicability of second moment methods. As a matter of fact if we choose $h > -\beta^2/2$ ($h_c(0) = 0$), as usual with $\delta = h + \beta^2/2$ we can write

$$\frac{\mathrm{var}_{\mathbb{P}}\left(Z_{N,\omega}^{\mathtt{f}}\right)}{\left(\mathbb{E}Z_{N,\omega}^{\mathtt{f}}\right)^2} = \mathbf{E}_\delta^{\otimes 2}\left[\exp\left(\beta^2 \sum_{n=1}^{N} \mathbf{1}_{n\in\tau\cap\tau'}\right) - 1\right]. \tag{71}$$

Note the analogy with (58) formula, in which $\delta = 0$: this time, since the underlying measure is $\mathbf{P}_h$ (see the beginning of Section 2 or Theorem 2.1), $\tau \cap \tau'$ is a positive persistent renewal and the expression in (71) is growing exponentially as $N \to \infty$ for every $\beta > 0$ (unlike for the $\delta = 0$ case, in which the exponential growth sets up only for β larger than a positive constant β_0). As we have pointed out, this was to be expected: can one still extract from (71) some interesting information? The answer is positive, as shown by K. Alexander in [4].

The crucial point is not to take the limit in N, but rather exploit (71) up to the scale of the correlation length of the annealed system, which is just a homogeneous system with pinning potential δ (see Remark 2.3). The idea is to establish that the quenched partition function is close to the annealed one with large probability up to the correlation length scale. Note that, as pointed out in Remark 2.3, on such a scale the annealed partition function starts exhibiting exponential growth and one feels the size of the free energy (we will actually need to choose the size of the system, call it N_0, to be a large, but finite, multiple of the correlation length, the relative error ε in (70) leaves the room to make such an estimate). Once such an estimate on a system of length N_0 is achieved, it is a matter of chopping the polymer into N/N_0 portions: some work at the boundary of these regions is needed and for that we refer to [4], while we focus on explaining why the second moment method works up to the correlation length scale.

Let us therefore go back to (71) and let us set $N_0 := q/\,\mathsf{F}(0,\delta)$ (and assume that it is an integer number). With such a choice, by exploiting the estimates outlined in Remark 2.2, it is not difficult to see that for every $K(\cdot)$ and $q > 0$ there exists $c_K(q) > 0$ such that

$$\mathbb{E}Z^{\mathtt{f}}_{N_0,\omega} = \mathbf{E}\exp\left(\delta\sum_{n=1}^{N}\mathbf{1}_{n\in\tau}\right) \leq c_K(q). \tag{72}$$

This is because, as long as q is finite and δ tends to zero, the system is in the *critical window* (as a matter of fact, in [50] the limit of $\mathbb{E}Z^{\mathtt{f}}_{N_0,\omega}$ as δ tends to zero, q kept fixed, is computed and the notion of critical window is further elaborated). The constant $c_K(q)$ of course diverges as $q \nearrow \infty$. On the other hand $\mathbb{E}Z^{\mathtt{f}}_{N,\omega} \geq 1$ so that in order to estimate the quantity in (71) it suffices to estimate

$$\mathbf{E}^{\otimes 2}\left[\exp\left(\delta\sum_{n=1}^{N}(\mathbf{1}_{n\in\tau}+\mathbf{1}_{n\in\tau'})\right)\left(\exp\left(\beta^2\sum_{n=1}^{N}\mathbf{1}_{n\in\tau\cap\tau'}\right)-1\right)\right]. \tag{73}$$

We now use the Cauchy-Schwarz inequality and the fact that $(\exp(x)-1)^2 \leq \exp(2x)-1$ for $x \geq 0$ to bound the expression in (73) (and therefore the expression in (71)) by

$$\mathbf{E}\left[\exp\left(2\delta\sum_{n=1}^{N}\mathbf{1}_{n\in\tau}\right)\right]\mathbf{E}\left[\left(\exp\left(2\beta^2\sum_{n=1}^{N}\mathbf{1}_{n\in\tau\cap\tau'}\right)-1\right)\right]^{1/2} =: T_1\cdot T_2. \tag{74}$$

But, by (72), T_1 is bounded by a constant (which depends on q). On the other hand, T_2 has been already estimated in (58)–(62), and it is $O(\beta)$, thanks to the fact that the renewal $\tau\cap\tau'$ is terminating (since $\alpha < 1/2$). Therefore, by Chebychev inequality, for every $\epsilon > 0$

$$\mathbb{P}\left(Z^{\mathtt{f}}_{N_0,\omega} \geq (1-\epsilon)\mathbf{E}Z^{\mathtt{f}}_{N_0,\omega}\right) \leq \frac{C_K(q)}{\epsilon^2}\sqrt{\widetilde{c}}\,\beta, \tag{75}$$

where $C_K(q)$ is a constant depending on $K(\cdot)$ and q (it is just the constant $c_K(q)$ of (72) when δ is replaced by 2δ) and $\widetilde{c}$ is taken form (62). Since $\mathbb{E}Z^{\mathtt{f}}_{N_0,\omega}$ is bounded from below by $\exp(-\delta)\exp(\mathsf{F}(0,\delta)N_0) = \exp(-\delta+q)$ (see (34)) we see that on the scale of correlation length the quenched partition function grows (almost) like the annealed one with large probability if β is small enough.

4.2. Lower bounds on the free energy beyond the irrelevant disorder regime

The technique for lower bounds on the free energy that we have outlined, as well as the technique in [54, 56], lead to upper bounds on $h_c(\beta)$ also in the case $\alpha \in [1/2, 1)$, see (67) and Remark 3.5. But a look at Theorem 3.3(2) suffices to see that the case $\alpha > 1$ is somewhat different, because it is no longer true that $h_c(\beta) - h_c^{\mathrm{ann}}(\beta) = o(\beta^2)$. So, in this case, the easy bound $h_c(\beta) \le h_c(0)$ (which is just a consequence of convexity) is *optimal* in the sense that $h_c(\beta) - h_c^{\mathrm{ann}}(\beta) \le \beta^2/2 = O(\beta^2)$.

Let us therefore explain why the second moment method yields also (67) when $\alpha \in (1/2, 1)$. For this we go back to (71) and (73) (we are still placing ourselves on the scale of the correlation length). The term T_1 is still bounded by a constant, that of course depends on $K(\cdot)$ and q, just as in the $\alpha < 1/2$ case. The term T_2 this time grows exponentially in N, because this time $\tau \cap \tau'$ is persistent, and we have to worry about the size of N also for this term. But let us quickly estimate the growth rate of T_2 and for which values of N we can expect this term to be small for β small. A necessary condition, that with some careful work one can show also to be sufficient, is that the expectation of the term in the exponent is small, namely that (*cf.* Proposition 1.1)

$$\beta^2 \sum_{n=1}^{N} \mathbf{P}(n \in \tau)^2 \overset{N\to\infty}{\sim} c_\alpha \beta^2 N^{2\alpha-1}, \tag{76}$$

has to be chosen small. However we still keep $N = N_0 = q/\,\mathsf{F}(0,\delta)$, that is N of the order of $\delta^{-1/\alpha}$ (by Theorem 2.1), times a constant which is large if q is large. Plugging such a value of N in (76) we see that we are asking $\beta^2\delta^{(1-2\alpha)/(2\alpha)}$ to be small. Therefore in this regime we expect the second moment method to work, leading to localization and also to the fact that the quenched free energy is fairly close to the annealed one, if

$$\delta \ge c\,\beta^{2\alpha/(2\alpha-1)}, \tag{77}$$

with c a small (fixed) constant. But this what is claimed in (67).

5. Relevant disorder estimates: critical point shift

Annealing is the standard procedure to get upper bounds on disordered partition functions. One can go beyond by partial annealing procedures, like the *constrained* annealing procedure [44], and this does give some results, see, *e.g.*, [4], but for pinning models constrained annealing, in the infinite volume limit, yields nothing beyond the annealed bound if we are concerned with identifying the critical point [15]. There is therefore the need for a different idea.

5.1. Fractional moment estimates

A tool that allows to go beyond the annealed bound $h_c(\beta) \ge h_c^{\mathrm{ann}}(\beta)$ (we set $h_c(0) = 0$ also in this section) turns out to be estimating $A_N := \mathbb{E}[(Z_{N,\omega}^{\mathrm{c}})^\gamma]$ for

$\gamma \in (0,1)$ by means of the basic inequality

$$\Big(\sum_j a_j\Big)^\gamma \le \sum_j a_j^\gamma, \tag{78}$$

that holds whenever $a_j \ge 0$ for every j. This has been pointed out by F.L. Toninelli in [55]. Inequality (78) has been exploited also in other contexts, notably in [22, 14], to get upper bounds on the partition function of the directed polymer in random environment, and in [2] to establish localization of eigenfunctions for random operators, in particular in the Anderson localization context.

For pinning models one applies (78) to the renewal identity

$$Z^{\mathrm{c}}_{N,\omega} = \sum_{n=0}^{N-1} Z^{\mathrm{c}}_{n,\omega} K(N-n)\xi_N, \quad \text{with} \quad \xi_N := \exp(\beta\omega_N + h), \tag{79}$$

and, by taking the expectation, one gets to

$$A_N \le \mathbb{E}[\xi_1^\gamma] \sum_{n=0}^{N-1} A_n K(N-n)^\gamma = \sum_{n=1}^{N} A_{N-n} Q(n), \tag{80}$$

where $Q_n := \mathbb{E}[\xi_1^\gamma] K(n)^\gamma$. Now the point is that (80) implies

$$A_N \le \left(\sum_{n=1}^{\infty} Q(n)\right) \max_{n=0,1,\ldots,N-1} A_n, \tag{81}$$

so that, if $\sum_n Q(n) \le 1$ we have $A_N \le A_0 = 1$ for every N. Summing everything up

$$\mathbb{E}[\xi_1^\gamma] \sum_{n=1}^{\infty} K(n)^\gamma \le 1 \implies \sup_N A_N \le 1. \tag{82}$$

And of course if A_N has sub-exponential growth the free energy is zero since

$$\frac{1}{N}\mathbb{E}\log Z^{\mathrm{c}}_{N,\omega} = \frac{1}{\alpha N}\mathbb{E}\log\left(Z^{\mathrm{c}}_{N,\omega}\right)^\alpha \le \frac{1}{\alpha N}\log A_N. \tag{83}$$

Remark 5.1. The discrete convolution inequality (82) can actually be exploited more. Observe in fact that the solution to the renewal equation

$$u_0 = 1 \quad \text{and} \quad u_N = \sum_{n=1}^{N} Q(n) u_{N-n} \quad \text{for } N = 1, 2, \ldots, \tag{84}$$

dominates A_N. But if $Q(\cdot)$ is a probability distribution (possibly adding $Q(\infty)$), then $u_\cdot$ is the renewal function of the $Q(\cdot)$-renewal and if $\sum_n Q(n) < 1$ then $u_N \sim cQ(N)$ for $N \to \infty$ (c is an explicit constant, see Theorem 1.1). Therefore

$$\mathbb{E}\left[\xi_1^\gamma\right] \sum_{n=1}^{\infty} K(n)^\gamma < 1 \implies \text{there exists } C > 0 \text{ such that } A_N \le CK(N)^\gamma. \tag{85}$$

We are now left with verifying for which values of β and h we can find γ such that $\mathbb{E}[\xi_1^\gamma]\sum_{n=1}^\infty K(n)^\gamma \le 1$. Let us consider the case $\sum_n K(n) = 1$: in this case $\sum_n K(n)^\gamma > 1$, so the question is for which values of β and h the pre-factor $\mathbb{E}\xi_1^\gamma$ is sufficiently small. This is a straightforward computation ($\delta = h + \beta^2/2$):

$$\mathbb{E}\xi_1^\gamma = \exp\left(-\frac{\beta^2}{2}\gamma(1-\gamma) + \delta\gamma\right), \tag{86}$$

which is small for β sufficiently large, for every fixed value of δ. This result is therefore saying that (it may be helpful to keep in mind Figure 4):

1. $h_c(\beta) > h_c^{\mathrm{ann}}(\beta)$ if β is sufficiently large;
2. the gap between $h_c(\beta)$ and $h_c^{\mathrm{ann}}(\beta)$ becomes arbitrarily large as β tends to infinity: in fact it is of the order of β^2.

Note that this approach yields very explicit bounds, but it does not give results for small values of β.

5.2. Iterated fractional moment estimates

To go beyond the estimate we have just presented, in [19] another renewal identity has been exploited, namely: for every fixed k and every $N \ge k$

$$Z_{N,\omega}^{c} = \sum_{n=k}^{N} Z_{N-n,\omega}^{c} \sum_{j=0}^{k-1} K(n-j)\,\xi_{N-j} Z_{j,\theta^{N-j}\omega}^{c}. \tag{87}$$

This is simply obtained by decomposing the constrained partition function according to the value $N-n$ of the last point of τ before or at $N-k$ ($0 \le N-n \le N-k$ in the sum), and to the value $N-j$ of the first point of τ to the right of $N-k$ (so that $N-k < N-j \le N$). Of course $Z_{j,\theta^{N-j}\omega}^{c}$ has the same law as $Z_{j,\omega}^{c}$ and the three random variables $Z_{N-n,\omega}^{c}$, ξ_{N-j} and $Z_{j,\theta^{N-j}\omega}^{c}$ are independent, if $n \ge k$ and $j < k$.

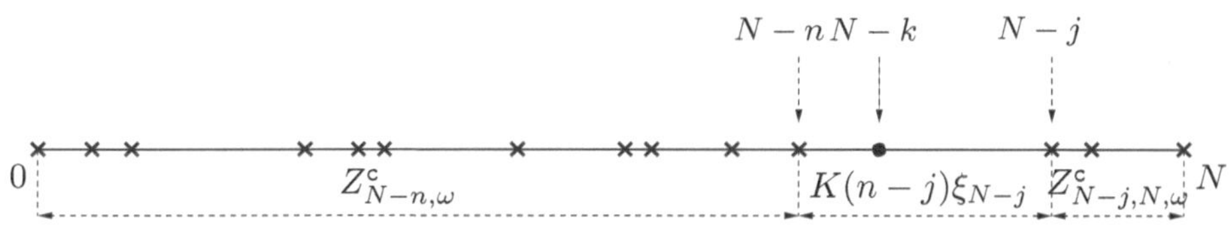

FIGURE 5. The renewal identity (87) is obtained by fixing a value of k and summing over the values of the last contact before $N-k$ (the large dot in the figure) and the first contact after $N-k$. The two contacts are respectively $N-n$ and $N-j$ (crosses are contacts in the figure).

Let $0 < \gamma < 1$, set once again $A_N := \mathbb{E}[(Z_{N,\omega}^{c})^\gamma]$ and use (78) and (87) to get for $N \ge k$

$$A_N \le \mathbb{E}[\xi_1^\gamma] \sum_{n=k}^{N} A_{N-n} \sum_{j=0}^{k-1} K(n-j)^\gamma A_j. \tag{88}$$

This is still a renewal type inequality since it can be rewritten as

$$A_N \le \sum_{n=1}^{N} A_{N-n} Q_k(n), \tag{89}$$

with $Q_k(n) := \mathbb{E}[\xi_1^\gamma] \sum_{j=0}^{k-1} K(n-j)^\gamma A_j$ if $n \ge k$ and $Q_k(n) := 0$ for $n < k$. In particular if for given β and h one can find $k \in \mathbb{N}$ and $\gamma \in (0,1)$ such that

$$\rho := \sum_n Q_k(n) = \mathbb{E}[\xi_1^\gamma] \sum_{n=k}^{\infty} \sum_{j=0}^{k-1} K(n-j)^\gamma A_j \le 1, \tag{90}$$

then one directly extracts from (89) that

$$A_N \le \rho \max\{A_0, \ldots, A_{N-k}\}, \tag{91}$$

for $N \ge k$, which implies that $A_N \le \max\{A_0, \ldots, A_{k-1}\}$ and hence $\mathsf{F}(\beta, h) = 0$.

Remark 5.2. Like in Remark 5.1 one can be sharper by exploiting the renewal structure in (89). The difference with Remark 5.1 is that in this case $N \ge k$. In order to put (89) into a more customary *renewal form* we set $\widetilde{A}_N := A_N \mathbf{1}_{N \ge k}$, so that

$$\widetilde{A}_N \le \sum_{n=1}^{N-(k-1)} \widetilde{A}_{N-n} Q_k(n) + P_k(N), \quad \text{with} \quad P_k(N) = \sum_{n=0}^{k-1} A(n) Q_k(N-n), \tag{92}$$

and therefore there exists $c > 0$ (depending on k, $K(\cdot)$ and γ, besides of course β and h) such that $P_k(N) \le c Q_k(N)$. Let us now consider the standard renewal equation for the $Q_k(n)$-renewal: $u_0 = 1$ (but of course one can choose an arbitrary $u_0 > 0$) and

$$u_N = \sum_{n=1}^{N} u_{N-n} Q_k(n) \overset{N \ge k}{=} \sum_{n=1}^{N-(k-1)} u_{N-n} Q_k(n) + u_0 Q_k(N), \tag{93}$$

where the first equality holds for $N = 1, 2, \ldots$ and for the second one we have used that, since $Q_k(n) = 0$ up to $n = k-1$, we have $u_1 = u_2 = \cdots = u_{k-1} = 0$. Once again if $\sum_n Q_k(n) < 1$, that is if $\rho < 1$, we are dealing with a renewal equation of a terminating process and therefore u_N behaves asymptotically like (a constant times) $Q_k(N)$. Comparing (92) and (93) one obtains that there exists a constant $C = C(K(\cdot), k, \gamma, h, \beta)$ such that

$$A_N \le C K(N)^\gamma. \tag{94}$$

5.3. Finite size estimates by shifting

What the iterated fractional moment has done for us is reducing the problem of estimating the free energy from above to a finite volume estimate. Notice in fact that estimating ρ, *cf.* (90), amounts to estimating only (a fractional moment of) $Z^{\mathtt{c}}_{j,\omega}$, for $j < k$ (note the parallel with Remark 3.2!). This type of estimates demands a new ingredient, which is more easily explained when $\alpha > 1$. A preliminary observation is that ρ is bounded above by $\varepsilon^{-1} \sum_{j=0}^{k-1} (A_j/(k-j)^{(1+\alpha)\gamma - 1}$, with ε a constant that depends on $K(\cdot)$ and γ so that $\rho \leq 1$ is implied if for a given γ

$$\sum_{j=0}^{k-1} \frac{A_j}{(k-j)^{(1+\alpha)\gamma-1}} \leq \varepsilon. \tag{95}$$

Note that, unlike the case treated in §5.1, here the pre-factor $\mathbf{E}\xi_1^\gamma$, and therefore β and h, has only a marginal role: as long as β and h are chosen in a compact, which is what we are doing since we are focusing on the critical region of the annealed model at small or moderate values of β, ε can be chosen independent of the value of β and h. We shall see that the expression in (95) can be made small for example by choosing k large.

Let us start by observing that we know of course (Jensen inequality) that for $h = h_c^{\mathrm{ann}}(\beta) + \delta$

$$A_j \leq (\mathbb{E}Z_{j,\omega}^{c})^\gamma = \left(\mathbf{E}\left[\exp\left(\delta \sum_{n-1}^{j} \mathbf{1}_{n\in\tau}\right); j \in \tau\right]\right)^\gamma = \exp(\gamma \, \mathsf{F}(0,\delta)j)\mathbf{P}_\delta(j\in\tau)^\gamma. \tag{96}$$

We are of course interested in $\delta > 0$: by the Renewal Theorem $\mathbf{P}_\delta(j \in \tau)$ is bounded from below by a positive constant (even if δ were zero!), so this term cannot be of much help and we simply bound it above by one. On the other hand the exponentially growing term stays bounded for j up to the correlation length of the annealed system (*cf.* Remark 2.3): we therefore choose $k := 1/\,\mathsf{F}(0,\delta)$ (again, assume that it is in $\mathbb{N}$). At this point we observe that we can choose $\gamma \in (0,1)$ such that

$$(1+\alpha)\gamma > 2, \tag{97}$$

then the expression in (95) is bounded for k large, that is δ small. This is not yet what we want, but a more attentive analysis shows that one has

$$\sum_{j=0}^{k-1-R} \frac{A_j}{(k-j)^{(1+\alpha)\gamma-1}} \leq \exp(\gamma) \sum_{j>R} j^{-(1+\alpha)\gamma+1} \leq \varepsilon/2, \tag{98}$$

for any $k \geq R$ and R chosen sufficiently large (depending only on γ, α and ε). This has been achieved by using (96). We have therefore to show that

$$\sum_{j=k-R}^{k-1} \frac{A_j}{(k-j)^{(1+\alpha)\gamma-1}} \leq \varepsilon/2. \tag{99}$$

For this we set

$$\widehat{A}_k := \limsup_{\delta\searrow 0} \max_{j=k-R,\ldots,k-1} A_j. \tag{100}$$

If we are able to show that

$$\widehat{A}_k \sum_{i=1}^{R} i^{-((1+\alpha)\gamma-1)} \leq \varepsilon/3, \tag{101}$$

then (95) would be established (for δ small and $k = 1/\,\mathsf{F}(0,\delta)$). Of course in (101) one can replace R with ∞ obtaining thus a more stringent condition (but, in the end, equivalent, since we are not tracking the constants). For a proof of (101) one has to go beyond (96) and in doing so the size of β turns out to play a role.

In order to go beyond (96) the new idea is a tilting procedure (first proposed in [31]), that, given the Gaussian context, reduces to a shift. The idea is based on the following consequence of Hölder inequality

$$A_j = \mathbb{E}'\left[\left(Z^{\mathtt{c}}_{j,\omega}\right)^\gamma \frac{\mathrm{d}\mathbb{P}}{\mathrm{d}\mathbb{P}'}(\omega)\right] \le \mathbb{E}'[Z^{\mathtt{c}}_{j,\omega}]^\gamma \mathbb{E}'\left[\left(\frac{\mathrm{d}\mathbb{P}}{\mathrm{d}\mathbb{P}'}(\omega)\right)^{1/(1-\gamma)}\right]^{1-\gamma}, \tag{102}$$

where $\mathbb{P}'$ is a probability with respect to which $\mathbb{P}$ is absolutely continuous. In order to make the choice of $\mathbb{P}'$ let us fix $\delta = a\beta^2$, a a constant that we are going to choose along the way: $\mathbb{P}'$ is the law of the sequence

$$\omega_1 - \sqrt{a\beta^2}, \omega_2 - \sqrt{a\beta^2}, \ldots, \omega_k - \sqrt{a\beta^2}, \omega_{k+1}, \omega_{k+2}, \ldots \tag{103}$$

which is a sequence of independent (non identically distributed) variables. One then readily computes

$$\mathbb{E}'\left[\left(\frac{\mathrm{d}\mathbb{P}}{\mathrm{d}\mathbb{P}'}(\omega)\right)^{1/(1-\gamma)}\right]^{1-\gamma} = \exp\left(\frac{\gamma}{1-\gamma} a\beta^2 k\right), \tag{104}$$

but $a\beta^2 k = \delta / \mathsf{F}(0,\delta)$ and the ratio $\delta/\mathsf{F}(0,\delta)$ tends to a positive constant as $\delta \searrow 0$ since $\alpha > 1$, cf. Theorem 2.1. Let us now turn our attention to $\mathbb{E}' Z^{\mathtt{c}}_{j,\omega}$ which, for $j \le k$, coincides with $\mathbb{E}' Z^{\mathtt{c}}_{j,\omega-\sqrt{a\beta^2}}$. But this is just the partition function of a homogeneous model with negative pinning potential if we choose a small, namely (for conciseness we look only at the case $j = k$)

$$\begin{aligned} \mathbb{E}' Z^{\mathtt{c}}_{k,\omega} &= \mathbf{E}\left[\exp\left(-\beta^2(\sqrt{a}-a)\sum_{n=1}^k \mathbf{1}_{n\in\tau}\right); k \in \tau\right] \\ &= \mathbf{E}\left[\exp\left(-\left(\frac{1}{\sqrt{a}}-1\right) \times \frac{\delta}{\mathsf{F}(0,\delta)} \times \frac{1}{k}\sum_{n=1}^k \mathbf{1}_{n\in\tau}\right); k \in \tau\right]. \end{aligned} \tag{105}$$

But $\lim_{k\to\infty}(1/k)\sum_{n=1}^k \mathbf{1}_{n\in\tau} = 1/\mathbf{E}[\tau_1]$ $\mathbf{P}$-a.s. and this readily implies that $\mathbb{E}' Z^{\mathtt{c}}_{k,\omega}$ is bounded by a constant that can be chosen arbitrarily small, provided one chooses a sufficiently small (so $a^{-1/2}-1$ is large). The argument easily extends to j between $k-R$ and k so that (101) is proven since $\widehat{A}_k$ can be chosen arbitrarily small for a sufficiently small and every $\delta \le \delta_0$ (for some $\delta_0 > 0$). This concludes the argument for the case $\alpha > 1$, that is Theorem 3.3(2).

The case $\alpha \in (1/2, 1)$ (Theorem 3.3(1)) is conceptually not very different: the main difference lies in the fact that it is no longer sufficient to show that A_j is small for j close to k, one has actually to extract some decay in j. But the fact that A_j does decay with j, at least if $j \le 1/\mathsf{F}(0,\delta)$, is already rather evident from (96) from the fact that the term $\mathbf{P}_\delta(j \in \tau)$ is, at least till $j < k = 1/\mathsf{F}(0,\delta)$, close to $\mathbf{P}(j \in \tau)$ which behaves for j large as $j^{-(1-\alpha)}$ (times a constant, *cf.* Theorem 1.1). The argument is however somewhat technical and we refer to [19] for details.

6. Relevant disorder estimates: the critical exponent

The argument leading to Theorem 3.4 is based on the *rare stretch* strategy sketched in Figure 6. It is based on a one-step coarse graining of the environment on the scale $\ell \in \mathbb{N}$, $1 \ll \ell \ll N$. Actually one should think of ℓ as very large but finite. We assume $N/\ell \in \mathbb{N}$ and we look at the sequence of i.i.d. random variables defined as

$$Y_j := \mathbf{1}_{E_j}, \quad \text{with} \quad E_j = \left\{\omega : \log Z^{\mathtt{c}}_{\ell,\theta^{(j-1)\ell}\omega} \geq a\,\mathsf{F}(\beta, h+\delta)\right\}, \tag{106}$$

where $\delta > 0$, $a \in (0,1)$ (eventually $a \nearrow 1$) and $j = 1, 2, \ldots$, but of course only the j's up to N/ℓ are relevant to us. Note that the Y variables are Bernoulli random variables of parameter $p(\ell) := \mathbb{P}(E_1)$ and, since $a < 1$, $p(\ell)$ is small when ℓ is large, by the very definition of the free energy and its self-averaging property (*cf.* Theorem 3.1). One can actually show rather easily that

$$\liminf_{\ell\to\infty} \frac{1}{\ell} \log p(\ell) \geq -\frac{\delta^2}{2\beta^2}. \tag{107}$$

We give a proof of this inequality below, but the intuitive reason is that the probability of observing $\sum_{i=1}^{\ell} \omega_i \approx \ell\delta/\beta$ behaves like $\exp(-\ell\delta^2/2\beta^2)$ for ℓ large (this is the standard Cramer Large Deviation result). When such a Large Deviation event occurs, the environment in the ℓ-block will look like the original ω variables translated of δ/β, that is $\beta\omega_i + h$ looks like $\beta\omega_i + h + \delta$. And in that block the logarithm of the partition function will hence be close to $\ell\,\mathsf{F}(\beta, h+\delta)$. Shifting the mean is of course only one possible strategy to make the event E_1 typical and hence such an argument yields only a lower bound on $p(\ell)$.

We now make a lower bound on $Z^{\mathtt{f}}_{N,\omega}$ by considering only the τ-trajectories that visit all and only the ℓ blocks for which $Y_j(\omega) = 1$ (see Figure 6). The renewal property leads to a rather explicit lower bound, namely (with the notation of the figure)

$$\log Z^{\mathtt{f}}_{N,\omega} \geq \sum_{\substack{j \leq N/\ell: \\ Y_j = 1}} \log Z^{\mathtt{c}}_{\ell,\theta^{(j-1)\ell}\omega} + \sum_{n=1}^{\mathcal{N}_Y(\omega)} \log K(G_n\ell) + O(\log N), \tag{108}$$

where the $O(\log N)$ term comes from the last excursion. Let us now divide by N and take the limit, keeping into account that, by definition of the Y variables, we have a lower bound on the partition functions $\log Z^{\mathtt{c}}_{\ell,\theta^{(j-1)\ell}\omega}$ that appear in the right-hand side. We therefore obtain

$$\begin{aligned}\mathsf{F}(\beta, h) &\geq a\ell\,\mathsf{F}(\beta, h+\delta) \lim_{N\to\infty} \frac{\mathcal{N}_Y(\omega)}{N} + \limsup_{N\to\infty}\left(\frac{\mathcal{N}_Y(\omega)}{N} \frac{1}{\mathcal{N}_Y(\omega)} \sum_{n=1}^{\mathcal{N}_Y(\omega)} \log K(G_n\ell)\right) \\ &= ap(\ell)\,\mathsf{F}(\beta, h+\delta) + \frac{p(\ell)}{\ell} \limsup_{N\to\infty} \frac{1}{\mathcal{N}_Y(\omega)} \sum_{n=1}^{\mathcal{N}_Y(\omega)} \log K(G_n\ell),\end{aligned} \tag{109}$$

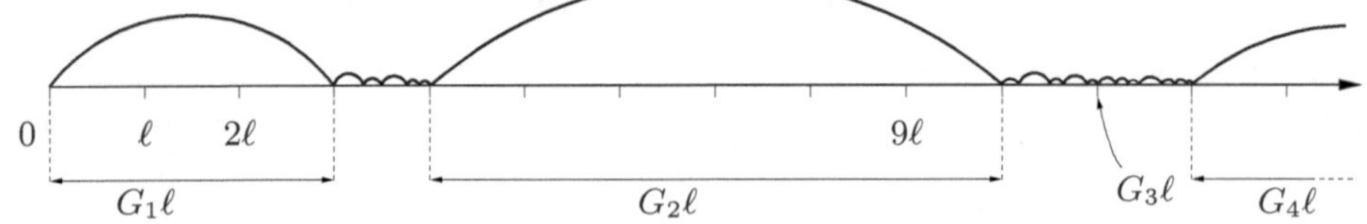

FIGURE 6. The rare stretch strategy is implemented by looking at blocks of ω variables of size ℓ. To block j is associated a Bernoulli random variable Y_j: such a random variable is a function of the ω variables in the block and it determines whether a system of the size of that block, with constrained boundary conditions and precisely with the ω variables of the block, has a sufficiently large (in fact, an atypically large) partition function. The precise definition of Y_j is in (106). In the figure $Y_4 = Y_{11} = Y_{12} = 1$, while the other Y variables are zero. The gaps between the success blocks are parametrized as $G_n\ell$, so that $\{G_n\}_n$ are i.i.d. Geometric random variables: $\mathbb{P}(G_1 = k) = (1-p)^k p$ ($k = 0, 1, 2, \ldots$ and $p = p(\ell)$, given in the text). The lower bound is then achieved by restricting the partition function to the τ trajectories that visit only success blocks and visit the first and the last point of the block (also if two success blocks are contiguous: this is the case of the second and third success blocks in the figure). Such a strategy is profoundly different from that employed in [3], that is based on the effect of typical fluctuations (on the Central Limit Theorem scale) and competition with boundary effects. Our strategy is instead a Large Deviation strategy and, in a sense, it exploits the *flexibility* of the polymer to target rare regions (the boundary conditions play no role).

where in the last step we have used the strong law of large numbers (the limits are in the $\mathbb{P}(\mathrm{d}\omega)$-a.s. sense) to estimate the leading behavior of the number of successes in an array of N/ℓ Bernoulli variables of parameter $p(\ell)$. Moreover, to be precise, (109) holds if we set $K(0) = 1$ (which we do only here). The (superior) limit that is left in the expression is also easily evaluated by using the strong law of large numbers after having observed that, since $\log K(x) \overset{x\to\infty}{\sim} -(1+\alpha)\log x$, when $G_n = 1, 2, \ldots$ we have $\log K(G_n\ell) \geq -a^{-1}(1+\alpha)(\log G_n + \log \ell)$ for ℓ sufficiently large (uniformly in the value of G_n: note that $1/a$ is once again just a number larger than 1). The net outcome is therefore

$$\begin{aligned}\mathsf{F}(\beta, h) \geq\, & ap(\ell)\,\mathsf{F}(\beta, h+\delta) \\ & - \frac{p(\ell)}{\ell} a^{-1}(1+\alpha)\big(\mathbb{E}[\log G_1; G_1 > 0] + \mathbb{P}(G_1 > 0)\log \ell\big). \end{aligned} \tag{110}$$

Since G_1 is a geometric variable of parameter $p(\ell)$ we directly compute

$$\mathbb{E}[\log G_1; G_1 > 0] = (1 + o_\ell(1))\log(1/p(\ell)),$$

which, by (107), is bounded above by $a^{-1}\ell\delta^2/(2\beta^2)$ (ℓ large: once again a^{-1} is just used as an arbitrary constant larger than one). Therefore

$$\mathsf{F}(\beta, h) \geq p(\ell)\left[a\,\mathsf{F}(\beta, h+\delta) - a^{-2}\frac{\delta^2(1+\alpha)}{2\beta^2} + o_\ell(1)\right], \tag{111}$$

where the term $o_\ell(1)$ is $-c(\log \ell)/\ell$ ($c > 0$) and this bound holds, given $a \in (0,1)$, for every ℓ larger than some ℓ_0.

Now we set $h = h_c(\beta)$ in (111), so that the left-hand side is zero and therefore

$$a\,\mathsf{F}(\beta, h_c(\beta) + \delta) - a^{-2}\frac{\delta^2(1+\alpha)}{2\beta^2} + o_\ell(1) \le 0, \tag{112}$$

for every $\ell > \ell_0$, so that $\mathsf{F}(\beta, h_c(\beta)+\delta) \le a^{-3}(\delta^2(1+\alpha)/(2\beta^2))$, and since $a \in (0,1)$ is arbitrary we can let $a \nearrow 1$ and we are done.

For completeness we give a proof of (107). We call $\widetilde{\mathbb{P}}_\ell$ the law of the sequence of random variables

$$\omega_1 + \delta/\beta, \omega_2 + \delta/\beta, \ldots, \omega_\ell + \delta/\beta, \omega_{\ell+1}, \omega_{\ell+2}, \ldots \tag{113}$$

Note that $\mathbb{P}_\ell(E_1)$ tends to one as ℓ becomes large, simply by definition of free energy and because the law of $\{\beta\omega_n + h\}_{n=1,\ldots,\ell}$, when ω is distributed according to $\widetilde{\mathbb{P}}_\ell$, coincides with the law of $\{\beta\omega_n + h + \delta\}_{n=1,\ldots,\ell}$, when ω is distributed according to $\mathbb{P}$. We compute the relative entropy

$$\mathcal{H}\left(\widetilde{\mathbb{P}}_\ell|\mathbb{P}\right) := \widetilde{\mathbb{E}}_\ell\left[\log\frac{\mathrm{d}\widetilde{\mathbb{P}}_\ell}{\mathrm{d}\mathbb{P}}(\omega)\right] = \frac{\delta^2}{2\beta^2}\ell, \tag{114}$$

and by a standard entropy inequality (see, *e.g.*, [29, § A.2])

$$\log\frac{\mathbb{P}(E_1)}{\widetilde{\mathbb{P}}_\ell(E_1)} \ge -\frac{1}{\widetilde{\mathbb{P}}_\ell(E_1)}\Big(\mathcal{H}(\widetilde{\mathbb{P}}_\ell|\mathbb{P}) + \frac{1}{\mathrm{e}}\Big) = -\frac{1}{\widetilde{\mathbb{P}}_\ell(E_1)}\Big(\frac{\delta^2}{2\beta^2}\ell + \frac{1}{\mathrm{e}}\Big), \tag{115}$$

and this yields (107).

Acknowledgement

The author is very grateful to the organizers of the Summer School on Spin Glasses for the invitation to lecture on disordered pinning models. He also acknowledges the support of ANR, grant POLINTBIO.

References

[1] D.B. Abraham, *Surface Structures and Phase Transitions, Exact Results*, in *Phase Transitions and Critical Phenomena* **10**, Academic Press, London (UK) (1986), 1–74.

[2] M. Aizenman and S. Molchanov, *Localization at large disorder and at extreme energies: an elementary derivation*, Commun. Math. Phys. **157** (1993), 245–278.

[3] M. Aizenman and J. Wehr, *Rounding effects of quenched randomness on first-order phase transitions*, Commun. Math. Phys. **130** (1990), 489–528.

[4] K.S. Alexander, *The effect of disorder on polymer depinning transitions*, Commun. Math. Phys. **279** (2008), 117–146.

[5] K.S. Alexander, *Ivy on the ceiling: first-order polymer depinning transitions with quenched disorder*, Markov Proc. Rel. Fields **13** (2007), 663–680.

[6] K.S. Alexander and V. Sidoravicius, *Pinning of polymers and interfaces by random potentials*, Ann. Appl. Probab. **16** (2006), 636–669.

[7] K.S. Alexander and N. Zygouras, *Quenched and annealed critical points in polymer pinning models*, arXiv:0805.1708 [math.PR]

[8] S. Asmussen, *Applied probability and queues*, Second Edition, Applications of Mathematics **51**, Springer-Verlag, New York (2003).

[9] N.H. Bingham, C.M. Goldie and J.L. Teugels, *Regular variation*, Cambridge University Press, Cambridge, 1987.

[10] M. Birkner and R. Sun, *Annealed vs quenched critical points for a random walk pinning model*, arXiv:0807.2752v1 [math.PR]

[11] R. Blossey and E. Carlon, *Reparametrizing the loop entropy weights: effect on DNA melting curves*, Phys. Rev. E **68** (2003), 061911 (8 pages).

[12] E. Bolthausen, F. Caravenna and B. de Tilière, *The quenched critical point of a diluted disordered polymer model*, arXiv:0711.0141v1 [math.PR]

[13] T.W. Burkhardt, *Localization-Delocalization transition in a solid-on-solid model with a pinning potential*, J. Phys. A: Math. Gen. **14** (1981), L63–L68.

[14] A. Camanes and P. Carmona, *Directed polymers, critical temperature and uniform integrability*, preprint (2007).

[15] F. Caravenna and G. Giacomin, *On constrained annealed bounds for pinning and wetting models*, Elect. Comm. Probab. **10** (2005), 179–189.

[16] F. Caravenna, G. Giacomin and L. Zambotti, *A renewal theory approach to periodic copolymers with adsorption*, Ann. Appl. Probab. **17** (2007), 1362–1398.

[17] D. Cule and T. Hwa, *Denaturation of Heterogeneous DNA*, Phys. Rev. Lett. **79** (1997), 2375–2378.

[18] B. Coluzzi and E. Yeramian, *Numerical evidence for relevance of disorder in a Poland–Scheraga DNA denaturation model with self-avoidance: scaling behavior of average quantities*, Eur. Phys. Journal B **56** (2007), 349–365.

[19] B. Derrida, G. Giacomin, H. Lacoin and F.L. Toninelli, *Fractional moment bounds and disorder relevance for pinning models*, Commun. Math. Phys. **287** (2009), 867–887.

[20] B. Derrida, V. Hakim and J. Vannimenus, *Effect of disorder on two-dimensional wetting*, J. Statist. Phys. **66** (1992), 1189–1213.

[21] R.A. Doney, *One-sided local large deviation and renewal theorems in the case of infinite mean*, Probab. Theory Rel. Fields **107** (1997), 451–465.

[22] M.R. Evans and B. Derrida, *Improved bounds for the transition temperature of directed polymers in a finite-dimensional random medium*, J. Statist. Phys. **69** (1992), 427–437.

[23] W. Feller, *An introduction to probability theory and its applications*, Vol. I, Third edition, John Wiley & Sons, Inc., New York–London–Sydney, 1968.

[24] M.E. Fisher, *Walks, walls, wetting, and melting*, J. Statist. Phys. **34** (1984), 667–729.

[25] G. Forgacs, J.M. Luck, Th.M. Nieuwenhuizen and H. Orland, *Wetting of a disordered substrate: exact critical behavior in two dimensions*, Phys. Rev. Lett. **57** (1986), 2184–2187.

[26] S. Galluccio and and R. Graber, *Depinning transition of a directed polymer by a periodic potential: a d-dimensional solution*, Phys. Rev. E **53** (1996), R5584-R5587.

[27] A. Garsia and J. Lamperti, *A discrete renewal theorem with infinite mean*, Comment. Math. Helv. **37** (1963), 221–234.

[28] P.G. de Gennes, *Scaling concepts in polymer physics*, Cornell University Press, Ithaca, NY (1979).

[29] G. Giacomin, *Random Polymer Models*, Imperial College Press, World Scientific (2007).

[30] G. Giacomin, *Renewal convergence rates and correlation decay for homogeneous pinning models*, Elect.J. Probab. **13** (2008), 513–529.

[31] G. Giacomin, H. Lacoin and F.L. Toninelli, *Hierarchical pinning models, quadratic maps and quenched disorder*, arXiv:0711.4649 [math.PR]

[32] G. Giacomin and F.L. Toninelli, *Estimates on path delocalization for copolymers at selective interfaces*, Probab. Theor. Rel. Fields **133** (2005), 464–482.

[33] G. Giacomin and F.L. Toninelli, *Smoothing of Depinning Transitions for Directed Polymers with Quenched Disorder*, Phys. Rev. Lett. **96** (2006), 060702.

[34] G. Giacomin and F.L. Toninelli, *Smoothing effect of quenched disorder on polymer depinning transitions*, Commun. Math. Phys. **266** (2006), 1–16.

[35] G. Giacomin and F.L. Toninelli, *The localized phase of disordered copolymers with adsorption*, ALEA **1** (2006),149–180.

[36] G. Giacomin and F.L. Toninelli, *On the irrelevant disorder regime of pinning models*, arXiv: 0707.3340v1 [math.PR]

[37] A.B. Harris, *Effect of Random Defects on the Critical Behaviour of Ising Models*, J. Phys. C **7** (1974), 1671–1692.

[38] Y. Imry and S.–K. Ma, *Random-field instability of the ordered state of continuous symmetry*, Phys. Rev. Lett. **35** (1975), 1399–1401.

[39] Y. Isozaki and N. Yoshida, *Weakly pinned random walk on the wall: pathwise descriptions of the phase Transition*, Stoch. Proc. Appl. **96**(2001), 261–284.

[40] N.C. Jain and W.E. Pruitt, *The range of random walk*, in *Proceedings of the Sixth Berkeley Symposium on Mathematical Statistics and Probability* (Univ. California, Berkeley, Calif., 1970/1971), Vol. III: Probability theory, 31–50. Univ. California Press, Berkeley, Calif., 1972.

[41] Y. Kafri, D. Mukamel and L. Peliti, *Why is the DNA denaturation transition first order?*, Phys. Rev. Lett. **85** (2000), 4988–4991.

[42] Y. Kafri, D.R. Nelson and A. Polkovnikov, *Unzipping flux lines from extended defects in type-II superconductors*, Europhys. Lett. **73** (2006), 253–259.

[43] J.F.C. Kingman, *Subadditive Ergodic Theory*, Ann. Probab. **1** (1973), 882–909.

[44] R. Kühn, *Equilibrium ensemble approach to disordered systems I: general theory, exact results*, Z. Phys. B **100** (1996), 231–242.

[45] M. Ledoux, *The concentration of measure phenomenon*, Mathematical Surveys and Monographs, **89**, American Mathematical Society, Providence, RI (2001).

[46] J.M.J. van Leeuwen and H.J. Hilhorst, *Pinning of rough interface by an external potential*, Phys. A **107** (1981), 319–329.

[47] D. Marenduzzo, A. Trovato and A. Maritan, *Phase diagram of force-induced DNA unzipping in exactly solvable models*, Phys. Rev. E **64** (2001), 031901 (12 pages).

[48] C. Monthus and T. Garel, *Distribution of pseudo-critical temperatures and lack of self-averaging in disordered Poland–Scheraga models with different loop exponents*, Eur. Phys. J. B **48** (2005), 393–403.

[49] N. Pétrélis, *Polymer Pinning at an Interface*, Stoch. Proc. Appl. **116** (2006), 1600–1621.

[50] J. Sohier, *Finite size scaling for homogeneous pinning models*, arXiv:0802.1040 [math.PR]

[51] M. Talagrand, *A new look at independence*, Ann. Probab. **24** (1996), 1–34.

[52] F.L. Toninelli, *Critical properties and finite-size estimates for the depinning transition of directed random polymers*, J. Statist. Phys. **126** (2007), 1025–1044.

[53] F.L. Toninelli, *Correlation lengths for random polymer models and for some renewal sequences*, Electron. J. Probab. **12** (2007), 613–636.

[54] F.L. Toninelli, *A replica-coupling approach to disordered pinning models*, Commun. Math. Phys. **280** (2008), 389–401.

[55] F.L. Toninelli, *Disordered pinning models and copolymers: beyond annealed bounds*, Ann. Appl. Probab. **18** (2008), 1569–1587.

[56] F.L. Toninelli, *Localization transition in disordered pinning models. Effect of randomness on the critical properties*, Lecture Notes from the 5th Prague Summer School on Mathematical Statistical Mechanics, September 2006, arXiv:0703912 [math.PR].

[57] S.G. Whittington, *A directed-walk model of copolymer adsorption*, J. Phys. A: Math. Gen. **31** (1998), 8797–8803.

Giambattista Giacomin
U.F.R. de Mathématiques, Case 7012 (site Chevaleret)
Laboratoire de Probabilités et Modèles Aléatoires (LPMA, CNRS U.M.R. 7599)
Université Paris Diderot (Paris 7)
F-75205 Paris Cedex 13, France
e-mail: giacomin@math.jussieu.fr

Progress in Probability, Vol. 62, 271–278

A Smoothing Inequality for Hierarchical Pinning Models

Hubert Lacoin and Fabio Lucio Toninelli

Abstract. We consider a hierarchical pinning model introduced by B. Derrida, V. Hakim and J. Vannimenus in [3], which undergoes a localization/delocalization phase transition. This depends on a parameter $B > 2$, related to the geometry of the hierarchical lattice. We prove that the phase transition is of second order in presence of disorder. This implies that disorder smoothes the transition in the so-called *relevant disorder case*, i.e., $B > B_c = 2 + \sqrt{2}$.

Mathematics Subject Classification (2000). 60K35, 82B44, 37H10.

Keywords. Hierarchical models, quadratic recurrence equations, pinning models, disorder, smoothing

1. Model

The model we are going to introduce can be interpreted either as an infinite-dimensional dynamical system or as a pinning model on a hierarchical lattice. We refer to [3], where the model was introduced, or to [7], for the latter interpretation, and we mention only that the parameter B in (2) is related to the geometry of the lattice. The mathematical understanding of disordered pinning models, hierarchical or not, has witnessed a remarkable progress lately, cf. in particular [1, 2] and [4]–[8].

Let $\{\omega_n\}_{n\in\mathbb{N}}$ be a sequence of i.i.d. random variables (with related probability distribution denoted by $\mathbb{P}$), with zero mean, unit variance and satisfying

$$M(\beta) := \mathbb{E}[\exp(\beta\omega_1)] < \infty \quad \forall \beta > 0.$$

In Section 3 we will state our result in the particular case of Gaussian random variables, $\omega_1 \sim \mathcal{N}(0,1)$.

We are interested in the dynamical system defined by the following initial condition and recurrence equation

$$R_0^{(i)} = \exp\left(\beta\omega_i - \log M(\beta) + h\right), \tag{1}$$

$$R_{n+1}^{(i)} = \frac{R_n^{(2i)} R_n^{(2i-1)} + (B-1)}{B}, \tag{2}$$

where $B > 2$, $\beta > 0$ and $h \in \mathbb{R}$ are fixed parameters.

One can easily note that for every n, the random variables $\{R_n^{(i)}\}_{i\geq 1}$, are i.i.d. We are mainly interested in the evolution of the probability law of $R_n^{(1)}$ (let us call it $\mathcal{L}_n$). $R_n^{(1)}$ is essentially the partition function of the pinning model at generation n [3]. One can re-interpret this system as a dynamical system on probability laws, $\mathcal{L}_{n+1}$ being the law of R_{n+1} given by

$$R_{n+1} = \frac{R_n^{(1)} R_n^{(2)} + (B-1)}{B}, \tag{3}$$

where $R_n^{(i)}$ $i = 1, 2$ are i.i.d. random variables with probability law $\mathcal{L}_n$.

One can also study this dynamical system in a non-random set up, choosing $\beta = 0$ or, equivalently, considering $r_n = \mathbb{E}R_n$ (respectively pure or annealed system). The recursion becomes then

$$r_{n+1} = \frac{r_n^2 + (B-1)}{B} \tag{4}$$

(with initial condition $r_0 = \exp(h)$), which is nothing but a particular case of the well-known *logistic map*.

In any case, there exists a (non-random) quantity $\mathsf{F}(\beta, h)$ such that

$$\lim_{n\to\infty} \frac{1}{2^n} \log R_n = \mathsf{F}(\beta, h) \quad \text{almost surely w.r.t. } \mathbb{P}.$$

The convergence also holds in $\mathbb{L}_1$, and $\mathsf{F}(\beta, h) \geq 0$. For a proof of these statements see [4, Theorem 1.1]. Observe that the non-negativity of F just follows from $R_n^{(i)} \geq (B-1)/B$ (cf. (2)). We call $\mathsf{F}(\beta, h)$ the *free energy* of the system, and we remark that $\mathsf{F}(\beta, \cdot)$ is increasing. We define the critical point

$$h_c(\beta) = \inf\{h \text{ such that } \mathsf{F}(\beta, h) > 0\}.$$

For an interpretation of $h_c(\beta)$ as the transition point between a delocalized and a localized phase ($h < h_c(\beta)$ and $h > h_c(\beta)$, respectively) we refer to [4] and [3].

Our aim in this paper is to describe the influence of the disorder on the shape of the curve $\mathsf{F}(\beta, \cdot)$ around $h_c(\beta)$, and in particular to obtain a bound on the critical exponent which governs the vanishing of F when $h_c(\beta)$ is approached from the localized phase.

2. Known results

In [4] various results have been obtained about the influence of disorder on the free energy curve. We summarize the situation in the two following theorems. The first one describes the behavior of the free energy around h_c and the value of h_c for the pure system ($\beta = 0$). The second one concerns the behavior of $h_c(\beta)$ for small values of β, making a distinction between two different situations: irrelevant disorder for $B < 2 + \sqrt{2}$ (critical properties of the system are unchanged, for β small, with respect to $\beta = 0$) and relevant disorder for $B > 2 + \sqrt{2}$ (the disordered system behaves very differently from the pure one, for every $\beta > 0$). For this reason we will refer to $B_c = 2 + \sqrt{2}$ as the critical value of B.

Theorem 2.1 (Annealed system estimates, [4]**).** *The function $h \mapsto \mathsf{F}(0, h)$ is real analytic except at $h = h_c := h_c(0)$. Moreover $h_c = \log(B-1)$ and there exists $c = c(B) > 0$ such that for all $h \in (h_c, h_c + 1)$*

$$c(B)^{-1}(h - h_c)^{1/\alpha} \leq \mathsf{F}(0, h) \leq c(B)(h - h_c)^{1/\alpha}, \tag{5}$$

where

$$\alpha := \frac{\log(2(B-1)/B)}{\log 2}. \tag{6}$$

Theorem 2.2 ([4]**).** *When $B < 2 + \sqrt{2} = B_c$, there exists β_0 such that for all $0 \leq \beta \leq \beta_0$,*

$$h_c(\beta) = h_c(0).$$

Moreover, for any $\varepsilon > 0$, one can find $h_\varepsilon > h_c(0)$ such that for any $h \in (h_c(0), h_\varepsilon)$ we have

$$(1 - \varepsilon)\,\mathsf{F}(0, h) \leq \mathsf{F}(\beta, h) \leq \mathsf{F}(0, h).$$

When $B > 2 + \sqrt{2}$, we have for any $\beta > 0$, $h_c(\beta) > h_c(0)$. Moreover we can estimate the difference: there exists $C(B) > 0$ such that for all $\beta \leq 1$

$$C(B)^{-1}\beta^{\frac{2\alpha}{2\alpha-1}} \leq h_c(\beta) - h_c(0) \leq C(B)\beta^{\frac{2\alpha}{2\alpha-1}}.$$

Finally, if $B = B_c$ then

$$0 \leq h_c(\beta) - h_c(0) \leq \mathrm{e}^{-C/\beta^2} \tag{7}$$

for some positive constant C.

It is a very interesting open problem to close the gap between the upper and lower bound in (7). Let us mention that the arguments in [3] suggest that the upper bound is the correct one, with an explicit prediction for the constant C.

When $B \geq B_c$ the previous theorem says nothing on the shape of the free energy around $h_c(\beta)$; in particular, one may wonder if it is different from the one of the pure system. In this spirit, we prove a general theorem on the effect of disorder on the phase transition of the system. This theorem is the analog of what has been proved in [5, 6] for the non-hierarchical pinning model based on a renewal process.

3. Smoothness of the phase transition in presence of disorder

Assume that $\omega_1 \sim \mathcal{N}(0,1)$. We have:

Theorem 3.1. *For every $B > 2$ there exists $c(B) < \infty$ such that for every $\beta > 0$ and $\delta > 0$ one has*

$$\mathsf{F}(\beta, h_c(\beta) + \delta) \le \frac{\delta^2}{\beta^2} c(B). \tag{8}$$

Remark 3.1. In view of Theorem 2.1 and the definition of α, this shows that if $B > B_c$ (which corresponds to $\alpha > 1/2$) and $\beta > 0$ the critical exponent of the transition is different from that of the pure model, i.e., we have

$$\liminf_{h \to h_c(\beta)+} \frac{\log \mathsf{F}(\beta, h)}{\log(h - h_c(\beta))} \ge 2 > \lim_{h \to h_c(0)+} \frac{\log \mathsf{F}(0, h)}{\log(h - h_c(0))} = \frac{1}{\alpha}.$$

Proof of Theorem 3.1. Fix $\beta > 0$ and let $h = h_c(\beta)$. Let $N \in \mathbb{N}$, $\ell \in \mathbb{N}$ with $\ell < N$ and

$$\mathcal{I}_N(\omega) := \{1 \le j \le 2^{N-\ell} \;:\; R_\ell^{(j)} \ge \exp(2^{\ell-1} \mathsf{F}(\beta, h_c(\beta) + \delta))\}. \tag{9}$$

Defining

$$p_\ell := \mathbb{P}(1 \in \mathcal{I}_N(\omega)), \tag{10}$$

one has $\mathbb{P}(\mathrm{d}\omega)$-a.s.

$$\lim_{N \to \infty} \frac{|\mathcal{I}_N(\omega)|}{2^{N-\ell}} = p_\ell \tag{11}$$

from the strong law of large numbers and, for ℓ sufficiently large,

$$p_\ell \ge \mathrm{e}^{-2^\ell \delta^2/\beta^2}. \tag{12}$$

To prove the latter estimate we proceed as in [5] and we use the classical entropic inequality

$$\mathbb{P}(A) \ge \tilde{\mathbb{P}}(A) \exp\left(-\frac{1}{\tilde{\mathbb{P}}(A)} \left(\mathrm{H}(\tilde{\mathbb{P}}|\mathbb{P}) + \mathrm{e}^{-1}\right)\right) \tag{13}$$

which holds for every event A if the laws $\mathbb{P}$ and $\tilde{\mathbb{P}}$ are mutually absolutely continuous, and

$$\mathrm{H}(\tilde{\mathbb{P}}|\mathbb{P}) := \mathbb{E}\left(\frac{\mathrm{d}\tilde{\mathbb{P}}}{\mathrm{d}\mathbb{P}} \log \frac{\mathrm{d}\tilde{\mathbb{P}}}{\mathrm{d}\mathbb{P}}\right)$$

denotes the relative entropy. Equation (12) then easily follows if we apply (13) with $A = \{1 \in \mathcal{I}_N(\omega)\}$ and $\tilde{\mathbb{P}} = \tilde{\mathbb{P}}_\ell$ being the law under which $\{\omega_n\}_{n \in \mathbb{N}}$ are independent Gaussian variables of unit variance and mean $\tilde{\mathbb{E}}\,\omega_n = (\delta/\beta)$ for $n \le 2^\ell$ and $\tilde{\mathbb{E}}\omega_n = 0$ otherwise. Indeed, in that case $\tilde{\mathbb{P}}(A) \ge (3/4)$ for ℓ sufficiently large by the fact that $2^{-\ell} \log R_\ell$ converges to $\mathsf{F}(\beta, h_c(\beta) + \delta)$ in $\tilde{\mathbb{P}}$-probability for $\ell \to \infty$ and the relative entropy is immediately computed: $\mathrm{H}(\tilde{\mathbb{P}}|\mathbb{P}) = 2^\ell (\delta/\beta)^2/2$.

Consider the binary tree $\mathcal{T}_N$ with $N + 1$ levels (see Figure 1): at level 0 are the leaves, at level N is the root. Pairs (n, j) with $0 \le n \le N$ and $1 \le j \le 2^{N-n}$ are called *nodes* (leaves are also considered to be nodes). The number of nodes in

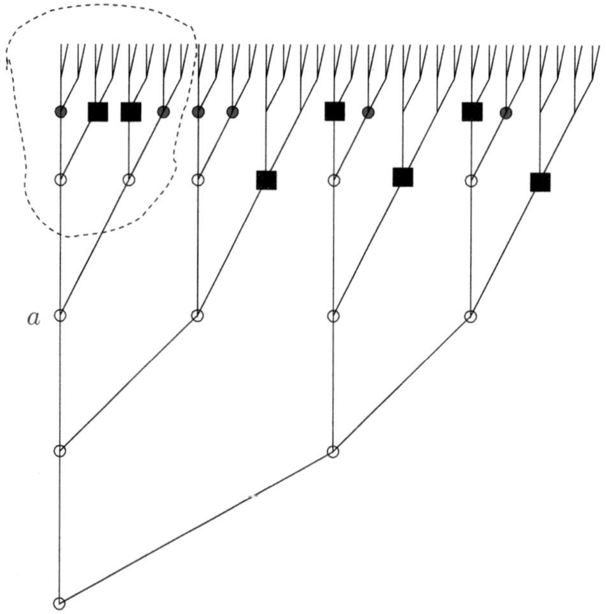

FIGURE 1. The binary tree $\mathcal{T}_N$ for $N=6$. At level 0 are the leaves. Nodes (ℓ, j) with $j \in \mathcal{I}_N(\omega)$ are marked by full dots (here we have taken $\ell=2$), good nodes by empty dots and bad nodes by squares. The dashed line includes all the descendents of the node a.

$\mathcal{T}_N$ is $2^{N+1}-1$. Given a node $(n,j) \in \mathcal{T}_N$ with $n<N$ we call $(n+1, \lceil j/2 \rceil)$ its father. The descendents of a node $(n,j) \in \mathcal{T}_N$ are defined in the natural way.

Given a node $(n,j) \in \mathcal{T}_N$ with $n>\ell$ we say that it is a *good node* if there exists $i \in \mathcal{I}_N(\omega)$ such that (ℓ, i) is a descendent of (n,j). A node (n,j) with $n \geq \ell$ will be called *bad* if its father is good but he himself is not good (cf. Figure 1). Note that descendents of bad nodes are neither good nor bad. If $\mathcal{I}_N(\omega)=\emptyset$, by convention we say that the root is a bad node, so that the root is always either bad or good. Let $g_N(\omega)$ and $b_N(\omega)$ be the number of good and bad nodes in $\mathcal{T}_N$.

Given ω, one has the inequality

$$R_N \geq \exp(\mathsf{F}(\beta, h_c(\beta)+\delta) 2^{\ell-1} |\mathcal{I}_N(\omega)|) B^{-g_N(\omega)} \left(\frac{B-1}{B}\right)^{b_N(\omega)}, \tag{14}$$

which will be proved below. As a consequence, $\mathbb{P}(\mathrm{d}\omega)$-almost surely from (11)

$$0 = \mathsf{F}(\beta, h_c(\beta)) \geq \frac{p_\ell}{2} \mathsf{F}(\beta, h_c(\beta)+\delta) - \liminf_{N\to\infty} 2^{-N} \Big(g_N(\omega) \log B + b_N(\omega) \log\Big(\frac{B}{B-1}\Big)\Big). \tag{15}$$

Now it is obvious that, if $\mathcal{I}_N(\omega)$ is not empty, $g_N(\omega)$ is bounded above by the total number of points of $\mathcal{T}_{\lceil \log_2 |\mathcal{I}_N(\omega)|\rceil}$, plus $|\mathcal{I}_N(\omega)|(N-\ell-\lfloor \log_2 |\mathcal{I}_N(\omega)|\rfloor)$. In formulas,

$$g_N(\omega) \le |\mathcal{I}_N(\omega)|(2+N-\lfloor \log_2 |\mathcal{I}_N(\omega)|\rfloor - \ell), \tag{16}$$

with the convention that $0\log 0 = 0$. Letting $N\to\infty$, almost surely one has via (11) and (12)

$$\liminf_{N\to\infty} 2^{-N} g_N(\omega) \le 2^{-\ell} p_\ell (2-\log_2 p_\ell) \le p_\ell \frac{\delta^2}{\beta^2 \log 2}(1+o_\ell(1)), \tag{17}$$

where $o_\ell(1)$ denotes a quantity which vanishes for $\ell\to\infty$.

As for the last term in (15), write $\mathcal{I}_N(\omega) = \{x_1(\omega), x_2(\omega), \dots,\}$ with $x_r(\omega) < x_{r+1}(\omega)$. Then we have

$$b_N(\omega) \le 2 \sum_{r=1}^{|\mathcal{I}_N(\omega)|+1} \log_2(x_r(\omega) - x_{r-1}(\omega)+1) \tag{18}$$

with the convention that $x_0(\omega) := 1$ and $x_{|\mathcal{I}_N(\omega)|+1} := 2^{N-\ell}$. This can be proven as follows (see Figure 2). Let $1 < r \le |\mathcal{I}_N(\omega)|$ and a_r be the first common ancestor of $(\ell, x_{r-1}(\omega))$ and $(\ell, x_r(\omega))$. Let π_1 (respectively π_2) denote the unique path without self-intersections which leads from a_r to $(\ell, x_{r-1}(\omega))$ (resp. from a_r to $(\ell, x_r(\omega))$). Let L be the number of times the walk π_1 makes a move to the left, and R the number of times π_2 makes a move to the right. Then the number of bad nodes which are enclosed by the paths π_1 and π_2, call it b_r, equals $b_r = L+R-2$ (cf. Figure 2). On the other hand,

$$x_r(\omega) - x_{r-1}(\omega) - 1 \ge 2^{L-1} + 2^{R-1} - 2 \ge 2^{(L+R)/2} - 2 \tag{19}$$

from which one deduces that $b_r \le 2\log_2(x_r(\omega) - x_{r-1}(\omega)+1)$. Similarly one proves that the number of bad nodes which are descendents of the first common ancestor of $(\ell,1)$ and $(\ell, x_1(\omega))$ (respectively, of $(\ell, x_{|\mathcal{I}_N(\omega)|})$ and $(\ell, 2^{N-\ell})$) is at most $2\log_2(x_1(\omega))$ (resp. at most $2\log_2(2^{N-\ell} - x_{|\mathcal{I}_N(\omega)|}+1)$) and Eq. (18) is proven.

Using Jensen's inequality for the logarithm, from Eqs. (18) and (12) one has almost surely

$$\liminf_{N\to\infty} 2^{-N} b_N(\omega) \le 2^{-\ell+1} p_\ell \log_2\left(\frac{1}{p_\ell}\right) \le 2p_\ell \frac{\delta^2}{\beta^2\log 2}. \tag{20}$$

Putting together Eqs. (15), (17) and (20) and taking ℓ large one obtains then (8) for a suitable $c(B)$.

It remains to prove (14), and we will do by induction on $N-\ell$. We need only consider the case where $\mathcal{I}_N(\omega)\neq\emptyset$ (i.e., the root is a good node) otherwise (14) reduces to $R_N \ge (B-1)/B$ which is evident from (3). For $N-\ell=1$, Eq. (14) is easily checked using the definition of good and bad sites and the basic recursion (3). Assume therefore that the statement holds for $N-\ell<k$, and let $N=\ell+k$.

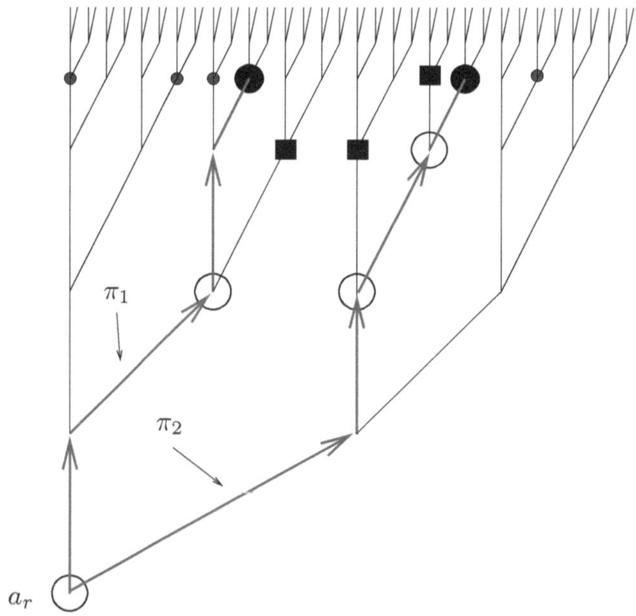

FIGURE 2. Again, the binary tree $\mathcal{T}_N$ for $N = 6$, $\ell = 2$ and nodes (ℓ, j) with $j \in \mathcal{I}_N(\omega)$ marked by full dots. The two big full dots denote nodes $x_{r-1}(\omega)$ and $x_r(\omega)$ ($r = 5$ in the figure) and their first common ancestor a_r is the root. The paths π_1, π_2 are marked by thicker lines and follow the arrows. Empty circles mark left turns in π_1, and right turns in π_2 (note that a_r contributes both to L and to R). Squares mark bad nodes which are between paths π_1 and π_2.

Call $\mathcal{T}_{N-1}^{(1)}$ and $\mathcal{T}_{N-1}^{(2)}$ the two trees of depth $N-1$ which originate from the root. Since the root is assumed to be good, the following statements are true:

- $g_N(\omega)$ equals the number $g_{N-1}^{(1)}(\omega)$ of good sites in $\mathcal{T}_{N-1}^{(1)}$ plus the number $g_{N-1}^{(2)}(\omega)$ of good sites in $\mathcal{T}_{N-1}^{(2)}$ plus one;
- $b_N(\omega)$ equals $b_{N-1}^{(1)}(\omega) + b_{N-1}^{(2)}(\omega)$ (with obvious notations).

From (3) we have $R_N \geq R_{N-1}^{(1)} R_{N-1}^{(2)}/B$ which, in view of the induction hypothesis, reads

$$R_N \geq \exp(F(\beta, h_c(\beta) + \delta)2^{\ell-1}|\mathcal{I}_N(\omega)|) \times \left(\frac{B-1}{B}\right)^{b_{N-1}^{(1)}(\omega)+b_{N-1}^{(2)}(\omega)} B^{-[1+g_{N-1}^{(1)}(\omega)+g_{N-1}^{(2)}]}. \tag{21}$$

Thanks to the two observations above, this coincides with inequality (14) and the proof is complete. □

Acknowledgement

We are very much indebted to Bernard Derrida and Giambattista Giacomin for several enlightening discussions. We acknowledge also the support of ANR grant *POLINTBIO*, and F.L.T. acknowledges the support of ANR grant *LHMSHE*.

References

[1] Alexander, K.S.: The effect of disorder on polymer depinning transitions. Comm. Math. Phys. **279**(1), 117–146 (2008).

[2] Derrida, B., Giacomin, G., Lacoin, H., Toninelli, F.L.: Fractional moment bounds and disorder relevance for pinning models. Commun. Math. Phys. **287**(3), 867–887 (2009).

[3] Derrida, B., Hakim, V., Vannimenus, J.: Effect of disorder on two-dimensional wetting. J. Statist. Phys. **66**(5-6), 1189–1213 (1992).

[4] Giacomin, G., Lacoin, H., Toninelli, F.L.: Hierarchical pinning models, quadratic maps and quenched disorder, Probab. Theory Relat. Fields **145**(1-2) (2009).

[5] Giacomin, G., Toninelli, F.L.: Smoothing effect of quenched disorder on polymer depinning transitions. Comm. Math. Phys. **266**(1), 1–16 (2006).

[6] Giacomin, G., Toninelli, F.L.: Smoothing of depinning transitions for directed polymers with quenched disorder. Physical Review Letters **96**(7), 070602 (2006).

[7] Monthus, C., Garel, T.: Critical points of quadratic renormalizations of random variables and phase transitions of disordered polymer models on diamond lattices. Physical Review E (Statistical, Nonlinear, and Soft Matter Physics) **77**(2), 021132 (2008).

[8] Toninelli, F.L.: A replica-coupling approach to disordered pinning models. Comm. Math. Phys. **280**(2), 389–401 (2008).

Hubert Lacoin
U.F.R. de Mathématiques
Case 7012 (site Chevaleret)
Laboratoire de Probabilités et Modèles Aléatoires (LPMA, CNRS U.M.R. 7599)
Université Paris Diderot (Paris 7)
F-75205 Paris Cedex 13, France
e-mail: `lacoin@math.jussieu.fr`

Fabio Lucio Toninelli
Laboratoire de Physique (CNRS U.M.R. 5672)
École Normale Supérieure de Lyon
46 Allée d'Italie
F-69364 Lyon Cedex 07, France
e-mail: `fltonine@ens-lyon.fr`

MIX
Papier aus verantwortungsvollen Quellen
Paper from responsible sources
FSC® C105338

If you have any concerns about our products,
you can contact us on
ProductSafety@springernature.com

In case Publisher is established outside the EU,
the EU authorized representative is:
Springer Nature Customer Service Center GmbH
Europaplatz 3, 69115 Heidelberg, Germany

Printed by Libri Plureos GmbH
in Hamburg, Germany